SEMIGROUPS OF MATRICES

SERIES IN ALGEBRA

Editors: J. M. Howie, D. J. Robinson, W. D. Munn

SERIES IN
ALGEBRA
VOLUME 6

SEMIGROUPS OF MATRICES

Jan Okniński

Warsaw University

World Scientific
Singapore • New Jersey • London • Hong Kong

Published by

World Scientific Publishing Co. Pte. Ltd.

5 Toh Tuck Link, Singapore 596224

USA office: 27 Warren Street, Suite 401-402, Hackensack, NJ 07601

UK office: 57 Shelton Street, Covent Garden, London WC2H 9HE

British Library Cataloguing-in-Publication Data
A catalogue record for this book is available from the British Library.

SEMIGROUPS OF MATRICES

ISBN-13 978-981-02-3445-4
ISBN-10 981-02-3445-7

To   Iwona,
Gosia, Asia and Adam

Preface

This book is concerned with the structure of linear semigroups, that is, subsemigroups of the multiplicative semigroup $M_n(K)$ of $n \times n$ matrices over a field K (or, more generally, skew linear semigroups – if K is allowed to be a division ring) and its applications to certain problems on associative algebras, semigroups and linear representations. It is motivated by several recent developments in the area of linear semigroups and their applications.

The theory of linear semigroups has been developed almost only in the context of linear algebraic (= Zariski closed) monoids $S \subseteq M_n(K)$, while it turned out really fruitful in the case where S is connected. The book of M.S.Putcha "Linear Algebraic Monoids" (Cambridge University Press, 1988) is the only monograph in this area. However, there is no strong link with arbitrary algebraic monoids and even less with arbitrary semigroups of matrices. In contrast to the theory of algebraic groups, the theory has lead only to a few applications to infinite dimensional associative algebras.

In this context, there has been a need for a structure theory of arbitrary linear semigroups that would enable us to use the powerful techniques of algebraic groups and to obtain applications, in particular to infinite dimensional associative algebras and their representations. Actually, some of the motivating ideas have arisen from the author's monograph "Semigroup Algebras" (Marcel Dekker, 1991).

The results obtained within the last ten years allow us to build such a general structural approach to semigroups of matrices. We summarize the state of knowledge in this area, presenting the approach and its consequences in a unified form. Since the object of our study seems to play a fundamental role in many fields, the presented material may be of interest not only to a broad audience in associative algebra and linear algebra, but also to research mathematicians working in areas where these theories are applicable, such as for example some aspects of theoretical computer science.

A general structure theorem for semigroups $S \subseteq M_n(K)$ is the main motive of our approach. It allows us to associate to S a collection of at most 2^n linear groups G_α and as many so called sandwich matrices P_α over these groups. This result has a strong flavour of Wedderburn's structure theorem for finite dimensional algebras. Moreover, the theorem yields a chain of ideals of S that often allows an inductive approach and suggests a reduction of problems concerning S to questions on the groups G_α and the associated matrices P_α. In the classical case of finite semigroups, P_α, G_α come from the principal factors of S. Our approach is based on the mutual action of the linear groups G_α and their action on the sandwich matrices P_α.

This settles foundations for developing a theory of linear semigroups that would make use of, and extend, the powerful techniques and results of the classical theory of linear groups, including the highlights of the theory of algebraic groups. An obvious step is then to describe connections of the structural theorem for $S \subseteq M_n(K)$ with the representation theory (irreducible representations) of S and with the topological structure (connected components) of the Zariski closure $\overline{S}$ of S.

Another basic ingredient of the approach is the so called generalised Tits alternative. Recall that a finitely generated linear group either has a free nonabelian subgroup or is almost solvable. For a finitely generated linear semigroup S the following equivalence is true: S has no free noncommutative subsemigroups if and only if every maximal cancellative subsemigroup of S is almost nilpotent (equivalently, the linear groups associated to S are almost nilpotent). Together with the structure theorem this result turns out to be a powerful tool.

Having established the generalised Tits alternative, it is natural to ask for a description of certain important special classes of linear semigroups with no free noncommutative subsemigroups. In this respect we discuss identities of linear semigroups, including Malcev nilpotent semigroups, and the Gelfand – Kirillov dimension of linear semigroups. The corresponding results for groups are now classical tools in group theory and its applications. The aim is to find their analogues and exploit their consequences for the structure of $S \subseteq M_n(K)$.

If $S \subseteq M_n(D)$ for a division ring D which is not a field, the main ingredients of our approach are still retained. This leads to applications to associative algebras that have 'many' skew linear representations.

While the main focus is on the structural approach and its consequences, combinatorial aspects of a very promising class of finite semigroups, called monoids of Lie type and including $M_n(\mathbf{F}_q)$ as the simplest example, are also discussed.

We start in Chapter 1 with recalling some fundamental notions, results and methods, frequently used in the book. These include basics of semigroup theory, semigroup algebras, methods of algebraic geometry (Zariski closure), multilinear methods (exterior power), and certain combinatorial tools often useful when dealing with sets of matrices.

Chapter 2 is devoted to a thorough analysis of the full linear monoid $M_n(K)$. First, the ideal structure of $M_n(K)$ and the egg-box pattern on each principal factor are described. Then, we discuss systems of idempotents and nilsemigroups arising from $M_n(K)$, as it turns out that these two topics are essentially connected. We continue with a combinatorial approach to $M_n(K)$, with an emphasis on the case of the finite field $K = \mathbf{F}_q$, modelled after classical combinatorics on the group $GL_n(\mathbf{F}_q)$. Finally, a proof of semisimplicity of the complex algebra $\mathbf{C}[M_n(\mathbf{F}_q)]$ is presented, followed by a description of irreducible representations.

In Chapter 3 our structural approach to arbitrary linear semigroups $S \subseteq M_n(K)$ is explained. It leads to a decomposition of S into so called uniform components U_α, each leading to a linear group $G_\alpha \subseteq GL_{n_\alpha}(K), n_\alpha \leq n$, and inheriting an egg-box pattern from a principal factor of $M_n(K)$. Also certain 'nilpotent components' of S come to the picture. A closure operation $S \mapsto \mathrm{cl}(S) \subseteq M_n(K)$, that makes each associated group G_α a subgroup of $\mathrm{cl}(S)$, is described. Its relation to S and to the Zariski closure $\overline{S}$ is set up in structural terms. This is followed by some technical results often useful when applying the structural approach.

Semigroups $S \subseteq M_n(K)$ which act irreducibly on the column vector space K^n are discussed in Chapter 4. As in the case of linear groups, this is an important class, which provides us with a general method of studying arbitrary linear semigroups. We continue with a discussion of irreducible representations, with an emphasis on connections with the structural approach. Finally, an application to triangularizability of linear semigroups, is given.

Linear semigroups satisfying identities of certain important types are discussed in Chapter 5. These include semigroup identities, with a special emphasis on Malcev nilpotent semigroups, and semigroups S

such that the semigroup algebra $K[S]$ satisfies a polynomial identity. In all cases, the groups G_α associated to S turn out to be almost nilpotent, that is, finite extensions of nilpotent groups.

An extension of the celebrated Tits alternative to the context of semigroups of matrices is obtained in Chapter 6. So, we discuss semigroups with no free noncommutative subsemigroups, the main difficulty being in the case where $S \subseteq GL_n(K)$. In particular, this condition is shown to lead to a semigroup identity on S. Consequences for the properties of the closure $cl(S)$ are presented.

The next natural step is to look at finitely generated semigroups of polynomial growth. This is motivated by the generalised Tits alternative obtained before and, on the other hand, by the known description of groups of polynomial growth (these are exactly almost nilpotent groups). So, in Chapter 7, we are looking for a structural description of this class of semigroups. Semigroups of linear growth and their relation to so called repetitive semigroups, are also treated.

Chapter 8 is independent of most of the foregoing material, except for Chapters 2 and 4. Namely, an exceptional class of finite monoids, with $M_n(\mathbf{F}_q)$ as the starting example, is discussed. The emphasis is on the combinatorics arising from Tits systems, and extending the powerful results on the structure and combinatorics of finite groups of Lie type. This is then used for a discussion of the complex and modular representations of such monoids and their connection with groups representations.

Chapter 9 provides examples of applications of the developed methods to infinite dimensional associative algebras. The problems considered here stem from the earlier study of semigroup algebras, and were in fact one of the main motivations for developing our structural approach to linear semigroups. Namely, graded rings R are treated via the semigroup $H(R)$ of homogeneous elements, in cases, where $H(R)$ can be effectively attacked via its linear (or, more generally, skew linear) representations. In particular, this applies to rings satisfying certain finiteness conditions, such as being Noetherian or satisfying a polynomial identity. In order to apply this idea, we first need to discuss an extension of the structural approach to the case of skew linear semigroups, that is, subsemigroups of $M_n(D)$ for a division ring D. Differences with the case where D is a field, and resulting invariants for D are discussed. This is followed by applications to some selected topics: prime Goldie semigroup algebras, principal ideal semigroup algebras, semilocal and perfect graded rings, and the homogeneity problem of the Jacobson radical of a graded ring.

While the first part of the book is reasonably self - contained, for the later chapters we need more background. This particularly applies to Chapters 6,8,9, where some deep classical results on algebraic groups, finite groups of Lie type, and associative algebras are needed.

This project was supported in part by a KBN research grant, Poland.

March, 1998   Jan Okniński

Contents

Chapter 1

General techniques

In this chapter we recall some basic facts about semigroups and we introduce notation. Standard results on semigroup algebras, linear groups and algebraic groups and semigroups are also given. The last section is devoted to certain combinatorial techniques, applicable to sequences of matrices. Only a few proofs are given, but standard references are provided.

1.1 Semigroups and their algebras

If A is a subset of a semigroup S, then by $\langle A \rangle$ we denote the subsemigroup of S generated by A. If $A = \{a_1, \ldots, a_k\}$, then we also write $\langle a_1, \ldots, a_k \rangle$. If G is a subgroup of S and $A \subseteq G$, then $\mathrm{gp}(A)$ stands for the subgroup of G generated by A.

If S is a semigroup that is not a monoid, then by S^1 we mean the monoid $S \cup \{1\}$ obtained by adjoining an identity element to S. Otherwise, we put $S^1 = S$. If S has a zero element, we often write θ for the zero of S. By S^0 we mean the semigroup S with zero adjoined.

Let $e = e^2 \in S$. Then eSe is a monoid and hence the set of units of eSe is a subgroup of S. Subgroups of this type are exactly the maximal subgroups of S. For example, if $S = M_n(K)$ the multiplicative semigroup of $n \times n$ matrices over a field K, then every $eM_n(K)e$ is isomorphic to $M_j(K)$, where j is the rank of e. Hence, the maximal subgroup of $M_n(K)$ containing e is isomorphic to $GL_j(K)$.

A nonempty subset I of S is a right (left) ideal of S if $IS \subseteq I$ ($SI \subseteq I$, respectively). Green's relations on S reflect the ideal structure.

1

They are defined as follows:

- $a\mathcal{R}b$ if $aS^1 = bS^1$

- $a\mathcal{L}b$ if $S^1a = S^1b$

- $a\mathcal{H}b$ if $a\mathcal{R}b$ and $a\mathcal{L}b$

- $a\mathcal{J}b$ if $S^1aS^1 = S^1bS^1$.

If I is a (two-sided) ideal of S, then one can form the Rees factor S/I, which is the semigroup with zero $(S \setminus I) \cup \theta$ with operation defined by $a \cdot b = ab$ if $ab \in S \setminus I$ and $ab = \theta$ otherwise. It is often convenient to put $S/I = S$ for $I = \emptyset$.

If $a \in S$, then the set I_a of nongenerators of the ideal S^1aS^1 is an ideal of S, if nonempty. The Rees factor S^1aS^1/I_a is called the principal factor of a in S. It is clear that, if $a \neq \theta$, then the set of nonzero elements of S^1aS^1/I_a can be identified with the $\mathcal{J}$-class of a in S. Principal factors are used to study the structure of S 'locally'. This approach is especially useful in the class of finite semigroups or (von Neumann) regular semigroups.

A semigroup S is called 0-simple if S has no ideals other than S and possibly θ if S has a zero element θ. If additionally S has a primitive idempotent, then S is completely 0-simple. (Note that, for practical reasons, we do not make any distinction between completely simple semigroups and completely 0-simple semigroups.)

It is known that every principal factor of a semigroup S is either completely 0-simple or a null semigroup (that is, a semigroup with zero multiplication).

Let X, Y be nonempty sets, G a group and $P = (p_{yx})$ a $Y \times X$ matrix with entries in G^0. (Clearly, by this we mean a function $(y, x) \mapsto p_{yx}$ from $Y \times X$ to G^0.) By $S = \mathcal{M}(G, X, Y, P)$ we denote the set of all triples $(g, x, y) \in G^0 \times X \times Y$ with $(g, x, y,) \cdot (g'x', y') = (gp_{yx'}g', x, y')$, where all triples (θ, x, y) are identified with the zero of S. In particular, θ also denotes the zero of S. S is called a semigroup of matrix type over the group G with sandwich matrix P. Then the sets $S_{(x)} = \{(g, x, y) \in S \mid g \in G^0, y \in Y\}$, $x \in X$, are called the rows of S, while $S^{(y)} = \{(g, x, y) \in S \mid g \in G^0, x \in X\}$, $y \in Y$, are the columns of S. If $|X| = k < \infty$, then we also write $S = \mathcal{M}(G, k, Y, P)$, identifying X with the set $\{1, \dots, k\}$. The same applies to the set Y of columns of S.

For any subsets $A \subseteq X, B \subseteq Y$ we put $S_{(A)} = \bigcup_{x \in A} S_{(x)}$, $S^{(B)} = \bigcup_{y \in B} S^{(y)}$, and $S_{(A)}^{(B)} = S_{(A)} \cap S^{(B)} = \bigcup_{x \in A, y \in B} S_{(x)}^{(y)}$. Clearly, $S_{(A)}^{(B)} = \mathcal{M}(G, A, B, P_{BA})$ is a semigroup of matrix type, where P_{BA} is the $B \times A$ submatrix of P.

S can be represented as the set of $X \times Y$ matrices over $G \cup \{0\}$ with at most one nonzero entry (under the identification of every (g, x, y) with the matrix with g in position (x, y) and zeros elsewhere) with $a \cdot b = aPb$, where the latter is the usual multiplication of matrices.

It is known that a semigroup S with zero is completely 0-simple if and only if it is isomorphic to a semigroup of matrix type $\mathcal{M}(G, X, Y, P)$ such that every row and every column of the matrix P has a nonzero entry. If S is a semigroup with no zero element, then it is completely 0-simple if and only if S^0 is of the latter type (in particular, the sandwich matrix P has no zero entries). The basic information on the properties of semigroups of the above type is given below.

Lemma 1.1 *Let $S = \mathcal{M}(G, X, Y, P)$ be a semigroup of matrix type. Then*

1. *for every nonempty subsets $A \subseteq X, B \subseteq Y$, $S_{(A)}$ is a right ideal of S and $S^{(B)}$ is a left ideal of S,*

2. *$E = \{(g, x, y) \mid p_{yx} \in G, g = p_{yx}^{-1}\}$ is the set of nonzero idempotents of S. If $e = (g, x, y) \in E$, then $S_{(x)}^{(y)} \simeq G^0$ via the map $(h, x, y) \mapsto h p_{yx}$,*

3. *if $a, b, c \in S$ and $ab \neq \theta, bc \neq \theta$, then $abc \neq \theta$; if $a \neq \theta$, then any of the conditions $ab = a, ba = a$ implies that b is an idempotent,*

4. *if $p_{yx} = 0$, then $S_{(x)}^{(y)}$ is a semigroup with zero multiplication,*

5. *if S is completely 0-simple, then for every nonzero $a = (g, x, y) \in S$ there exist $e, f \in E$ such that $a = ea = af$ and there exist $b, c \in S$ such that bac lies in a maximal subgroup of S. Moreover, $aS = S_{(x)}$ and $Sa = S^{(y)}$.*

Note that the last assertion identifies the $\mathcal{R}, \mathcal{L}$ and $\mathcal{H}$-classes of a completely 0-simple S. In particular, we get an egg-box pattern on $S \setminus$

$\{\theta\}$ (the boxes being the nonzero $\mathcal{H}$-classes $S^{(y)}_{(x)} \setminus \{\theta\}$), which will be crucial for the general structural approach developed in Chapter 3.

Let $S = \mathcal{M}(G, X, Y, P)$ be a semigroup of matrix type over a group G. If H is a normal subgroup of G, then $\phi_H((g, x, y)) = (gH, x, y)$ determines a homomorphism $\phi_H : S \longrightarrow S_H$, where $S_H = \mathcal{M}(G/H, X, Y, P_H)$ is the semigroup of matrix type over G/H with sandwich matrix $P_H = (p_{yH,xH})$ defined by $p_{xH,yH} = p_{yx}H$.

It is well known that $\mathcal{M}(G, X, Y, P) \simeq \mathcal{M}(G', X', Y', P')$ if and only if $|X| = |X'|, |Y| = |Y'|$ and there exists an isomorphism $\phi : G \longrightarrow G'$, such that $\phi(P) = Q_1 P' Q_2$ for some $Y \times Y'$ and $X' \times X$ matrices Q_1, Q_2 respectively, over $G' \cup \{0\}$ with exactly one nonzero entry in each row and each column. Here $\phi(P) = (\phi(p_{yx}))$.

Using this fact, it is often useful to normalize the given sandwich matrix P. Namely, if we fix $x \in X$ and $y \in Y$, then the above allows to change the presentation of S so that all entries of the row y and of the column x of P lie in the set $\{0, 1\}$.

More generally, we shall sometimes consider a semigroup of matrix type $\mathcal{M}(T, X, Y, P)$ over a semigroup T (here p_{yx} are simply chosen from T^0).

If K is a field, then let $K[S]$ be the semigroup algebra of S over K. That is, $K[S]$ is the K-space with basis S and multiplication on $K[S]$ extends that in S.

If S has a zero θ, then $K\theta$ is an ideal of $K[S]$. We put $K_0[S] = K[S]/K\theta$ and call it the contracted semigroup algebra of S over K. Thus, we identify the zeros of S and of $K[S]$. If S has no zero, then we put $K_0[S] = K[S]$. The elements of $K_0[S]$ are written as finite linear combinations $a = \sum \lambda_s s$ of elements of $s \in S \setminus \{\theta\}$. By supp$(a)$ we denote the set $\{s \in S \mid \lambda_s \neq 0\}$. It is easy to see that $K[S] \simeq K_0[S] \oplus K\theta$ as K-algebras. More generally, if I is any ideal of S, then $K_0[S/I]$ is isomorphic to, and shall be identified with, $K[S]/K[I]$.

If J is an ideal of $K[S]$, then $\sim_J$ defined by: $a \sim_J b$ if $a - b \in J$, is a congruence on S. Moreover, the homomorphism $K[S] \longrightarrow K[S]/J$ factors through $K[S/\sim_J]$. The image of S can be identified with $S/\sim_J$ and the kernel of the map $K[S] \longrightarrow K[S/\sim_J]$ is the linear span of the set $\{s - t \mid s \sim_J t, s, t \in S\}$. If A is a subset of a K-algebra R, then we write $K\{A\}$ for the subspace of R which is the linear span of S, if unambiguous. In particular, this will be applied to the case $R = M_n(K)$ and $A = S$ is a subsemigroup of R. Clearly, $K\{S\}$ is a subalgebra in this case.

With every semigroup of matrix type $S = \mathcal{M}(G, X, Y, P)$ we can associate the contracted semigroup algebra $K_0[S]$. It is called the Munn algebra of S. We shall see in Section 4.2 that $K_0[S]$ can be identified with the algebra $\mathcal{M}(K[G], X, Y, P)$ of $X \times Y$ matrices over the group algebra $K[G]$ with finitely many nonzero entries and multiplication given by $a \cdot b = aPb$.

Our basic references for the above material are [15], [38], [87].

1.2 Zariski topology and closed semigroups

In what follows K denotes a field. $\overline{K}$ is the algebraic closure of K. If $n \geq 1$, then we write $M_n(K)$ for the multiplicative monoid of $n \times n$ matrices over K. By a linear semigroup over K we mean a subsemigroup of some $M_n(K)$. If $a \in M_n(K)$, then rank(a) denotes the rank of the matrix a. As usual, $GL_n(K)$ is the group of invertible matrices. $M_n(K)$ acts on the column vector space K^n by left multiplication. By ker(a), Im$(a) \subseteq K^n$ we mean the kernel and the image of $a \in M_n(K)$ under this action.

A subset $X \subseteq K^n$ is closed if it is the zero set of a collection of polynomials in $K[x_1, \ldots, x_n]$. This defines the Zariski topology on K^n. Closed subsets of K^n are in one-to-one correspondence with radical ideals of $K[x_1, \ldots, x_n]$. In particular, since $K[x_1, \ldots, x_n]$ is Noetherian, we have d.c.c. on closed subsets of K^n.

When the K-space $M_n(K)$ is identified with K^{n^2}, the polynomial ring $K[x_{ij}| \, i, j = 1, 2, \ldots, n]$ can be used to introduce the Zariski topology on $M_n(K)$. Here the matrix units e_{ij} form the standard basis of $M_n(K)$ and they correspond to the variables x_{ij}.

The closure of a subset A of $M_n(K)$ is denoted by $\overline{A}$. It is well known that right (and left) multiplication by any $a \in M_n(K)$ is a continuous map in Zariski topology. Topological tools can be applied to linear semigroups because of the following observation.

Theorem 1.2 *Let $S \subseteq M_n(K)$ be a semigroup. Then $\overline{S}$ is a subsemigroup of $M_n(K)$ containing S.*

Proof. Let $a \in S$. Consider the set $L_a = \{x \in M_n(K)| \, ax \in \overline{S}\}$. Clearly $S \subseteq L_a$. Moreover L_a is a closed subset of $M_n(K)$ because $\overline{S}$ is closed and the left multiplication by a is a continuous map. Therefore $\overline{S} \subseteq L_a$. Since $a \in S$ is arbitrary, $S\overline{S} \subseteq \overline{S}$. Now, let $b \in \overline{S}$ and $R_b = \{x \in M_n(K)| xb \subseteq \overline{S}\}$. Again, R_b is a closed set and $S \subseteq R_b$. This implies that $\overline{S} \subseteq R_b$. The assertion follows. $\square$

If A is a subset of $GL_n(K)$, then by the closure of A in $GL_n(K)$ we mean $\overline{A} \cap GL_n(K)$. This defines the Zariski topology on $GL_n(K)$. More generally, if $A \subseteq B$ are subsets of $M_n(K)$, then A is closed in B if it is of the form $X \cap B$ for a closed subset X of $M_n(K)$.

A semigroup S is called π-regular if a power of every element of S lies in a subgroup of S. Then every homomorphic image of S also is of this type. Clearly, every periodic semigroup is π-regular. Other important examples of π-regular semigroups are given below. Recall that S is a (von Neumann) regular semigroup if for every $a \in S$ there exists $x \in S$ such that $axa = a$.

Proposition 1.3 *$M_n(K)$ is π-regular. Moreover, if $a \in M_n(K)$, then the set $A_a = \{a^k| k \geq n\}$ is contained in a subgroup of $M_n(K)$. More generally, if $S \subseteq M_n(K)$ is a regular semigroup, then, for every $a \in S$, A_a is contained in a subgroup of S.*

Proof. First, consider the case $R = M_n(K)$. If $a \in R$, then by Fitting's lemma there exist subspaces V, W of K^n such that $K^n = V \oplus W$, for every $k \geq n$ we have $a^k V = 0$ and a^k maps isomorphically W onto W. Let $\phi : W \longrightarrow W$ be the inverse of the latter map. By the Cayley-Hamilton theorem, a linear combination b of some powers of a^k acts on W as ϕ. Therefore $f = a^k b = ba^k$ is the idempotent such that $fV = 0$ and $fw = w$ for $w \in W$. It is easy to see that $\{x \in R| xW = W, xV = 0\}$ is a monoid with identity f and b is the inverse of a^k in this monoid. Hence, the assertion follows.

Now, assume that $S \subseteq M_n(K)$ is regular. Let $a \in S$. If $k \geq n$, then have seen that $c = a^k \in G$ for a subgroup G of $M_n(K)$. By the hypothesis there exists $x \in S$ such that $c^3 x c^3 = c^3$. Then $\text{rank}(c^2 x c) = \text{rank}(c)$ and $c^2 x c \in e M_n(K) e$, where $e = e^2 \in G$. This implies that $c^2 x c = e$ and cxc is the inverse of c in the maximal subgroup of $M_n(K)$ containing e. Hence, also in the maximal subgroup of S containing e. The assertion follows. $\square$

The basic property of $\overline{S}$ reads as follows.

Theorem 1.4 *If $S \subseteq M_n(K)$ is a closed subsemigroup, then S is π-regular.*

Proof. Let $a \in S$. It is enough to show that $b = a^n$ lies in a subgroup of S. From Proposition 1.3 we know that $b \in G$ for a maximal subgroup G of $M_n(K)$. Let $e = e^2 \in G$. Then $bc = cb = e$ for some $c \in G$. Let $Z = \{x \in M_n(K) | ex = xe = x\}$. Then Z is a closed subset of $M_n(K)$. Hence $T = S \cap Z$ is closed in S. Since $b \in T$, we get a chain

$$bT \supseteq b^2T \supseteq b^3T \supseteq \cdots .$$

It is easy to see that $b^jT = \{x \in T | c^jx \in T\}$. Therefore b^jT is closed in S for every $j \geq 1$. D.c.c. on closed subsets implies now that $b^iT = b^{i+1}T$ for some $i \geq 1$. Therefore

$$T = eT = c^ib^iT = c^ib^{i+1}T = ebT = bT.$$

Similarly one shows that $T = Tb$. There exists $x \in T$ such that $b = bx$. Hence $x = ex = cbx = cb = e$. Consequently $e \in T$. Finally, there exist $y, z \in T$ such that $by = e = zb$. This means that b is invertible in $eTe \subseteq S$. The assertion follows. $\square$

Corollary 1.5 *For every semigroup $S \subseteq M_n(K)$ there exists the smallest π-regular semigroup $\mathrm{cl}(S) \subseteq M_n(K)$ containing S. We have $\mathrm{cl}(S) \subseteq \overline{S} \subseteq K\{S\}$ and $K\{S\}$ is a closed subset of $M_n(K)$. Moreover, for every maximal subgroup G of $M_n(K)$ intersecting $\mathrm{cl}(S)$, $\mathrm{cl}(S) \cap G$ is a maximal subgroup of $\mathrm{cl}(S)$ and, for every $a \in \mathrm{cl}(S)$, a^n lies in a maximal subgroup of $\mathrm{cl}(S)$.*

Proof. First, we show that $K\{S\}$ is a closed subset of $M_n(K)$. Let $e_1, \ldots, e_{n^2}$ be a basis of $M_n(K)$ such that $e_1, \ldots, e_r$, for some $r \leq n$, is a basis of $K\{S\}$. Every $a = (a_{ij}) \in M_n(K)$ is uniquely written as $a = \sum_{k=1}^{n^2} \lambda_k e_k$. Then $a \in K\{S\}$ if and only if $\lambda_k = 0$ for $k = r+1, \ldots, n^2$. Hence $K\{S\}$ is described by a system of $n^2 - r$ linear equations in the indeterminates $x_{ij}, i, j = 1, \ldots, n$. In particular, it is a closed subset of $M_n(K)$. This implies that $\overline{S} \subseteq K\{S\}$.

We know that $M_n(K)$ is π-regular. Let A be a nonempty intersection of a family $\mathcal{A}$ of π-regular subsemigroups of $M_n(K)$. We claim that

$A \cap G$ is a maximal subgroup of A whenever G is a maximal subgroup of $M_n(K)$ intersecting A.

Let $a \in A \cap G$. If $T \in \mathcal{A}$, then there exists $n_T \geq 1$ such that $a^{n_T} \in G_T$ for a maximal subgroup G_T of T. Since G_T is a subgroup of $M_n(K)$, we must have $G_T \subseteq G$. Thus $a^{n_T} b_T = e = e^2 \in G$ for some $b_T \in G_T \subseteq T$. Then $d = a^{n_T-1} b_T \in T$ is the inverse of a in G, and it follows that $d \in \bigcap_{T \in \mathcal{A}} T = A$. Therefore $A \cap G$ is a subgroup of A. It is a maximal subgroup because every subgroup of $M_n(K)$ containing e is contained in G. This proves the claim.

From Theorem 1.4 it follows that $\mathrm{cl}(S) \subseteq \overline{S}$. Then the remaining assertion is a consequence of Proposition 1.3. $\square$

The semigroup $\mathrm{cl}(S)$ will be called the π-regular closure of S. As we shall see in Chapter 3, it can be viewed as a group approximation of S. In the simplest case, where $S \subseteq GL_n(K)$, $\mathrm{cl}(S)$ is the group generated by S. In general, $\mathrm{cl}(S)$ is much closer to S than $\overline{S}$, so it is more useful when studying algebraic properties of S.

If S is a closed subsemigroup of $M_n(K)$, then S is uniquely expressible as a finite union of irreducible closed sets (that is, sets that are not a union of two proper closed subsets). These are called the irreducible (or connected) components of S. A connected semigroup $S \subseteq M_n(K)$ is a semigroup that has only one component. Such semigroups have many nice properties, that generalize deep structural properties of connected linear groups.

In general, two different components can intersect nontrivially. However, if $G \subseteq GL_n(K)$ is a group and we consider the topology induced on $GL_n(K)$, then the component G^c containing the identity is a normal subgroup of finite index, and the set of components coincides with the set of cosets of G^c in G.

If $G \subseteq GL_n(K)$ is a linear group, then G has the largest normal subgroup consisting of unipotent matrices. This is called the unipotent radical of G. If G is brought by conjugation in $M_n(K)$ to a block diagonal form with irreducible diagonal blocks, then the unipotent radical of G is the kernel of the projection onto the diagonal blocks.

If $G \subseteq GL_n(K)$ is solvable and $K = \overline{K}$, then G has a normal subgroup of finite index (bounded by a function of n), which is triangularizable in $GL_n(K)$. Moreover, the connected component $\overline{G}^c$ of the Zariski closure $\overline{G} \cap GL_n(K)$ of G in $GL_n(K)$ is triangularizable. It is known that solvable subgroups of $GL_n(K)$ are solvable of bounded solvability class (not exceeding $2n$.)

If K is a finitely generated field or $\mathrm{ch}(K) = 0$, then $G \subseteq GL_n(K)$ either has a free nonabelian subgroup or it is almost solvable (that is, a finite extension of a solvable group). If $\mathrm{ch}(K) > 0$ and G has no free nonabelian subgroups, then G has a solvable normal subgroup H such that G/N is periodic. This fundamental result is called Tits alternative.

Let $G \subseteq GL_n(K)$ be a nilpotent group, $K = \overline{K}$. Then the connected component $\overline{G}^c$ of the Zariski closure $\overline{G} \cap GL_n(K)$ of G in $GL_n(K)$ is of the form $G_d \times G_u$, where G_d is a diagonalizable group and G_u is a unipotent group.

Other results on closed linear groups, which will be needed later, are stated in Section 6.1. Our standard references for linear groups, algebraic groups and algebraic semigroups are [7], [83], [108], [134].

1.3 Exterior power

Let $\Lambda^j V$ be the j-th exterior power of the column vector space $V = K^n$. Recall that, if $e_1, \ldots, e_n$ is a basis of K^n, then $e_{i_1} \wedge \cdots \wedge e_{i_j}, 1 \le i_1 < \ldots < i_j \le n$, is a basis of $\Lambda^j V$. If W is a vector space, then every linear map $\phi : V \longrightarrow W$ uniquely extends to a linear map $\Lambda^j \phi : \Lambda^j V \longrightarrow \Lambda^j W$. The exterior power map $\Lambda^j : \mathrm{End}(V) \longrightarrow \mathrm{End}(\Lambda^j V)$ is defined by $\Lambda^j(\phi) = \Lambda^j \phi$. Therefore, Λ^j can be viewed as a map $M_n(K) \longrightarrow M_{\binom{n}{j}}(K)$. More precisely, we have

$$\Lambda^j(a)(a_1 \wedge \cdots \wedge a_j) = a(a_1) \wedge \cdots \wedge a(a_j)$$

for every $a_1, \ldots, a_j \in K^n$.

The following well-known observations will be very useful, cf.[9].

Lemma 1.6 $\Lambda^j : M_n(K) \longrightarrow M_{\binom{n}{j}}(K)$ *is a semigroup homomorphism and*

1. $\mathrm{rank}(\Lambda^j(a)) = \binom{r}{j}$ *if* $\mathrm{rank}(a) = r \ge j$, *and* $\mathrm{rank}(\Lambda^j(a)) = 0$ *otherwise.*

2. *For every vectors* $a_1, \ldots, a_j \in K^n$ *and every* $a_{ik} \in K, i, k = 1, \ldots, j$, $(\sum_{k=1}^j a_{1k} a_k) \wedge \cdots \wedge (\sum_{k=1}^j a_{jk} a_k) = \det(a_{ik})(a_1 \wedge \cdots \wedge a_j)$.

3. If $V \subseteq K^n$ is a subspace with basis $f_1, \ldots, f_j$, and $a \in M_n(K)$ is a matrix of rank j which determines an isomorphism $a_{|V} : V \longrightarrow V$, then $\Lambda^j(a)(f_1 \wedge \cdots \wedge f_j) = \det(a_{|V})(f_1 \wedge \cdots \wedge f_j)$.

Let ΛV be the exterior algebra of V. If $a_1, \ldots, a_j$ is a basis of a subspace $W \subseteq V$, then the vector $W' = a_1 \wedge \cdots \wedge a_j \in \Lambda V$ is determined uniquely up to a scalar multiple. On the other hand, W' determines the subspace spanned by the vectors $a_1, \ldots, a_j$ (if they are K-independent). Moreover, $W_1 \cap W_2 = 0$ if and only if $W_1' \wedge W_2' \neq 0$.

Exterior power provides a convenient tool to approach linear semigroups inductively. This is a consequence of Lemma 1.8 below. Here $M_j = \{a \in M_n | \operatorname{rank}(a) \leq j\}$ for $j = 0, 1, \ldots, n$, is easily seen to be an ideal of the monoid $M_n(K)$. First we recall the following classical observation, which can be found in [17], the proof of Theorem 36.2, or [36], the proof of Lemma 2.3.4.

Lemma 1.7 If K is a finitely generated field and $n \geq 1$, then the set of roots of unity that are roots of polynomials over K of degree at most n is finite.

Lemma 1.8 Let $1 \leq j \leq n$. For every $r > 1$ there exists a homomorphism $\phi : M_n(K)/M_{j-1} \longrightarrow M_t(K)$, $t = \binom{rn}{r(j-1)+1}$, such that $\phi(a) = \phi(b)$ if and only if $a = \lambda b$ for an $r(j-1)+1$-th root of unity $\lambda \in K$. Moreover, if the field K is finitely generated, then r can be chosen so that ϕ is an embedding.

Proof. Let $r > 1$. Consider the homomorphism $\varphi : M_n(K) \longrightarrow M_{rn}(K)$ defined by

$$
\varphi(a) = \begin{pmatrix} a & 0 & \cdots & 0 \\ 0 & a & \cdots & 0 \\ & & \ddots & \\ 0 & \cdots & 0 & a \end{pmatrix}.
$$

Let $s = r(j-1)+1$ and $\eta = \Lambda^s \varphi$. Clearly, $\eta(M_{j-1}) = 0$. Moreover $\varphi(M_n(K))$ has no nonzero matrices of rank less than rj.

To prove the first assertion, assume that $b, c \in M_{rn}(K)$ are of rank $\geq rj$ and such that $\Lambda^s(b) = \Lambda^s(c)$. Let $k = \operatorname{rank}(b) = \operatorname{rank}(c)$. Note that $s < rj \leq k$. Then

$$
\begin{aligned}
b(v_1) \wedge \cdots \wedge b(v_s) &= \Lambda^s(b)(v_1 \wedge \cdots \wedge v_s) \\
&= \Lambda^s(c)(v_1 \wedge \cdots \wedge v_s) \\
&= c(v_1) \wedge \cdots \wedge c(v_s)
\end{aligned}
$$

for every $v_1, \ldots, v_s \in K^{rn}$. This implies that $b(v_1), \ldots, b(v_s)$ span the same subspace as $c(v_1), \ldots, c(v_s)$, whenever they are K-independent. Consequently, $\operatorname{Im}(b) = \operatorname{Im}(c)$ and $\ker(b) = \ker(c)$. Let $e \in M_{rn}(K)$ be a projection onto $\operatorname{Im}(b)$. Then there exists $x \in GL_{rn}(K)$ such that $bx, cx \in D$ for the maximal subgroup D of $M_{rn}(K)$ containing e (choose x such that $x(\ker(e)) = \ker(b)$ and $bx(\operatorname{Im}(b)) = \operatorname{Im}(b)$). Then, for $d = (bx)(cx)^{-1} \in D$, $\Lambda^s(d)$ is an idempotent (here $(cx)^{-1}$ denotes the inverse of cx in D). If $u_1, \ldots, u_s \in eK^{rn}$ are independent and U is the subspace they span, then $d(u_1) \wedge \cdots \wedge d(u_s) = d^2(u_1) \wedge \cdots \wedge d^2(u_s)$ implies that $d(U)$ is a d-invariant subspace. Hence, every s-dimensional subspace of eK^{rn} is d-invariant. Since $s < k = \operatorname{rank}(e)$, it follows that $(bx)(cx)^{-1} = d = \lambda e$ for a scalar $\lambda \in K$. This implies that $bx = \lambda cx$ and so $b = \lambda c$. Hence, the displayed formula implies also that λ is an s-th root of unity in K.

Finally, if K is finitely generated, then it contains finitely many roots of unity by Lemma 1.7. Therefore, ϕ is an embedding if r is chosen big enough. $\square$

Corollary 1.9 *Let* $\eta = \Lambda^s\varphi : M_n(K) \longrightarrow M_n(K)/M_{j-1} \longrightarrow M_t(K)$ *be the homomorphism found above. If* $S \subseteq M_n(K)$ *is a semigroup, then* $\operatorname{cl}(\phi(S)) = \phi(\operatorname{cl}(S))$.

Proof. Let $T \supseteq \eta(S)$ be a π-regular subsemigroup of $M_t(K)$. Then $T' = T \cap \eta(M_n(K))$ is also π-regular because so is $\eta(M_n(K))$ (see the proof of Corollary 1.5). Clearly $T' \supseteq \eta(S)$. Let $\eta(a) \in T', a \in M_n(K)$. Then, for some $k \geq 1$, $\eta(a)^k$ lies in a subgroup H of T'. So $\eta(a)^k\eta(x) = \eta(e)$ for some $x, e \in M_n(K)$ such that $\eta(e)$ is the identity of H and $\eta(x) \in H$. By Lemma 1.8, $\eta(e) = \eta(e^2)$ implies that $e^2 = \lambda e$ for an s-th root of unity λ. Hence, replacing e by e^s, we may assume that $e = e^2$. Let G be the maximal subgroup of $M_n(K)$ containing e. Since $\eta(x) =$

$\eta(exe)$, we come to $x^s = (exe)^s \in eM_n(K)e$. Similarly, $a^{ks} \in eM_n(K)e$. Now $\eta(a^{ks}x^s) = \eta(e)$ implies that $(a^{ks}x^s)^s = e$. Therefore, the inverse $x^s(a^{ks}x^s)^{s-1}$ of a^{ks} in G is contained in $\eta^{-1}(T')$. We have thus shown that $\eta^{-1}(T') \subseteq M_n(K)$ is a π-regular semigroup which contains S. Therefore $\mathrm{cl}(S) \subseteq \eta^{-1}(T')$, so $\eta(\mathrm{cl}(S)) \subseteq \eta\eta^{-1}(T') \subseteq T' \subseteq T$. It follows that $\eta(\mathrm{cl}(S)) \subseteq \mathrm{cl}(\eta(S))$. Since $\eta(\mathrm{cl}(S))$ is π-regular and contains $\eta(S)$, we also have $\eta(\mathrm{cl}(S)) \supseteq \mathrm{cl}(\eta(S))$. This completes the proof. $\square$

1.4   Combinatorial techniques

For a given semigroup $S \subseteq M_n(K)$ the nonempty intersections $S \cap D$ with maximal subgroups of $M_n(K)$ will play a crucial role in the approach developed in Chapter 3. Here, we present a combinatorial technique which indicates the role of these intersections. The first result of this type was obtained in [43]. We follow the improved approach of [116], (compare also [70], Chapter 10, and so called repetitive mappings [73] in connection with Lemma 1.11 below).

Lemma 1.10 *The following conditions are equivalent for a matrix* $a \in M_n(K)$

   1. $a \in D$ *for a maximal subgroup* D *of* $M_n(K)$,

   2. $\mathrm{rank}(a^2) = \mathrm{rank}(a)$,

   3. K^n *is the direct sum of the subspaces* $\ker(a)$ *and* $\mathrm{Im}(a)$,

   4. *the minimal polynomial* $\chi_a(x)$ *of* a *is not divisible by* x^2,

   5. a *is conjugate to a matrix* $\begin{pmatrix} g & 0 \\ 0 & 0 \end{pmatrix}$ *for some* $g \in GL_j(K)$ *and some* j, $0 \le j \le n$.

Proof. If $a \in D$, then $a^2 \in D$, so that $\mathrm{rank}(a^2) = \mathrm{rank}(a)$. The latter implies that $\ker(a) \cap \mathrm{Im}(a) = 0$. Hence 1) implies 2) and 2) implies 3). If 3) holds, then the restriction of a to $\mathrm{Im}(a)$ is an automorphism. Changing the basis of K^n we derive 5). Assume that 5) holds. If $a = 0$ or $\mathrm{rank}(a) = n$, 4) is clear. Otherwise, $\chi_a(x) = x \cdot \chi_g(x)$, and $\chi_g(0) \ne 0$. Hence 4) follows. If 4) holds, then the Jordan form of a in $M_n(\overline{K})$ is as in 5). Therefore $a \in G'$ for a maximal subgroup G' of $M_n(\overline{K})$. Since $M_n(K) \cap G'$ is a subgroup of $M_n(K)$, 1) follows. This proves the lemma. $\square$

The basic combinatorial lemma reads as follows.

Lemma 1.11 *Let X be a nonempty set, $n, k_0, k_1, \ldots, k_n$ natural numbers, and $r : X^* \longrightarrow \{0, 1, \ldots, n\}$ a mapping such that $r(tuv) \leq r(u)$ for all t, u, v in the free monoid X^* generated by X. Then, for every $w \in X^*$ of length at least $k_0 \cdots k_n$ in X there exists l, $0 \leq l \leq n$, such that w has a factor of the form $w_1 \cdots w_{k_l}$ with all $w_i \in X^* \setminus \{1\}$, and $r(w_i \cdots w_j) = l$ for all i, j with $1 \leq i \leq j \leq k_l$.*

Proof. Let

$$r_n = k_n, \; r_{n-1} = k_n k_{n-1}, \; \ldots, \; r_0 = k_n k_{n-1} \cdots k_0.$$

Assume first that $w \in X^*$ is of length at least k_n and $r(w) = n$. Then w has a factor of the form $x_1 \cdots x_{k_n}$ with all $x_i \in X$. Hence, for all $i, j \leq k_n$, $n \geq r(x_i \cdots x_j) \geq r(w) = n$ implies that $r(x_i \cdots x_j) = n$ and the assertion follows.

Thus assume that $l < n$ and the assertion is true for all elements $x \in X^*$ with $r(x) > l$. Let $w \in X^*$ be of length $r_l = k_l r_{l+1}$ and $r(w) = l$. Then w has a factor $w_1 \cdots w_{k_l}$ with every w_i of length r_{l+1}. If $l' = r(w_i) > l$ for some i, then the assertion follows by induction, applied to w_i because $r_{l+1} \geq r_{l'}$. Otherwise, $r(w_i) \leq l$ for every i. Therefore, for all $i, j \leq k_l$ with $i \leq j$ we have

$$l = r(w) \leq r(w_i \cdots w_j) \leq r(w_i) \leq l,$$

which implies that $r(w_i \cdots w_j) = l$. This completes the proof. $\square$

The following result shows that sufficiently long products of elements of $M_n(K)$ contain subproducts which lie in maximal subgroups of $M_n(K)$.

Theorem 1.12 *There exists a natural number $N = N(n)$ such that for all elements $a_1, \ldots, a_N$ of $M_n(K)$ there exist $i < j$, $1 \leq i, j \leq N$, with $a_i \cdots a_j \in D$ for a maximal subgroup D of $M_n(K)$.*

Proof. Let $k_0 = 1$, $k_i = \binom{n}{i} + 1$ for $i = 1, 2, \ldots, n-1$, and $k_n = 1$. Define $N = k_0 \cdots k_n$. Let $w = a_1 \cdots a_N$. We apply the preceding lemma with the function r defined by $r(v) = \mathrm{rank}(v)$. It follows that there exists $l \le n$ such that $w = uw_1 \cdots w_{k_l}v$ is a factorization of w as a word in $a_1, \ldots, a_N$, and $\mathrm{rank}(w_i \cdots w_j) = l$ for every $1 \le i \le j \le k_l$.

If $l = 0$, then $r(w_1) = 0$, so that $w_1 = 0$ lies in a maximal subgroup of $M_n(K)$. If $l = n$, then $\mathrm{rank}(w_1) = n$, so that $w_1 \in GL_n(K)$. Hence, assume that $1 < l < n$. Suppose that none of $w_i \cdots w_j$ lies in a maximal subgroup of $M_n(K)$. Let $V_i = \ker(w_i)$ and $W_i = \mathrm{Im}(w_i)$ for every i. Since $\mathrm{rank}(w_i w_{i+1}) = \mathrm{rank}(w_i)$, we must have $V_i \cap W_{i+1} = 0$. On the other hand $V_j \cap W_i \neq 0$ for $i \le j$, because by Lemma 1.10 $\mathrm{rank}(w_i \cdots w_j) > \mathrm{rank}((w_i \cdots w_j)^2)$. For every subspace Z of V put $Z' = z_1 \wedge \cdots \wedge z_r \in \Lambda^r V$, where $z_1, \ldots, z_r$ is a fixed basis of Z. Then $V_i' \wedge W_{i+1}' \neq 0$ and $V_j' \wedge W_i' = 0$ for $i \le j$. Since $k_l > \binom{n}{l}$, and the latter is the dimension of $\Lambda^l(K^n)$, it follows that there exists j, $1 < j \le k_l$, such that $W_j' = \alpha_1 W_1' + \cdots + \alpha_{j-1} W_{j-1}'$ for some $\alpha_k \in K$. Therefore

$$V_{j-1}' \wedge W_j' = \sum_{i=1}^{j-1} \alpha_i V_{j-1}' \wedge W_i' = 0,$$

a contradiction. This completes the proof. $\square$

Remark The above proof shows also that, if $w_1, \ldots, w_r \in M_n(K)$ are matrices of rank k for $r > \binom{n}{k}$ and such that $\mathrm{rank}(w_1 \cdots w_r) = k$, then $w_i \cdots w_j$ is contained in a maximal subgroup of $M_n(K)$ for some i, j with $1 \le i \le j \le r$. We shall see in Section 2.2 that it is enough to assume $r \ge \binom{n}{k}$ here.

We shall need another result on factors of sufficiently long words in a free semigroup. First, we state the following well-known observation.

Lemma 1.13 *Let P be a subset of a finitely generated free semigroup Z on a finite set X. Then the following conditions are equivalent*

1. *there exists $r \ge 1$ such every $z \in Z$ which is a word of length $\ge r$ has a factor in P,*

2. *every infinite word in the generators from X has a factor in P.*

Proof. Clearly 2) is a consequence of 1). So, suppose 1) does not hold. Then there exists an infinite set $F \subseteq Z$ of words with no factors in P. Then F has an infinite subset F_1 all of whose elements start with the same generator $x_1 \in X$. Next, there exists an infinite subset F_2 of F_1 whose all elements have equal initial factors of length two, say $x_1 x_2$, where $x_2 \in X$. Continuing this way, we come to an infinite word $x = x_1 x_2 \cdots$ which has no factors in P, because each finite factor of x is a factor of a word from F. The result follows. $\square$

The above lemma is often used in the context of the following extension of van der Waerden's theorem on arithmetical progressions (cf.[73]), obtained in [11], see also [73], Problem 4.1.1, [70], Lemma 10.1.6.

Theorem 1.14 *Let $\alpha : \mathbf{N} \longrightarrow X$ be a function into a finite set X. Then there exist $p \geq 1$ and an element $x \in X$ such that for every $q \geq 1$ there exist $i_1 < i_2 < \cdots < i_q$ with $\alpha(i_1) = \cdots = \alpha(i_q) = x$ and $i_{j+1} - i_j \leq p$ for $j = 1, \ldots, q - 1$.*

Our next result comes from [53]. We give a proof only in the case where S is a group, the case which will be needed later.

Corollary 1.15 *Let $x = x_1 x_2 \cdots$ be an infinite word in elements of a free semigroup Z on a set X. Assume that $\phi : Z \longrightarrow S$ is a homomorphism into a finite semigroup S. Then there exists an idempotent $e \in S$ and $p \geq 1$ such that for every $q \geq 1$ there are q consecutive factors $u_1, \ldots, u_q$ in x of length $\leq p$ with $\phi(u_i) = e$ for $i = 1, \ldots, q$.*

Proof. Let S be a group. Consider the sequence $y_n = x_1 \cdots x_n$ for $n = 1, 2, \ldots$. Let $\alpha : \mathbf{N} \longrightarrow S$ be defined by $\alpha(n) = \phi(y_n)$. From Theorem 1.14 it follows that there exists $p \geq 1$ and $e \in S$ such that for every $q \geq 1$ there exist $i_1 < i_2 < \cdots < i_{q+1}$ such that $\phi(y_{i_j}) = e$ and $i_{j+1} - i_j \leq p$ for $j = 1, \ldots, q$. Since S is a group, the consecutive factors $u_j = \phi(x_{i_j+1} \cdots x_{i_{j+1}}), j = 1, \ldots, q$, are all equal to the identity of S. $\square$

The last combinatorial technique discussed in this section is concerned with linear recursions. Here we follow [33].

Let $u_1, u_2, \ldots$ be a sequence of elements of a field K. Suppose this sequence satisfies a linear recurrence of order r

$$u_{m+r} = a_1 u_{m+r-1} + a_2 u_{m+r-2} + \cdots + a_r u_m$$

for $m = 0, 1, 2, \ldots$, where $a_i \in K$ are constants. Let $k(x) = 1 - a_1 x - a_2 x^2 - \cdots - a_r x^r \in K[x]$. The formal power series $g(x) = \sum_{i=0}^{\infty} u_i x^i \in K[[x]]$ is called the generating function for this sequence. Then $C(x) = g(x)k(x)$ is a polynomial of degree at most $r - 1$.

By the characteristic polynomial of the recurrence we mean

$$f(x) = x^r - a_1 x^{r-1} - \cdots - a_r \in K[x].$$

We can assume that $a_r \neq 0$, since otherwise this recurrence is of smaller order. Write

$$f(x) = (x - \alpha_1)^{e_1} \cdots (x - \alpha_s)^{e_s}, \ e_1 + \cdots + e_s = r$$

where $\alpha_i \in \overline{K}$. The rational function $g(x) = C(x)/k(x)$ may be written in terms of partial fractions $g(x) = \sum_{k=1}^{s} \sum_{i=1}^{e_k} \beta_{ki}/(1 - \alpha_k x)^i$. Let F denote the ring of integers if $ck(K) = 0$ and let F be the prime subfield of K if $ch(K) > 0$. Every $(1 - \alpha_k x)^i$ is invertible in $F(\alpha_k)[[x]]$ with inverse $\sum_{m=0}^{\infty} \binom{m+i-1}{i-1} \alpha_k^m x^m$. Then $P_k(m) = \sum_{i=1}^{e_k} \beta_{ki} \binom{m+i-1}{i-1}$ is on F equal to a polynomial in m.

Theorem 1.16 *Let $u_m \in K$, $m = 0, 1, \ldots$, be a recurrent sequence of order r. Then for all $m \geq 0$ we have $u_m = \sum_{k=1}^{s} P_k(m)\alpha_k^m$, where every $P_k(x) \in F(\alpha_k)[x]$ is a polynomial of degree at most $e_k - 1$.*

Conversely, assume that the elements of a sequence $u_m \in K$, $m \geq 0$, satisfy $u_m = \sum_{k=1}^{s} P_k(m)\alpha_k^m$ for some polynomials $P_k(x) \in \overline{K}[x]$ and some $\alpha_k \in \overline{K}$. One can proceed as in [33], but in the reverse order. First, since every polynomial $h(x) \in F[x]$ satisfies $h(m) = \sum_i \gamma_i \binom{m+i-1}{i-1}$ for some fixed $\gamma_i \in \overline{K}$ we see that the generating function for this sequence can be written in the form

$$g(x) = \sum_{m=0}^{\infty} \sum_{k=1}^{s} \sum_{i=1}^{e_k} \beta_{ki} \binom{m+i-1}{i-1} \alpha_k^m x^m$$

for some scalars $\beta_{ki} \in \overline{K}$ and some natural numbers e_k. This implies that $g(x) = \sum_{k=1}^{s} \sum_{i=1}^{e_k} (\beta_{ki}/(1 - \alpha_k x)^i)$. Therefore $g(x)k(x) = C(x)$ for a polynomial $C(x)$, where $k(x) = \prod_k (1 - \alpha_k x)^{e_k}$. Finally, it follows that the sequence u_m satisfies a linear recurrence of order $r = \sum_{k=1}^{s} e_k$ with characteristic polynomial $f(x) = x^r k(1/x)$.

The above theorem applies also to every sequence $u_m = az^m b$, $m = 0, 1, \dots$, where $a, b, z \in M_n(K)$. Namely, the minimal polynomial for the matrix z yields a linear recurrence $z^{m+r} = a_1 z^{m+r-1} + a_2 z^{m+r-2} + \cdots + a_r z^m$, $a_i \in K$, of order $r \leq n$, which leads to

$$u_{m+r} = a_1 u_{m+r-1} + a_2 u_{m+r-2} + \cdots + a_r u_m \text{ for } m = 0, 1, \dots.$$

Therefore, if $u_m = (u_{ij}^{(m)})_{i,j=1,\dots,n}$, then for every $i, j = 1, \dots, n$ we have $u_{ij}^{(m)} = \sum_{k=1}^s P_{ij}^k(m) \alpha_k^m$ for some polynomials $P_{ij}^k(m)$ and some $\alpha_k \in \overline{K}$.

We conclude with another useful observation on sequences of the form $az^m b$ for $a, b, z \in M_n(K)$.

Lemma 1.17 *Let $a, b, z \in M_n(K)$. Assume that $z \in D$ for a maximal subgroup D of $M_n(K)$. If $\operatorname{rank}(az^m b) < j$ for some $j \leq n$ and every $m = 1, 2, \dots$, then $\operatorname{rank}(aeb) < j$ for the idempotent $e \in D$.*

Proof. From Lemma 1.6 we know that $\Lambda^j(az^m b)) = 0$ for $m \geq 1$. Moreover, $\Lambda^j(e)$ is the identity of the subgroup $\Lambda^j(D)$ of $M_{\binom{n}{j}}(K)$ and $\Lambda^j(z) \in \Lambda^j(D)$. Let r be the rank of $\Lambda^j(e)$. Then $\lambda_0 \Lambda^j(e) + \lambda_1 \Lambda^j(z) + \cdots + \lambda_r (\Lambda^j(z))^r = 0$ for some $\lambda_i \in K$ such that $\lambda_0 \neq 0$. Multiplying by $\Lambda^j(a)$ on the left and by $\Lambda^j(b)$ on the right we come to $\Lambda^j(aeb) = \Lambda^j(a)\Lambda^j(e)\Lambda^j(b) = 0$. This implies that $\operatorname{rank}(aeb) < j$, as desired. $\square$

Chapter 2

Full linear monoid

In this chapter we present the basic information on the structure of the full linear monoid $M_n(K)$ over a field K. This includes a discussion of the ideal structure and the Rees presentations of the principal factors of $M_n(K)$, a useful analogue of the Bruhat decomposition for $GL_n(K)$, and the semigroup interpretation of the exterior power mappings on $M_n(K)$. Certain auxiliary results on nilsubsemigroups and nil subfactors of $M_n(K)$ are given in Section 2. Their connections with the properties of sets of idempotents are presented. An analogue of the classical combinatorial approach to $GL_n(K)$, leading in particular to a Tits system on $M_n(K)$, is presented in Section 3. The case of a finite field $K = \mathbf{F}_q$ is treated in detail in Section 4, with an emphasis on the representation theory of $M_n(\mathbf{F}_q)$.

If not stated otherwise, elements of $M_n(K)$ will be identified with endomorphisms of K^n (written as column vectors), acting on K^n by left multiplication.

2.1 Structure

We start with a description of $\mathcal{J}, \mathcal{R}$, and $\mathcal{L}$-classes of $M_n(K)$.

Lemma 2.1 *Green's relations on $M_n(K)$ are given by*

1. *$a\mathcal{J}b$ if and only if $GL_n(K)bGL_n(K) = GL_n(K)aGL_n(K)$, which is also equivalent to* rank$(a) = $ rank(b),

2. $a\mathcal{R}b$ if and only if $aGL_n(K) = bGL_n(K)$, which is also equivalent to $\mathrm{Im}(a) = \mathrm{Im}(b)$,

3. $a\mathcal{L}b$ if and only if $GL_n(K)a = GL_n(K)b$, which is also equivalent to $\ker(a) = \ker(b)$.

Proof. Let $a, b \in M_n(K)$ be matrices of rank j. Since aK^n, bK^n are isomorphic as K-spaces, there exists an invertible matrix $g \in M_n(K)$ such that $gaK^n = bK^n$. Let $e_1, \ldots, e_n$ be a basis of K^n such that $b(e_1), \ldots, b(e_j)$ is a basis of bK^n. Choose $f_i \in K^n$ such that $ga(f_i) = b(e_i)$ for $i = 1, \ldots, j$. For any basis $f_1, \ldots, f_n$ of K^n we have $gah = b$, where $h \in GL_n(K)$ is such that $h(e_i) = f_i$ for every i. Hence $b \in GL_n(K)aGL_n(K)$. Therefore $GL_n(K)aGL_n(K) = GL_n(K)bGL_n(K)$. Since the latter implies that $a\mathcal{J}b$, which in turn implies that the ranks of a and b are equal, assertion 1) follows.

Assume that $aK^n = bK^n$. A reasoning as above, with $g = 1$, shows that $ah = b$ for some $h \in GL_n(K)$. Consequently $aGL_n(K) = bGL_n(K)$. On the other hand, if $aM_n(K) = bM_n(K)$, then $\mathrm{Im}(a) = aM_n(K)K^n = bM_n(K)K^n = \mathrm{Im}(b)$, which proves 2).

Assume that $\ker(a) = \ker(b)$. Choose a basis $u_1, \ldots, u_n$ of K^n such that $u_{j+1}, \ldots, u_n$ is a basis of $\ker(a)$. Then $b(u_1), \ldots, b(u_j)$ are linearly independent, so there exists $g \in GL_n(K)$ such that $g(b(u_i)) = a(u_i)$ for $i = 1, \ldots, j$. Therefore $gb = a$. It follows that $GL_n(K)a = GL_n(K)b$. On the other hand, if $M_n(K)a = M_n(K)b$, then $b(\ker(a)) = 0$. Since $\mathrm{rank}(a) = \mathrm{rank}(b)$ and $\ker(a) \subseteq \ker(b)$, we must have $\ker(a) = \ker(b)$. This implies that 3) holds. $\square$

It is now clear that every $\mathcal{R}$-class R of $M_n(K)$ contains an idempotent, namely any projection e onto the common image of the endomorphisms from R. Moreover, conjugating by an element $g \in GL_n(K)$, we can assume that e is of the form $e = \begin{pmatrix} I & 0 \\ 0 & 0 \end{pmatrix}$. Thus, R consists of all matrices of rank equal to the rank of e which are of the form $\begin{pmatrix} a & b \\ 0 & 0 \end{pmatrix}$.

Similarly, the transpose R^t of R is the $\mathcal{L}$-class of e.

Recall that a matrix $a \in M_n(K)$ of rank $j \leq n$ is said to be in reduced row elementary form if the following conditions are satisfied:

1. the leading coefficient of every nonzero row of a is 1,

2. for every $i \leq j$, the $i + 1$-th row of a has more leading zero coefficients than the i-th row,

3. the leading coefficient of a nonzero row is the only nonzero coefficient in its column.

Let Y_j be the set of all matrices of rank j which are in the reduced row elementary form. Put $X_j = Y_j^t$, the transpose of Y_j. Assume that $x, x' \in X_j$. Suppose that $\mathrm{Im}(x) = \mathrm{Im}(x')$. By induction on j we show that $x = x'$. Let $v_1, \ldots, v_j$ and $w_1, \ldots, w_j$ be the subsequent nonzero columns of x, x' respectively. If $j = 1$, then the assertion is clear because $\mathrm{Lin}_K(v_1) = \mathrm{Lin}_K(w_1)$. Assume that $j > 1$. It is easy to see that $\mathrm{Lin}_K(v_1, \ldots, v_j) = \mathrm{Lin}_K(w_1, \ldots, w_j)$ implies that $v_k \in \mathrm{Lin}_K(w_2, \ldots, w_j)$ for $k = 2, \ldots, j$. This and a symmetric argument show that $\mathrm{Lin}_K(v_2, \ldots, v_j) = \mathrm{Lin}_K(w_2, \ldots, w_j)$. Deleting the first row and the first column of the considered matrices we see that the induction hypothesis implies that $v_k = w_k$ for $k = 2, \ldots, j$. Moreover

$$v_1 = \sum_{i=1}^{j} \alpha_i w_i \text{ for some } \alpha_i \in K.$$

The rows of x, x' containing the leading 1's of the columns v_i, w_i have all other entries equal to 0. Comparing the coordinates of the vectors in the displayed equality, corresponding to these rows, we come to $\alpha_2 = 0, \ldots, \alpha_j = 0$. Consequently $\alpha_1 = 1$ and $v_1 = w_1$. Therefore $x = x'$, as desired.

It is well known that every matrix a of rank j is column equivalent to a matrix $x \in X_j$ (that is, a can be brought to x by a sequence of elementary column operations). Moreover, if two matrices a, b are column equivalent, then $a = bg$ for some $g \in GL_n(K)$, so $\mathrm{Im}(a) = \mathrm{Im}(b)$. It follows that X_j is a set of representatives of the set of $\mathcal{R}$-classes of $M_n(K)$ consisting of matrices of rank j. Similarly, Y_j can be interpreted as the set of $\mathcal{L}$-classes.

Lemma 2.2 *Let* $0 < j < n$. *If* $e \in Y_j$ *is the diagonal idempotent, then let* G_j *be the group of units of the monoid* $eM_n(K)e$. *Then* $G_j \simeq GL_j(K)$, *each* xgy, *where* $x \in X_j, y \in Y_j, g \in G_j$, *has rank* j *and every matrix of rank* j *has a unique presentation in this form.*

Proof. Assume that $a \in M_n(K)$ has rank j. Let $b \in M_n(K)$ be an endomorphism of rank j such that $\ker(b) = \ker(a)$ and $\mathrm{Im}(b) = eK^n$. Then $b = eb$ and there exists $y \in Y_j$ such that $y = gb$ for some $g \in G_j$ (apply Gauss elimination). Clearly, $\ker(y) = \ker(a)$. Therefore, there exists $c \in M_n(K)$ such that $cy = a$. Then $a = cgb = (ce)gb$ (elimination on columns). Choose $h \in G_j$ such that $ceh \in X_j$. Then $a = (ceh)(h'g)b$, where h' denotes the inverse of h in G_j. Therefore a is of the desired form.

For every xgy there exist $u, v \in M_n(K)$ such that $ux = e = yv$ (because $x\mathcal{L}e$ and $y\mathcal{R}e$). Hence $\mathrm{rank}(xgy) \geq \mathrm{rank}(uxgyv) = \mathrm{rank}(g) = j$, so that $\mathrm{rank}(xgy) = j$. Assume now that $xgy = x'g'y'$ for some $x' \in X_j, y' \in Y_j, g' \in G_j$. Since $\mathrm{rank}(xgy) = \mathrm{rank}(x) = j$, we must have $\mathrm{Im}(xgy) = \mathrm{Im}(x)$. Hence $\mathrm{Im}(x) = \mathrm{Im}(x')$, which implies that $x = x'$. A dual argument shows that $y = y'$. Now $x(g - g')y = 0$ implies that $g - g' = e(g - g')e = ux(g - g')yv = 0$, so that $g = g'$ and the result follows. $\square$

Let $M_j = \{a \in M_n(K) | \ \mathrm{rank}(a) \leq j\}$ for $j = 0, 1, \ldots, n$. We are now able to describe the semigroup structure of $M_n(K)$.

Theorem 2.3 *The sets $0 = M_0 \subset M_1 \subset \cdots \subset M_n = M_n(K)$ are the only ideals of the monoid $M_n(K)$. Each Rees factor M_j/M_{j-1} is isomorphic to the completely 0-simple semigroup $\mathcal{M}(GL_j(K), X_j, Y_j, Q_j)$, where the matrix $Q_j = (q_{yx})$ is defined for $x \in X_j, y \in Y_j$ by $q_{yx} = yx$ if yx is of rank j and θ otherwise.*

Proof. From Lemma 2.1 we know that M_j are the only ideals of $M_n(K)$. Lemma 2.2 implies that $(g, x, y) \mapsto xgy$ determines an isomorphism of the semigroup of matrix type $\mathcal{M}(GL_j(K), X_j, Y_j, Q_j)$ onto M_j/M_{j-1} because $(xgy) \cdot (x'g'y') = x(gyx'g')y$ in M_j/M_{j-1} if $\mathrm{rank}(yx') = j$ and it is zero otherwise. Finally, by Lemma 2.1, each nonzero $\mathcal{R}$-class ($\mathcal{L}$-class, respectively) of the latter semigroup contains an idempotent - a projection on the subspace of K^n that is the common image of all transformations in this $\mathcal{R}$-class (respectively, a projection whose kernel is the kernel of all transformations in this $\mathcal{L}$-class). Therefore, this is a completely 0-simple semigroup. $\square$

The ideal $M_1 \simeq M_1/M_0$ of $M_n(K)$ admits also another description. Let X be a set of vectors in K^n representing all different 1-dimensional subspaces. Let Y be a set of representatives of 1-dimensional subspaces of the space dual to K^n. It is asy to check that $M_1 \simeq \mathcal{M}(K^*, X, Y, Q)$, where the yx-entry of Q is the 'scalar product' $\langle y, x \rangle$, [29].

Every principal factor M_j/M_{j-1} will be approached via its egg-box pattern arising from Theorem 2.3. The following observation will be often used for M equal to one of these factors.

Proposition 2.4 *Let S be a completely 0-simple subsemigroup of a completely 0-simple semigroup M. If $\mathcal{S}, \mathcal{S}_S$ is one of the Green relations $\mathcal{R}, \mathcal{L}, \mathcal{H}$, on M, and on S, respectively, then we have $\mathcal{S}_{|S} = \mathcal{S}_S$.*

Proof. Assume $a\mathcal{R}_S b$ for some $a, b \in S$. Clearly, $a\mathcal{R}b$. On the other hand, if $a\mathcal{R}b$, then choose $e = e^2 \in S$ such that $aS = eS$. Then $a\mathcal{R}e$, so that $b\mathcal{R}e$. Thus $b = eb \in eS$ and consequently $b \in aS$. Similarly $a \in bS$, which implies that $a\mathcal{S}_S b$. A symmetric argument works for the $\mathcal{L}$-classes. $\square$

The Rees presentation of M_j/M_{j-1} in Theorem 2.3 is given in terms of subspaces of K^n. Sometimes, the following group theoretic description is more useful, especially because it reflects the representation theory of the monoid $M_n(K)$ and allows to develop combinatorics of $M_n(K)$ in the case of a finite base field K. For an idempotent e of rank j let $P_j = \{a \in GL_n(K) \,|\, ae = eae\}$, $P_j^- = \{a \in GL_n(K) \,|\, ea = eae\}$, let U_j, U_j^- be the unipotent radicals of P_j, P_j^-, respectively, and $H_j = \{a \in GL_n(K) \,|\, (1 - e)a = a(1 - e) = 1 - e\}$, $H_j^* = \{a \in GL_n(K) \,|\, ea = ae = e\}$, $L_j = \{a \in GL_n(K) \,|\, ea = ae\}$. If $e = \begin{pmatrix} I & 0 \\ 0 & 0 \end{pmatrix}$, then these subgroups of $GL_n(K)$ are of the form

$$P_j = \begin{pmatrix} * & * \\ 0 & * \end{pmatrix}, \; P_j^- = \begin{pmatrix} * & 0 \\ * & * \end{pmatrix}, \; U_j = \begin{pmatrix} I & * \\ 0 & I \end{pmatrix}, \; U_j^- = \begin{pmatrix} I & 0 \\ * & I \end{pmatrix}$$

$$H_j = \begin{pmatrix} * & 0 \\ 0 & I \end{pmatrix}, \; H_j^* = \begin{pmatrix} I & 0 \\ 0 & * \end{pmatrix}, \; L_j = \begin{pmatrix} * & 0 \\ 0 & * \end{pmatrix}.$$

Note that P_j is a semidirect product of L_j and U_j.

It is easy to see that $P_j^- P_j$ coincides with the set of nonsingular matrices a such that $\text{rank}(eae) = j$. (If $a \in GL_n(K)$ satisfies the latter condition, row elementary operations allow us to find a matrix $g \in P_j^-$

such that $ga \in P_j$, so that $a \in P_j^- P_j$. The converse is clear.) Moreover $P_j^- P_j = U_j^- L_j U_j$, so that every matrix $z \in P_j^- P_j$ can be written in the form $z = u_- lu$ with $u \in U_j, u_- \in U_j^-, l \in L_j$. This presentation is unique. In fact, if $vlw = l'$ for some $v \in U_j^-, w \in U_j, l \in L_j$, then $vl = l'w^{-1} = w'l'$ for some $w' \in U_j$. It follows easily that $l = l'$, whence $v = w' = 1$ and consequently $w = 1$.

Let $GL_n(K)/P_j, GL_n(K)/P_j^-$ denote the sets of left, respectively right, coset representatives of the respective subgroups of $GL_n(K)$. We are ready to give a description of the factors M_j/M_{j-1} entirely in terms of the structure of the group $GL_n(K)$.

Proposition 2.5 *Let $j < n$. Then M_j/M_{j-1} is isomorphic to the semigroup $\mathcal{M}(H_j, GL_n(K)/P_j, GL_n(K)/P_j^-, N_j)$, where the matrix $N_j = (n_{yx})$ is defined for $x \in GL_n(K)/P_j, y \in GL_n(K)/P_j^-$ by: $n_{yx} = h$ if $yx \in U_j^- lU_j$ for some $h \in H_j, l \in L_j$ such that $h^{-1}l \in H_j^*$, and $n_{yx} = \theta$ if $yx \notin P_j^- P_j$.*

Proof. $GL_n(K)$ acts transitively on the set of $\mathcal{R}$-classes of $M_n(K)$ contained in $M_j \setminus M_{j-1}$ by left multiplication. The stabilizer of this action is the group P_j. Therefore, the set of nonzero $\mathcal{R}$-classes of M_j/M_{j-1} can be identified with a set of left coset representatives of P_j in $GL_n(K)$. A similar argument, applied to the set of $\mathcal{L}$-classes under right multiplication by $GL_n(K)$, allows us to identify it with $GL_n(K)/P_j^-$.

We have shown that every $\mathcal{H}$-class contained in $M_j \setminus M_{j-1}$ is of the form $xP_j eP_j^- y$ for some $x \in GL_n(K)/P_j, y \in GL_n(K)/P_j^-$. Hence, it is equal to $xeH_j ey$. Therefore $M_j \setminus M_{j-1} = GL_n(K)eGL_n(K) = \bigcup_{x,y} xeH_j ey$, where the summation runs over all possible coset representatives x, y. It follows that

$$\phi : \mathcal{M}(H_j, GL_n(K)/P_j, GL_n(K)/P_j^-, N_j) \longrightarrow M_j \setminus M_{j-1}$$

given by $(g, x, y) \mapsto xegey$ is a bijection. We know that $(xegey)(x'eg'ey')$ is of rank j if and only if $\text{rank}(eyx'e) = j$, which is equivalent to $yx' \in P_j^- P_j$. Moreover, in this case $eyx'e = en_{yx'}e$ and so

$$(xegey)(x'eg'ey') = x(egn_{yx'}g'e)y'$$

$$= \phi((gn_{yx'}g', x, y')) = \phi((g, x, y))\phi((g', x', y')).$$

This means that our map is an isomorphism. $\square$

Define $R = \{a = (a_{ij}) \in M_n(K) |\ a_{ij} \in \{0,1\}$ and a has at most one nonzero entry in each row and in each column $\}$. It is clear that R is the semigroup of all one-to-one partial transformations of the set $\{1, \dots, n\}$. R is called the symmetric inverse monoid. It is the semigroup analogue of the symmetric group. The structure of R is particularly nice.

Lemma 2.6 *The sets $R_0 \subset R_1 \subset \cdots \subset R_n = R$, $R_j = R \cap M_j$, are the only ideals of R. The Rees factors R_j/R_{j-1} are completely 0-simple semigroups with presentations $\mathcal{M}(W_j, \binom{n}{j}, \binom{n}{j}, I_j)$, where W_j is the group of $j \times j$ permutation matrices and I_j is the identity matrix. Moreover, Green's relations on R are the restrictions of the corresponding relations on $M_n(K)$ and $a\mathcal{R}b$ if and only if $aW_n = bW_n$, $a\mathcal{L}b$ if and only if $W_na = W_nb$, $a\mathcal{J}b$ if and only if $W_naW_n = W_nbW_n$ for $a, b \in R$.*

Proof. Let $a \in R_j \setminus R_{j-1}$. Permuting rows and columns of a we can bring this matrix to the form $e = \begin{pmatrix} I & 0 \\ 0 & 0 \end{pmatrix}$, where I is the $j \times j$ identity matrix. Therefore $a = (ue)(ev)$ for some $u, v \in W_n$. Adjusting the order of the nonzero columns of ue and of the nonzero rows of ev we come to $a = xgy$ for some $x \in X_j \cap R, y \in Y_j \cap R$ and $g \in G_j \cap R$. It follows that $R_j \setminus R_{j-1} = (X_j \cap R)(G_j \cap R)(Y_j \cap R)$. The above shows also that R_j/R_{j-1} has no nonzero ideals, so it is completely 0-simple. It is clear that $G_j \cap R \simeq W_j$ and $|X_j \cap R| = |Y_j \cap R| = \binom{n}{j}$ (the number of choices for the columns in which the nonzero entries of matrices in $Y_j \cap R$ are located). It is easy to see that either $yx = e$ or $\text{rank}(yx) < j$ for $x \in X_j \cap R, y \in Y_j \cap R$. The proof of Theorem 2.3 implies that R_j/R_{j-1} has the desired presentation. By Proposition 2.4 Green's relations on R are induced from $M_n(K)$. They come from the action of W_n because a permutation of rows (of columns, respectively) brings any $a \in R$ to a uniquely determined diagonal idempotent. $\square$

Let $B \subseteq GL_n(K)$ be the group of upper triangular matrices. B is a Borel subgroup of $GL_n(K)$. The corresponding Weyl group is $W = W_n$ - the group of permutation matrices. The following analogue of the Bruhat decomposition for $GL_n(K)$ was established in [115]. In this context R is called the Renner monoid of $M_n(K)$.

Theorem 2.7 $M_n(K) = \bigcup_{\sigma \in R} B\sigma B$ *and if $B\sigma B = B\sigma'B$, then $\sigma = \sigma'$. If $a \in M_n(K)$, then we have $GL_n(K)aGL_n(K) = GL_n(K)eGL_n(K) = \bigcup_{\sigma \in WeW} B\sigma B$, where $e = e^2 \in R$ is such that $\text{rank}(a) = \text{rank}(e)$.*

Proof. For $i, j \in \{1, \dots, n\}$ and $t \in K$ let e_{ij} be the corresponding matrix unit and $x_{ij}(t) = I + te_{ij}$. Let $a = (a_{ij}) \in M_n(K)$. The operation $a \to x_{ij}(t)a$ adds t times row j to row i, while $a \to ax_{ij}(t)$ acts similarly on columns of a. We will use $x_{ij}(t)$ for $i < j$ only because in this case $x_{ij}(t) \in B$. Thus, we allow addition of rows from below to above and addition of columns from left to right. If all entries of the first column of a are zero then we go to the second column. If the first column has a nonzero entry, then let j_1 be the largest integer such that $a_{j_1 1} \neq 0$. Using the $(j_1 1)$ entry of a we eliminate the remaining nonzero entries in row j_1 and in column 1. Thus, we find $u, v \in B$ such that $a' = uav$ has these properties. Multiplying by a diagonal matrix we can assume that $a_{j_1 1} = 1$. If all entries of the second column are zero, move to the third column. Otherwise, let j_2 be largest with $a'_{j_2 2} \neq 0$. Then $j_2 \neq j_1$. Using the $(j_2 2)$ entry of a' we eliminate the nonzero entries in rows j_1, j_2 and columns $1, 2$, except perhaps for the entries $(j_1 1), (j_2 2)$. Again, $(j_2 2)$ can be chosen 1 if it is nonzero. Proceeding in this way, we come to an element of R. This proves the first assertion.

Suppose now that $B\sigma'B = B\sigma B$ for some σ, σ'. Then $\sigma' = b\sigma c$ for some $b, c \in B$. It is clear that b can be brought to the identity matrix by a sequence of elementary row operations of the above type ($i < j$.) The same holds for c with respect to the operations on columns. Hence σ' is obtained from σ by a sequence of operations in which addition of rows is done from below to above and addition of columns is done from left to right. If the first column of σ consists of zeros, the same is true for σ'. If the first column of σ contains 1 in position $(j_1 1)$, then σ' has a nonzero entry in this position. Hence, the first colums of σ and σ' are equal. A similar argument works for the remaining columns. Therefore $\sigma = \sigma'$.

Since WeW contains all matrices in R of rank equal to the rank of a by Lemma 2.6, the assertion on $GL_n(K)aGL_n(K)$ follows. $\square$

Recall from Section 1.1 that the j-th exterior power Λ^j can be viewed as a semigroup homomorphism $M_n(K) \longrightarrow M_{\binom{n}{j}}(K)$. The following observation is of interest when Lemma 1.6 is applied. Here, if $g \in G_j$, then we write $\det(g)$ for the determinant of the $j \times j$ invertible submatrix of g.

Lemma 2.8 *The homomorphism* $\Lambda^j : M_n(K) \longrightarrow M_{\binom{n}{j}}(K)$ *satisfies* *the following conditions*

1. *if $a, b \in M_n(K) \setminus M_{j-1}$, then we have $a\mathcal{S}b$ in $M_n(K)$ if and only if $\Lambda^j(a)\mathcal{S}\Lambda^j(b)$ in $M_{\binom{n}{j}}(K)$ for each of Green's relations $\mathcal{S} = \mathcal{R}, \mathcal{L}, \mathcal{J}$,*

2. *if M_j/M_{j-1} is identified with $\mathcal{M}(G_j, X_j, Y_j, Q_j)$, then $\Lambda^j(M_j)$ can be identified with $\mathcal{M}(K^*, X'_j, Y'_j, Q'_j)$, where $X'_j = \Lambda^j(X_j), Y'_j = \Lambda^j(Y_j)$, $q'_{\Lambda^j(y)\Lambda^j(x)} = \Lambda^j(q_{yx}) = \det(q_{yx})$ and Λ^j is given by the formula $\Lambda^j((g, x, y)) = (\det(g), \Lambda^j(x), \Lambda^j(y))$.*

Proof. If $a\mathcal{L}b$ in $M_n(K)$ then clearly $\Lambda^j(a)\mathcal{L}\Lambda^j(b)$ in $M_{\binom{n}{j}}(K)$. On the other hand, if the latter holds, then $\operatorname{rank}(a) = \operatorname{rank}(b)$ by Lemma 1.6. Suppose that $M_n(K)a \neq M_n(K)b$. Then $\ker(a) \neq \ker(b)$. Hence there exists $s \in M_j$ such that $\operatorname{rank}(as) = j$ and $\operatorname{rank}(bs) < j$. Thus $\Lambda^j(as) \neq 0$, while $\Lambda^j(bs) = 0$. Lemma 2.1 implies that $\Lambda^j(a)$ and $\Lambda^j(b)$ are not in the same $\mathcal{L}$-class of $M_{\binom{n}{j}}(K)$. A dual argument works for the relation $\mathcal{R}$, so that 1) follows.

It is clear that $\Lambda^j(M_j)$ is a completely 0-simple semigroup over K^* with the sets of rows and columns indexed by X'_j, Y'_j, respectively. If $xgy \in M_j \setminus M_{j-1}$, where $x \in X_j, y \in Y_j, g \in G_j$, then from Lemma 1.6 we know that $\Lambda^j(xgy) = \Lambda^j(x)\det(g)\Lambda^j(y)$. Therefore 2) follows. $\square$

2.2 Idempotents and nil semigroups

We start by introducing the notion of a triangular set of idempotents. This notion arises from a technical condition, which is used in Chapter 3 in the proof of the structural theorem for arbitrary subsemigroups of $M_n(K)$ (Theorem 3.5). Apparently, it is also connected with nilpotency of nil semigroups. Recall that a semigroup with zero θ is a nil semigroup if for every $s \in S$ there exists $k \geq 1$ with $s^k = \theta$. If $S^k = \{\theta\}$ for some k, then S is called power nilpotent, or just nilpotent. (The latter notion will have a more general meaning, for historical reasons, in Chapter 5 only.) The key features of triangular sets of idempotents show up already in the context of subsemigroups of completely 0-simple semigroups. The defining condition was first considered in [127]. We follow [119], where it was introduced in full generality and used to study the more general case of skew linear semigroups, see Chapter 9.

First, we need some preparatory results.

Lemma 2.9 *Let* $\Gamma = (V, E)$ *be an oriented graph, where* V *is the set of vertices and* $E \subseteq V \times V$ *is the set of edges. Then the following conditions are equivalent:*

1. *for any* $n \geq 1$ *and any sequence* $v_1, \dots, v_n$ *of vertices of* Γ *there exists* $i \in \{1, \dots, n\}$ *such that* $(v_i, v_{i+1}) \in E$, *with indices taken modulo* n,

2. *there exists a linear order* $\preceq$ *on* V *such that* $(v, w) \in E$ *whenever* $v \preceq w$.

Proof. Assume that Γ satisfies condition 1). By transfinite induction, we shall define a sequence of subgraphs Γ_α of Γ with the same set of vertices and the set of edges $E_\alpha \subseteq E$ such that:
(i) each graph $\Gamma_\alpha = (V, E_\alpha)$ satisfies condition 1),
(ii) $E_\beta \subseteq E_\alpha$ for $\alpha \leq \beta$,
(iii) $(v, v) \in E_\alpha$ for every $v \in V$,
(iv) the last graph Γ_{α_0} does not contain double edges (there are no $v, w \in V$ with $(v, w) \in E_{\alpha_0}$ and $(w, v) \in E_{\alpha_0}$).
Put $E_1 = E$. Then for any $v \in V$ we have $(v, v) \in E_1$ (because condition i) may be applied to the sequence v).

Assume that E_β has been defined for $\beta < \alpha$, where $\alpha = \alpha' + 1$. Let v, w be distinct vertices of V such that $(v, w), (w, v) \in E_{\alpha'}$. Suppose that none of the graphs $(V, E_{\alpha'} \setminus \{(v, w)\}), (V, E_{\alpha'} \setminus \{(w, v)\})$ satisfies i). There exist $v_1, \dots, v_n \in V$ and $w_1, \dots, w_m \in V$ such that $(v_i, v_{i+1}) \notin E_{\alpha'} \setminus \{(v, w)\}$ and $(w_j, w_{j+1}) \notin E_{\alpha'} \setminus \{(w, v)\}$ for $i = 1, \dots, n$ and $j = 1, \dots, m$. Assume that there are two equal vertices in the sequence $v_1, \dots, v_n$. Thus there exist $i, j \in \{1, \dots, n\}, i < j$, such that $v_i = v_j$ and $v_i, v_{i+1}, \dots, v_{j-1}$ are all different ($i \neq j - 1$ because otherwise $v_i = v_{i+1}$, so (iii) implies that $(v_i, v_{i+1}) \in E_{\alpha'} \setminus \{(v, W)\}$, contradicting the choice of $v_1, \dots, v_n$). Then the sequence $v_1, \dots, v_n$ may be replaced by $v_i, \dots, v_{j-1}$. Therefore, we may assume that $v_1, \dots, v_n$ are all different, and similarly that $w_1, \dots, w_m$ are different. Since $(V, E_{\alpha'})$ satisfies i), there exists $i_0 \in \{1, \dots, n\}$ such that $(v_{i_0}, v_{i_0+1}) \in E_{\alpha'}$. But $(v_{i_0}, v_{i_0+1}) \notin E_{\alpha'} \setminus \{(v, w)\}$, so that $(v_{i_0}, v_{i_0+1}) = (v, w)$. Similarly, there exists $j_0 \in$

$\{1, \ldots, m\}$ such that $(w_{j_0}, w_{j_0+1}) = (w, v)$. Consider the sequence:

$$(*) \qquad v_1, v_2, \ldots, v_{i_0-1}, v_{i_0} = v$$
$$= w_{j_0+1}, \ldots, w_m, w_1, w_2, \ldots, w_{j_0-1}, w_{j_0}$$
$$= w = v_{i_0+1}, \ldots, v_n$$

Suppose that $(v_i, v_{i+1}) \in E_{\alpha'}$ for some $i \in \{i_0+1, \ldots, n, 1, \ldots, i_0-1\}$. Since $(v_i, v_{i+1}) \notin E_{\alpha'} \setminus \{(v, w)\}$, we get $(v_i, v_{i+1}) = (v, w)$. This implies that $v_i = v_{i_0}, i \neq i_0$, contradicting the fact that $v_1, \ldots, v_n$ are different. Hence $(v_i, v_{i+1}) \notin E_{\alpha'}$ for every $i \in \{i_0+1, \ldots, n, 1, \ldots, i-1\}$. Similarly, we prove that $(w_j, w_{j+1}) \notin E_{\alpha'}$ for every $j \in \{j_0+1, \ldots, m, 1, \ldots, j_0-1\}$. This shows that i) is not satisfied for the sequence $(*)$, a contradiction. Therefore one of the graphs $(V, E_{\alpha'} \setminus \{(v, w)\}), (V, E_{\alpha'} \setminus \{(w, v)\})$, say the former one satisfies i). Define $E_\alpha = E_{\alpha'} \setminus \{(v, w)\}$. It is obvious that (i),(ii),(iii) are satisfied by E_α.

If α is a limit ordinal, define $E_\alpha = \bigcap_{\beta < \alpha} E_\beta$. We will prove that $\Gamma_\alpha = (V, E_\alpha)$ satisfies i). Let $v_1, \ldots, v_n$ be any sequence of elements of V. If there exists $i \in \{1, \ldots, n\}$ such that $(v_i, v_{i+1}) \in E_\alpha$ for all $\beta < \alpha$, then $(v_i, v_{i+1}) \in E_\alpha$. Therefore i) is satisfied for Γ_α. Thus, suppose that for every $i \in \{1, \ldots, n\}$ there exists $\beta_i < \alpha$ such that $(v_i, v_{i+1}) \notin E_{\beta_i}$. Let $\beta = \max\{\beta_1, \ldots, \beta_n\}$. Then, for every $i \in \{1, \ldots, n\}$ we have $(v_i, v_{i+1}) \notin E_\beta$. This contradicts the fact that i) is satisfied for $\Gamma_\beta, \beta < \alpha$, showing that the latter case is impossible. It is now clear that Γ_α satisfies (i),(ii),(iii).

An ordinal number α_0 may be found such that E_{α_0} has no double edges. Define an order $\preceq$ on V by: $v \preceq w$ if and only if $(v, w) \in E_{\alpha_0}$. We will check that $\preceq$ is a linear order.

First, $v \preceq v$ because $(v, v) \in E_{\alpha_0}$.

If $v_1 \preceq v_2$ and $v_2 \preceq v_3$, then consider the sequence v_3, v_2, v_1. Since Γ_{α_0} satisfies i) and $(v_3, v_2), (v_2, v_1) \notin E_{\alpha_0}$, it follows that $(v_1, v_3) \in E_{\alpha_0}$. Hence $v_1 \preceq v_3$.

If $v_1 \preceq v_2$ and $v_2 \preceq v_1$, then $v_1 = v_2$ because Γ_{α_0} has no double edges. For any $v_1, v_2 \in V$ we have either $v_1 \preceq v_2$ or $v_2 \preceq v_1$. This may be obtained by applying i) to the sequence v_1, v_2.

It is clear that $\preceq$ satisfies condition 2) of the lemma.

Now, assume that $\preceq$ is a linear order on V that satisfies 2). Let $v_1, \ldots, v_n$ be a sequence of elements of V. If 1) is not satisfied, then $v_{i+1} \prec v_i$ for $i = 1, \ldots, n$ (with indices modulo n). Therefore $v_1 \prec v_1$, a contradiction. This completes the proof of the lemma. $\square$

A linearly ordered set T of nonzero idempotents of a semigroup S with zero θ is called a triangular set of idempotents if $e \prec f$ implies that $ef = \theta$ for $e, f \in T$.

Recall that, in a semigroup $S = \mathcal{M}(G, X, Y, P)$ of matrix type over a group G, the equality $s_1 \cdots s_n = \theta$ for some $s_i \in S, n > 1$, implies that there exists $j < n$ such that $s_j s_{j+1} = \theta$. This will be frequently used without further reference.

Lemma 2.10 *Let $M = \mathcal{M}(G, X, Y, P)$ be a semigroup of matrix type. Assume that every triangular set of idempotents in M has at most n elements. If S is a nil subsemigroup of M, then $S^{n+2} = \theta$, and $S^n = \theta$ whenever M is a completely 0-simple semigroup.*

Proof. Assume that S is not nilpotent of index $n + 2$, (respectively, of index n if M is completely 0-simple). Then there exist $s_0, \ldots, s_{n+1} \in S$ (respectively $s_1, \ldots, s_n \in S$), such that $s_0 \cdots s_{n+1} \neq \theta$ (respectively $s_1 \cdots s_n \neq \theta$). Thus $s_i s_{i+1} \neq \theta$ for every $i \in \{0, \ldots, n\}$, ($i \in \{1, \ldots, n - 1\}$ respectively). By the multiplication rule in M there exist idempotents $e_i, i = 0, 1, \ldots, n$ ($i = 1, 2, \ldots, n-1$,) such that $s_i e_i = s_i, e_i s_{i+1} = s_{i+1}$. If M is completely 0-simple, then we also have idempotents e_0, e_n such that $s_1 = e_0 s_1$ and $s_n = s_n e_n$. Since S is a nil semigroup, for every i, j with $1 \leq i \leq j \leq n$ we have $(s_i s_{i+1} \cdots s_j)^2 = \theta$. Then one of the elements $s_i s_{i+1}, \ldots, s_{j-1} s_j, s_j s_i$ is equal to θ. But $s_k s_{k+1} \neq \theta$ for $k = 1, 2, \ldots, n - 1$. Therefore $s_j s_i = \theta$ for $i \leq j$. Let i, j be such that $0 \leq i < j \leq n$. We have $s_j e_j e_i s_{i+1} = s_j s_{i+1} = \theta$. Since also $s_j e_j \neq \theta$ and $e_i s_{i+1} \neq \theta$, it follows that $e_j e_i = \theta$. We have proved that $e_0, \ldots, e_n$ form a triangular set of idempotents in M. This is impossible by the hypothesis on M. Therefore $S^{n+2} = \theta$ ($S^n = \theta$ respectively). $\square$

Remark A converse of this lemma is also true. Assume that $T \subseteq M = \mathcal{M}(G, X, Y, P)$ is a triangular set of idempotents such that $|T| = n < \infty$. Let $S = \{s \in M \mid se = s, fs = s \text{ for some } e, f \in T, e \preceq f\}$. We will first prove that S is a subsemigroup of M. Let $s, t \in S$. By definition there exist $e_1, e_2, f_1, f_2 \in T$ such that $e_1 \preceq f_1, e_2 \preceq f_2$ and $s = se_1, s = f_1 s, t = te_2, t = f_2 t$. Consider the following two cases:
i) $e_1 \prec f_2$
Then $st = (se_1)(f_2 t) = s\theta t = \theta$. Therefore $st \in S$.
ii) $f_2 \preceq e_1$

Assume that $st \neq \theta$. Then $st = ste_2, st = f_1 st$ and $e_2 \preceq f_2 \preceq e_1 \preceq f_1$. This means that $st \in S$.

Next, we will prove the following useful fact:

$(*)$ If $st = gstg \neq \theta$ for some $g \in T$, $s, t \in S$, then $s = gsg$ and $t = gtg$.

Let $s, t \in S$ and let $e_1, e_2, f_1, f_2 \in T$ be defined as above. Since $st \neq \theta$, case ii) holds. We know that $st = ste_2$ and $st = gstg$. It follows easily that $e_2 \mathcal{L} g$ in M. Thus $e_2 = g$ ($e_2 \neq g$ implies $e_2 g = \theta$ or $g e_2 = \theta$, a contradiction). Similarly $f_1 = g$. By case ii) we have $e_2 \preceq f_2 \preceq e_1 \preceq f_1$. Since also $e_2 = g = f_1$, this yields $e_2 = f_2 = e_1 = f_1 = g$. Thus $s = gsg$ and $t = gtg$, as desired.

Let $N = \{s \in M \mid s = se, s = fs$ for some $e, f \in T, e \prec f\} \cup \{\theta\}$. $S \setminus N$ is composed of maximal subgroups of M intersecting T. By $(*)$ N is an ideal of S. Moreover N is a nil semigroup (because $s^2 = (se)(fs) = s(ef)s = s\theta s = \theta$).

We will show that N is not nilpotent of index $< n$. Indeed, there exist $e_1, \ldots, e_n \in T$ with $e_1 \prec e_2 \prec \cdots \prec e_n$. Choose $s_i \in N$ as follows: $s_i = s_i e_i, s_i = e_{i+1} s_i$ for $i = 1, \ldots, n-1$. If $i \geq 2$, then we have $s_i s_{i-1} = (s_i e_i)(e_i s_{i-1}) \neq \theta$ since $s_i e_i \neq \theta$ and $e_i s_{i-1} \neq \theta$. Thus also $s_{n-1} s_{n-2} \cdots s_1 \neq \theta$, which shows that $N^{n-1} \neq \theta$.

We note that $S \setminus N$ a disjoint union of n groups which are $\mathcal{J}$-classes of S. It is clear that S is a π-regular semigroup.

The above shows that if there exists a triangular set of idempotents of cardinality n in M, then there exists a π-regular semigroup $S \subseteq M$ which has n $\mathcal{J}$-classes containing idempotents and a nilideal N which is not nilpotent of index $< n$.

We now investigate the notion of a triangular set of idempotents in $M_n(K)$. The following key observation is based on an idea similar to that used in the proof of Theorem 1.12. In view of Corollary 2.12 it also leads to an improvement of the bound obtained there.

Lemma 2.11 *Let $T \subseteq M_n(K)$ be a set of elements of $M_j \setminus M_{j-1}$ whose images in M_j / M_{j-1} form a triangular set of idempotents. Then $|T| \leq \binom{n}{j}$.*

Proof. First assume that $T \subseteq M_1$ is a triangular set of idempotents of $M_n(K)$. Suppose that $e_1, \ldots, e_{n+1} \in T$ are such that $e_i e_j = 0$ if $i < j$.

Let $0 \neq v_i \in \text{Im}(e_i) \subseteq K^n$. Since $v_1, \ldots, v_{n+1}$ are linearly dependent, there exist k and $\alpha_{k+1}, \ldots, \alpha_{n+1} \in K$ such that

$$v_k = \alpha_{k+1} v_{k+1} + \cdots + \alpha_{n+1} v_{n+1}.$$

Thus $e_k v_k = \sum_{i=k+1}^{n+1} \alpha_i e_k v_i = 0$ (note that $e_k v_i = 0$ for $i > k$). This contradicts the fact that $e_k v_k = v_k \neq 0$, establishing the assertion in the case $j = 1$.

Assume now that $T \subseteq M_j \setminus M_{j-1}$ has cardinality exceeding $t = \binom{n}{j}$ and satisfies the conditions of the lemma. Apply the exterior power map Λ^j. From Lemma 1.6 it follows that $\Lambda^j(T)$ inherits the assumptions on T. Moreover, it has the same cardinality as T and consists of matrices of rank one in $M_t(K)$. This contradicts the first paragraph of the proof, completing the argument in the general case. $\square$

Corollary 2.12 *Assume that* $s_1, \ldots, s_{\binom{n}{j}} \in M_n(K)$ *are matrices of rank* j *such that* $\text{rank}(s_1 \cdots s_{\binom{n}{j}}) = j$. *Then there exist* $i \leq k$ *such that* $s_i \cdots s_k$ *lies in a maximal subgroup of* $M_n(K)$.

Proof. The proof of Lemma 2.10 shows that one can construct a triangular set of idempotents in M_j/M_{j-1} of cardinality $\binom{n}{j} + 1$ if the assertion does not hold. This contradicts Lemma 2.11. $\square$

Corollary 2.13 *Let* $T \subset M_n(K)$ *be a set of idempotents of rank* j. *Assume that* $(exf)^2 \in M_{j-1}$ *for every* $x \in \langle T \rangle^1$ *and every* $e, f \in T$ *with* $e \neq f$. *Then there exists a linear order* $\preceq$ *on* T *such that* $ef \in M_{j-1}$ *whenever* $e \prec f$. *In particular,* $|T| \leq \binom{n}{j}$.

Proof. Let $V = T$, and $E \subseteq V \times V$ be defined by the rule $(e, f) \in E$ if and only if $ef \in M_{j-1}$ or $e = f$. We will prove that $\Gamma = (V, E)$ satisfies condition 1) of Lemma 2.9. Let $e_1, \ldots, e_n \in E$ and $e_1 \neq e_n$. Then, by the hypothesis $(e_1(e_2 e_3 \cdots e_{n-1}) e_n)^2 \in M_{j-1}$. Hence, there exists i such that $(e_i e_{i+1}) \in E$ (with indices taken modulo n). If $e_1 = e_n$, then $(e_n, e_1) \in E$. By Lemma 2.9 there exists a linear order on V satisfying the desired condition. From Lemma 2.11 it follows that $|T| \leq \binom{n}{j}$. $\square$

Now, we turn to nil semigroups arising from $M_n(K)$.

Proposition 2.14 *Let* $S \subseteq M_n(K)$ *be a semigroup. Then*

1. *If S has a zero θ, then $S \subseteq (I - \theta)M_n(K)(I - \theta) + \theta \simeq M_t(K)$, where $t = n - \text{rank}(\theta)$. Moreover, if S is a nil semigroup, then $S^t = \theta$,*

2. *If J is an ideal of S and $S \setminus J$ does not intersect maximal subgroups of $M_n(K)$, then S/J is nilpotent of index not exceeding $m = \prod_{k=1}^{n} \binom{n}{k}$. In particular, this holds if S is π-regular and S/J is a nil semigroup.*

Proof. 1) Since $\theta s = s\theta = \theta$ for every $s \in S$, $s = (I - \theta)s(I - \theta) + \theta$ and the first assertion follows via the map $s \mapsto (I - \theta)s(I - \theta)$. To prove the second, it is then enough to consider the case where $\theta = 0$ and $t = n$. We will use standard facts about irreducible semigroups, presented in Section 4.1. For another proof we refer to [24], 17.19. Conjugating in $M_n(K)$ we can assume that S is in a block triangular form of Proposition 4.4 with irreducible or zero diagonal blocks $T^{(i)}, i = 1, \dots, r$. It is enough to show that there are no irreducible blocks, because in this case S is upper triangular with zero diagonal and $S^n = 0$ follows. If some $T^{(i)}$ is irreducible, then the identity e of the K-linear span of $T^{(i)}$ can be written in the form $\sum_{i=1}^{k} \alpha_i s_i$ for some $s_i \in T^{(i)}$ and $\alpha_i \in K$ (see Lemma 4.1). But the trace of each s_i is zero because S is nil. Therefore $\text{tr}(e) = 0$, a contradiction. Hence 1) follows.

2) Consider the chain $S_0 \subseteq S_1 \subseteq \cdots \subseteq S_n$, where $S_j = S \cap M_j$. If $S_j \neq \emptyset$ and $a_1, \dots, a_{\binom{n}{j}} \in S_j$, then from Corollary 2.12 it follows that $a_1 \cdots a_{\binom{n}{j}} \in J \cup M_{j-1}$. Hence $S_j^{\binom{n}{j}} \subseteq J \cup S_{j-1}$. This easily implies that $S^m \subseteq J$.

Assume that S is π-regular and S/J is nil. Then every intersection of S with a maximal subgroup of $M_n(K)$ is contained in J by Corollary 1.5, so the above applies. $\square$

It is clear that the last assertion of 2) is not true without the assumption that S is π-regular. In fact, $S = \mathbf{Z} \subseteq \mathbf{Q}$ with the ideal $I = \bigcup_{i=1}^{\infty} p_i^i S$, where $p_1, p_2, \dots$ are the subsequent prime numbers, is a counterexample.

We conclude with the following direct consequence of Lemma 2.10 and Lemma 2.11, or of Proposition 2.14 and Lemma 1.6.

Corollary 2.15 *Assume that N is a nil subsemigroup of a principal factor M_j/M_{j-1}, $j \leq n$, of the monoid $M_n(K)$. Then $N^{\binom{n}{j}} = \theta$.*

2.3   Combinatorics on $M_n(K)$.

In this section we present more detailed information on the structure of the full linear monoid $M_n(K)$. We concentrate on the combinatorial aspects of this monoid. This is justified by the role of the analogous results for the full linear group $GL_n(K)$, cf. [13],[18]. As before, $M_j, j = 0, 1, \ldots, n$, will denote the ideals of $M_n(K)$. Also, B, R stand for the Borel subgroup consisting of upper triangular matrices in $GL_n(K)$ and for the Renner monoid, respectively. By $T, U \subseteq B$ we mean the subgroups of diagonal matrices and of unipotent matrices, respectively.

To get a deeper insight into the combinatorics of $M_n(K)$ one needs, as in the case of $GL_n(K)$ (and also of more general classes of groups cf.[13],[18]), a notion of the length function. Our aim is to prove several technical results on the nature of this function, culminating in Proposition 2.27, and then to use it in Theorem 2.30. The latter shows that Theorem 2.7 leads to a nice analogue of the powerful Tits system for $GL_n(K)$. In our presentation we closely follow Solomon, [129], whose work is in part modelled after the classical papers of Chevalley and Iwahori, [14],[41]. Note that another approach to the length function was taken by Renner in [115].

Let $W \subseteq M_n(K)$ be the group of permutation matrices and let $S = \{(1,2), (2,3), \ldots, (n-1,n)\}$ be the set of its 'simple reflections' - a distinguished set of generators of W. By E_{ij} we denote the matrix unit with 1 in position (i,j) and 0 elsewhere. If $r \in \{1, \ldots, n\}$, then every element $a \in R^r = R \cap (M_r \setminus M_{r-1})$ is of the form $a = \sum_{k=1}^{j} E_{i_k j_k}$, where $I(a) = \{i_1, \ldots, i_r\}, J(a) = \{j_1, \ldots, j_r\}$ are subsets of $\{1, \ldots, n\}$ of cardinality r. Write $i_k a = j_k$ and $a j_k = i_k$. Thus

$$\sum_{i \in I(a)} E_{i,ia} = a = \sum_{j \in J(a)} E_{aj,j}. \tag{2.1}$$

Note that $i \mapsto ia, j \mapsto aj$ are bijective maps from $I(a)$ to $J(a)$, and from $J(a)$ to $I(a)$, respectively. If $w \in W$, then $I(w) = \{1, \ldots, n\} = J(w)$ and $wi = iw^{-1}$ for all $i \in \{1, \ldots, n\}$. Let x^* denote the transpose of

$x \in M_n(K)$. Since $E_{ij}^* = E_{ji}$, we have $I(a) = J(a^*)$ and $J(a) = I(a^*)$. Also $(ia)a^* = i$ for $i \in I(a)$ and $a^*(aj) = j$ for $j \in J(a)$. The group $W \times W$ acts on R by

$$(w, v)a = wav^{-1} \text{ for } a \in R \text{ and } w, v \in W. \tag{2.2}$$

Define a graph with vertex set R^r as follows. Two vertices a, b are adjacent if there exists $s \in S$ such that $a = sb$ or there exists $s \in S$ with $a = bs$. The graph is connected because S generates W and R^r is a $W \times W$-orbit of R by Lemma 2.6. For $a, b \in R^r$ define the distance $\delta(a, b)$ from a to b

$$\delta(a, b) = \min\{l(w) + l(w') \mid w, w' \in W \text{ and } a = wbw'\} \tag{2.3}$$

where $l(w)$ is the length of w in the generators from S. We will define $l(a) = \delta(a, z)$ for a suitably chosen $z \in R^r$. Let

$$m = E_{12} + E_{23} + \cdots + E_{n-1,n}. \tag{2.4}$$

Then

$$m_r = m^{n-r} = E_{1,n-r+1} + E_{2,n-r+2} + \cdots + E_{r,n} \tag{2.5}$$

is a matrix of rank r. We define the length of $a \in R^r$ by the rule

$$l(a) = \min\{l(w) + l(w') \mid w, w' \in W \text{ and } a = wm_r w'\}. \tag{2.6}$$

Then m_r is the only element in R^r of length zero. Moreover, $|l(sa) - l(a)| \leq 1$ and $|l(as) - l(a)| \leq 1$ for every $a \in R$ and $s \in S$.

Our first aim is to give a combinatorial description of $l(a)$ for $a \in R$. This will be accomplished in Proposition 2.27. Let

$$\Delta = \{(i, j) \mid 1 \leq i \neq j \leq n\} \tag{2.7}$$

and

$$\Delta^+ = \{(i, j) \in \Delta \mid i < j\}, \ \Delta^- = \{(i, j) \in \Delta \mid i > j\}. \tag{2.8}$$

W acts on Δ by $w(i, j) = (wi, wj)$ for $w \in W$. For $s \in S$ which is the transposition of k and $k + 1$ let $\alpha_s = (k, k + 1) \in \Delta^+$. Then

$$s(\Delta^+ \setminus \{\alpha_s\}) = \Delta^+ \setminus \{\alpha_s\}. \tag{2.9}$$

If $w \in W$, then let

$$\Psi'(w) = \{\alpha \in \Delta^+ \,|\, w^{-1}\alpha \in \Delta^+\}, \ \ \Psi''(w) = \{\alpha \in \Delta^+ \,|\, w^{-1}\alpha \in \Delta^-\} \quad (2.10)$$

Thus

$$\Delta^+ = \Psi'(w) \sqcup \Psi''(w) \tag{2.11}$$

where the symbol $\sqcup$ is used to emphasize that the union is disjoint. Note that $(i,j) \in \Psi''(w)$ if and only if (j,i) is an inversion of the permutation $k \mapsto wk$ of the set $\{1, \ldots, n\}$. Thus $n(w) = |\Psi''(w)|$, where $n(w)$ denotes the number of inversions of w. From (2.9) it follows that the function $w \mapsto \Psi''(w)$ satisfies the 'cocycle condition'

$$\Psi''(sw) = s\Psi''(w) \cup \{\alpha_s\} \text{ if } \alpha_s \in \Psi'(w)$$
$$\Psi''(w) = s\Psi''(sw) \cup \{\alpha_s\} \text{ if } \alpha_s \in \Psi''(w) \tag{2.12}$$

where the unions are disjoint and thus

$$n(sw) = \begin{cases} n(w) + 1 & \text{if } \alpha_s \in \Psi'(w) \\ n(w) - 1 & \text{if } \alpha_s \in \Psi''(w) \end{cases} \tag{2.13}$$

Our aim is to prove several formulas analogous to (2.13) with W replaced by R and to use them to derive a formula for $l(a), a \in R^r$, in terms of Δ.

For a subset $A \subseteq \{1, \ldots, n\}$ define the following partition of Δ

$$\Delta_{00}(A) = \{(i,j) \in \Delta \,|\, i \notin A, j \notin A\} \tag{2.14}$$
$$\Delta_{01}(A) = \{(i,j) \in \Delta \,|\, i \notin A, j \in A\}$$
$$\Delta_{10}(A) = \{(i,j) \in \Delta \,|\, i \in A, j \notin A\}$$
$$\Delta_{11}(A) = \{(i,j) \in \Delta \,|\, i \in A, j \in A\}$$

If $x, y \in \{0,1\}$ and $a \in R$, define subsets $\Psi_{xy}(a)$ and $\Phi_{xy}(a)$ of Δ by

$$\Psi_{xy}(a) = \Delta_{xy}(I(a)), \ \ \Phi_{xy}(a) = \Delta_{xy}(J(a)). \tag{2.15}$$

For any subset Γ of Δ we write $\Gamma^+ = \Gamma \cap \Delta^+$. Next, define

$$\Psi'(a) = \{(i,j) \in \Psi_{11}^+(a) \,|\, (ia, ja) \in \Delta^+\} \tag{2.16}$$
$$\Psi''(a) = \{(i,j) \in \Psi_{11}^+(a) \,|\, (ia, ja) \in \Delta^-\}$$
$$\Phi'(a) = \{(i,j) \in \Phi_{11}^+(a) \,|\, (ai, aj) \in \Delta^+\}$$
$$\Phi''(a) = \{(i,j) \in \Phi_{11}^+(a) \,|\, (ai, aj) \in \Delta^-\}$$

Since $J(a) = I(a^*)$, we have

$$\Phi_{xy}(a) = \Psi_{xy}(a^*), \quad \Phi^+_{xy}(a) = \Psi^+_{xy}(a^*) \tag{2.17}$$

for $x, y \in \{0, 1\}$. Also

$$\Phi'(a) = \Psi'(a^*), \quad \Phi''(a) = \Psi''(a^*). \tag{2.18}$$

There is a duality between the pairs J, ψ and I, Φ. Namely, to each 'Ψ-statement' concerning left multiplication $a \mapsto sa$ there corresponds a dual 'Φ-statement' concerning right multiplication $a \mapsto as$, which may be deduced from it if we replace a by a^* and use the fact that $(sa)^* = a^*s$. For example (2.12) yields

$$\Phi''(ws) = s\Phi''(w) \cup \{\alpha_s\} \text{ if } \alpha_s \in \Phi'(w) \tag{2.19}$$
$$\Phi''(w) = s\Phi''(ws) \cup \{\alpha_s\} \text{ if } \alpha_s \in \Phi''(w)$$

Lemma 2.16 *The map* $(i, j) \mapsto (ja, ia)$ *is bijective from* $\Psi''(a)$ *to* $\Psi''(a^*)$.

Proof. Suppose that $(i, j) \in \Psi''(a)$. Then $i \in I(a), j \in J(a), i < j$ and $ia > ja$. Thus $ja \in J(a), ia \in J(a), ja < ia$ and $(ja)a^* = j > i = (ia)a^*$ so that $(ja, ia) \in \Psi''(a^*)$. Replacing a by a^* we see that, if $(i', j') \in \Psi''(a^*)$, then $(j'a^*, i'a^*) \in \Psi''(a^{**}) = \Psi''(a)$. $\square$

Define a function $n : R \longrightarrow \mathbf{N}$ by

$$n(a) = |\Psi''(a)|.$$

If $a = w \in W$ this agrees with $n(w)$ defined earlier. It follows from Lemma 2.16 that

$$n(a^*) = n(a). \tag{2.20}$$

If $a = m_r$, then $I(a) = \{1, \ldots, r\}$ and $ia = i + n - r$ for $i \in I(a)$, so $\psi''(a)$ is empty and thus $n(a) = 0$. If $w \in W$ and $n(w) = 0$, then $w = 1$. It is not true that if $a \in R^r$ and $n(a) = 0$, then $a = m_r$. To overcome this difficulty we introduce another function $m : R \longrightarrow \mathbf{N}$. For $A \subseteq \{1, \ldots, n\}$ define

$$m_{01}(A) = |\Delta^+_{01}(A)| \text{ and } m_{10}(A) = |\Delta^+_{10}(A)|. \tag{2.21}$$

Lemma 2.17 *If* $A \subseteq \{1, \ldots, n\}$ *has cardinality* r, *then*

$$m_{01}(A) = \sum_{k \in A}(k-1) - \frac{r(r-1)}{2}$$

$$m_{10}(A) = \sum_{k \in A}(n-k) - \frac{r(r-1)}{2}.$$

Proof. Write $A = \{k_1, \ldots, k_r\}$, where $k_1 < \cdots < k_r$. For $1 \leq q \leq r$ let

$$\Delta_{01}^q(A) = \{(i,j) \in \Delta_{01}^+(A)|\, j = k_q\}.$$

Since

$$\begin{aligned}\Delta_{01}^q(A) &= \{(1, k_q), (2, k_q), \ldots, (k_q - 1, k_q)\}\\ &\quad\backslash \{(k_1, k_q), (k_2, k_q), \ldots, (k_{q-1}, k_q)\}\end{aligned}$$

we have $|\Delta_{01}^q(A)| = (k_q - 1) - (q - 1)$. Since $\Delta_{01}(A) = \bigcup_{q=1}^r \Delta_{01}^q(A)$ is a disjoint union, the first formula follows. The second formula is proved similarly. $\square$

If $a \in R$ define

$$m_{01}(a) = m_{01}(I(a)) = |\Psi_{01}^+(a)|, \ m_{10}(a) = m_{10}(J(a)) = |\Phi_{10}^+(a)| \quad (2.22)$$

and put

$$m(a) = m_{01}(a) + m_{10}(a).$$

From the definitions and (2.17) it follows that, if $r \in R^r$, then

$$m(a) = \sum_{i \in I(a)}(i-1) + \sum_{j \in J(a)}(n-j) - r(r-1). \quad (2.23)$$

Since $J(a) = I(a^*)$, it follows from (2.17) and (2.22) that $m_{01}(a) + m_{10}(a^*) = r(n-r)$. Similarly, since $I(a) = J(a^*)$ we have $m_{10}(a) + m_{01}(a^*) = r(n-r)$. Thus, if $r \in R^r$, then

$$m(a) + m(a^*) = 2r(n-r). \quad (2.24)$$

Define $p : R \longrightarrow \mathbf{N}$ by $p(a) = m(a) + n(a)$. We will prove in Proposition 2.27 that $p(a) = l(a)$.

Lemma 2.18 *If* $a \in R^r$, *then* $p(a) \geq 0$ *with equality if and only if* $a = m_r$.

Proof. We have already noticed that both $\Psi_{01}^+(m_r)$ and $\Phi_{10}^+(m_r)$ are empty. So is $\Psi''(m_r)$. Thus $p(m_r) = 0$. Suppose now that $a \in R^r$ is such that $p(a) = 0$. Then $m(a) = 0 = n(a)$. The former equality shows that $|\Psi_{01}^+(a)| = 0 = |\Phi_{10}^+(a)|$. Since $I(a)$ and $J(a)$ have cardinality r, it follows from Lemma 2.17 that $I(a) = \{1, \ldots, r\}$ and $J(a) = \{n-r+1, \ldots, n\}$. Since $|\Psi''(a)| = n(a) = 0$ we have $ia < ja$ for all $1 \leq i < j \leq r$. Since $ia \in J(a) = \{n-r+1, \ldots, n\}$ and the map $i \mapsto ia$ is bijective from $I(a)$ to $J(a)$, we must have $ia = n - r + i$ for $1 \leq i \leq r$. Thus $a = m_r$. $\square$

In view of (2.14),(2.15) and (2.16) each $a \in R$ determines the following partition of Δ^+

$$\Delta^+ = \Psi_{00}^+(a) \sqcup \Psi_{01}^+(a) \sqcup \Psi_{10}^+(a) \sqcup \Psi'(a) \sqcup \Psi''(a). \tag{2.25}$$

This replaces the two part partition (2.11) corresponding to an element $w \in W$. We need analogues of (2.13) for the sets in this partition. These will be proved in Lemma 2.22. If $w \in W$, then $I(wa) = wI(a)$ and $I(aw) = I(a)$. It follows that if $w \in W$ and $x, y \in \{0,1\}$, then

$$\Psi_{xy}(wa) = w\Psi_{xy}(a) \text{ and } \Psi_{xy}(aw) = \Psi_{xy}(a). \tag{2.26}$$

Lemma 2.19 *Suppose that* $x, y \in \{0,1\}, a \in R$ *and* $s \in S$. *Then we have*

1. $s(\Psi_{xy}^+(a) \setminus \{\alpha_s\}) = \Psi_{xy}^+(sa) \setminus \{\alpha_s\}$,

2. $s(\Psi''(a) \setminus \{\alpha_s\}) = \Psi''(sa) \setminus \{\alpha_s\}$.

Proof. Since $\Psi_{xy}^+(a) \setminus \{\alpha_s\} = (\Delta^+ \setminus \{\alpha_s\}) \cap \Psi_{xy}(a)$, the first assertion follows from (2.9) and (2.26). Suppose that $(i, j) \in \Psi''(a) \setminus \{\alpha_s\}$. Then $i \in I(a), j \in I(a), i < j$ and $ia > ja$. Thus $si \in I(sa), sj \in I(sa)$ and $si < sj$ because $(i, j) \neq \alpha_s$. Since $(si)(sa) = (is)(sa) = ia > ja = (js)(sa) = (sj)(sa)$, it follows that $(si, sj) \in \Psi''(sa)$. Thus $s(\Psi''(a) \setminus \{\alpha_s\}) \subseteq \Psi''(sa) \setminus \{\alpha_s\}$. To get the reverse inclusion replace a by sa. $\square$

Lemma 2.20 *Suppose that* $x, y \in \{0,1\}, a \in R$ *and* $s \in S$. *Then*

1. $\alpha_s \in \Psi_{xy}(a)$ *if and only if* $\alpha_s \in \Psi_{yx}(sa)$,

2. $\alpha_s \in \Psi'(a)$ *if and only if* $\alpha_s \in \Psi''(sa)$.

Proof. To prove the first assertion suppose for example that $x = 0$ and $b = 1$. Write $\alpha_s = (k, k+1)$, where $1 \leq k \leq n-1$. It follows from (2.26) that we have the following equivalences: $\alpha_s \in \Psi_{01}(a) \Leftrightarrow k \notin I(a)$ and $k+1 \in I(a) \Leftrightarrow sk \in I(a)$ and $s(k+1) \notin I(a) \Leftrightarrow k \in I(sa)$ and $k+1 \notin I(sa) \Leftrightarrow \alpha_s \in \Psi_{10}(sa)$.

The proof of the second assertion is similar. $\square$

Lemma 2.21 *For any* $a \in R, s \in S$ *we have:*

1. *If* $\alpha_s \in \Psi_{00}(a)$ *then* $sa = a$.

2. *If* $\alpha_s \in \Psi_{01}(a)$ *then*
$$\Psi_{01}^+(a) = s\Psi_{01}^+(sa) \sqcup \{\alpha_s\}$$
$$\Psi_{10}^+(sa) = s\Psi_{10}^+(a) \sqcup \{\alpha_s\}$$
$$\Psi''(sa) = s\Psi''(a)$$

3. *If* $\alpha_s \in \Psi_{10}(a)$ *then*
$$\Psi_{10}^+(a) = s\Psi_{10}^+(sa) \sqcup \{\alpha_s\}$$
$$\Psi_{01}^+(sa) = s\Psi_{01}^+(a) \sqcup \{\alpha_s\}$$
$$\Psi''(sa) = s\Psi''(a)$$

4. *If* $\alpha_s \in \Psi_{11}(a)$ *then*
$$\Psi_{01}^+(sa) = s\Psi_{01}^+(a)$$
$$\Psi_{10}^+(sa) = s\Psi_{10}^+(a)$$
$$\Psi''(sa) = s\Psi''(a) \sqcup \{\alpha_s\} \text{ if } \alpha_s \in \Psi'(a)$$
$$\Psi''(a) = s\Psi''(sa) \sqcup \{\alpha_s\} \text{ if } \alpha_s \in \Psi''(a)$$

Proof. Write $\alpha_s = (k, k+1)$, where $1 \leq k \leq n-1$. To prove 1) suppose that $\alpha_s \in \Psi_{00}(a)$. Then $k \notin I(a)$ and $k+1 \notin I(a)$. Thus $si = i$ for all $i \in I(a)$. Since $sE_{ij} = E_{si,j}$ for all $i, j \in \{1, \ldots, n\}$, we have $sa = a$. This proves 1).

We will deduce 2),3),4) from Lemma 2.19, Lemma 2.20, and the fact that the union (2.25) is disjoint. Note that the unions in 2),3),4) are disjoint because $s\alpha_s \in \Delta^-$. To prove 2) suppose that $\alpha_s \in \Psi_{01}(a)$. Then $\alpha_s \in \Psi_{10}(sa)$ by Lemma 2.20. Thus $\alpha_s \notin \Psi_{10}(a)$ and thus $\alpha_s \notin \Psi_{01}(sa)$.

It follows from Lemma 2.19 that $s(\Psi_{01}^+(a) \setminus \{\alpha_s\}) = \Psi_{01}^+(sa) \setminus \{\alpha_s\} = \Psi_{01}^+(sa)$ and $\Psi_{10}^+(sa) \setminus \{\alpha_s\} = s(\Psi_{10}^+(sa) \setminus \{\alpha_s\}) = s\Psi_{10}^+(sa)$. This proves the first two assertions of 2). Since $\alpha_s \in \Psi_{01}(a)$, we have $\alpha_s \notin \Psi_{11}(a)$ and thus $\alpha_s \notin \Psi_{11}(sa)$. Therefore $\alpha_s \notin \Psi''(a)$ and $\alpha_s \notin \Psi''(sa)$. Now the remaining assertion of 2) follows from Lemma 2.19. To prove 3) suppose that $\alpha_s \in \Psi_{10}(a)$. Then $\alpha_s \in \Psi_{01}(sa)$ by Lemma 2.20. Thus we may apply 2) with sa in place of s. This proves 3). To prove 4) suppose that $\alpha_s \in \Psi_{11}(a)$. Then $\alpha_s \notin \Psi_{01}(a)$ and $\alpha_s \notin \Psi_{10}(a)$, so $\alpha_s \notin \Psi_{01}(sa)$ and $\alpha_s \notin \Psi_{10}(sa)$ by Lemma 2.20. Now the first two assertions of 4) follow from Lemma 2.19. If $\alpha_s \in \Psi'(a)$, then $\alpha_s \notin \Psi''(a)$ and also $\alpha_s \in \Psi''(sa)$ by Lemma 2.20. Now the third assertion of 4) follows from Lemma 2.19. If $\alpha_s \in \Psi''(a)$, then $\alpha_s \in \Psi'(sa)$ so the last assertion of 4) follows from the third by replacing a by sa. $\square$

Lemma 2.22 *For any* $a \in R, s \in S$ *we have*

1. *If* $\alpha_s \in \Psi_{00}(a)$ *then* $sa = a$.

2. *If* $\alpha_s \in \Psi_{01}(a)$ *then* $m(sa) = m(a) - 1$ *and* $n(sa) = n(a)$.

3. *If* $\alpha_s \in \Psi_{10}(a)$ *then* $m(sa) = m(a) + 1$ *and* $n(sa) = n(a)$.

4. *If* $\alpha_s \in \Psi_{11}(a)$ *then* $m(sa) = m(a)$ *and*

$$n(sa) = \begin{cases} n(a) + 1 & \text{if } \alpha_s \in \Psi'(a) \\ n(a) - 1 & \text{if } \alpha_s \in \Psi''(a) \end{cases}$$

Proof. It follows from (2.26) that $m_{10}(sa) = |\Psi_{10}^+(a^*s)| = |\Psi_{10}^+(a^*)| = m_{10}(a)$. Thus we may replace m by m_{01} in each of 2)-4). Now the assertions follow at once from Lemma 2.21. $\square$

Corollary 2.23 *If* $a \in R$ *and* $s \in S$, *then* $sa = a$ *or* $p(sa) = p(a) \pm 1$.

Note that the assertions in Lemma 2.22, which compare $m(sa)$ with $m(a)$, may be expressed in a single formula: if $x, y \in \{0, 1\}$, then

$$\alpha_s \in \Psi_{xy}(a) \text{ implies that } m(sa) - m(a) = x - y. \tag{2.27}$$

Recall that $\Phi_{xy}(a) = \Psi_{xy}(a^*)$. Since $sa^* = (as)^*$ and $\text{rank}(a) = \text{rank}(as)$, it follows from Lemma 2.22 that, if $x, y \in \{0, 1\}$, then

$$\alpha_s \in \Phi_{xy}(a) \text{ implies that } m(as) - m(a) = y - x. \tag{2.28}$$

Note that $n(a^*) = n(a)$ by (2.20). Also $\Phi'(a) = \Psi'(a^*)$ and $\Phi''(a) = \Psi''(a^*)$ by (2.18). Thus, the analogue of Lemma 2.22 for right multiplication is:

Lemma 2.24 *For any $a \in R, s \in S$ we have:*

1. *If $\alpha_s \in \Phi_{00}(a)$ then $as = a$.*

2. *If $\alpha_s \in \Phi_{01}(a)$ then $m(as) = m(a) - 1$ and $n(as) = n(a)$.*

3. *If $\alpha_s \in \Phi_{10}(a)$ then $m(as) = m(a) + 1$ and $n(as) = n(a)$.*

4. *If $\alpha_s \in \Phi_{11}(a)$ then $m(as) = m(a)$ and*

$$n(as) = \begin{cases} n(a) + 1 & \text{if } \alpha_s \in \Phi'(a) \\ n(a) - 1 & \text{if } \alpha_s \in \Phi''(a) \end{cases}$$

Corollary 2.25 *If $a \in R$ and $s \in S$, then $as = s$ or $p(as) = p(a) \pm 1$.*

Lemma 2.26 *If $a \in R^r$ and $a \neq m_r$, then there exists $s \in S$ such that $p(sa) = p(a) - 1$ or $p(as) = p(a) - 1$.*

Proof. Suppose first that $I(a) \neq \{1, \dots, r\}$. Write $I(a) = \{i_1, \dots, i_r\}$, where $i_1 < \cdots < i_r$. Then either (i) $i_1 > 1$ or (ii) there exists $q \in \{2, \dots, r\}$ such that $i_q - i_{q-1} > 1$. If (i) occurs, let $k = i_1 - 1$. If (ii) occurs, let $k = i_q - 1$. Then $k \notin I(a)$ and $k + 1 \in I(a)$ so $(k, k+1) \in \Psi_{01}(a)$. Define $s \in S$ by $\alpha_s = (k, k+1)$. From 2) in Lemma 2.22 it follows that $m(sa) = m(a) - 1$ and $n(sa) = n(a)$. Hence $p(sa) = p(a) - 1$. Thus we may assume that $I(a) = \{1, \dots, r\}$. If $J(a) \neq \{n - r + 1, \dots, n\}$ it follows by a similar argument using 3) of Lemma 2.24 that there exists $s \in S$ such that $p(as) = p(a) - 1$. Thus we may assume that $I(a) = \{1, \dots, r\}$ and $J(a) = \{n - r + 1, \dots, n\}$. Then $a = \sum_{i=1}^{r} E_{i,ia}$, where $\{1a, \dots, ra\} = \{n - r + 1, \dots, n\}$. Since $a \neq m_r$ there exists $p \in \{1, \dots, r - 1\}$ such that $pa > (p + 1)a$. Thus $(p, p + 1) \in \Psi''(a)$. Define $s \in S$ by $\alpha_s = (p, p + 1)$. It follows from 4) in Lemma 2.22 that $p(sa) = p(a) - 1$. $\square$

Proposition 2.27 *If $a \in R^r$ then $l(a) = m(a) + n(a)$.*

Proof. By induction on $l(a)$ we first show that $p(a) \le l(a)$. Let $a = wm_r w'$, where $l(w) + l(w') = l(a)$. If $l(a) = 0$, then $w = 1 = w'$ so that $a = m_r$. From Lemma 2.18 it follows that $p(a) = 0$. Suppose that $l(a) > 0$. Then $l(w) > 0$ or $l(w') > 0$. Assume that $l(w) > 0$ (the other case goes similarly). Write $w = sw''$, where $s \in S, w'' \in W$ and $l(w'') = l(w) - 1$. Let $b = sa = w'' m_r w'$. Then $l(b) < l(a)$. The induction hypothesis implies in view of Corollary 2.23 that $p(a) \le p(b) + 1 \le l(b) + 1 \le l(a)$, as claimed.

Next, we prove that $l(a) \le p(a)$ by induction on $p(a)$. If $p(a) = 0$, then $a = m_r$ by Lemma 2.18, so $l(a) = 0$. Suppose that $p(a) > 0$. Then $a \ne m_r$, so Lemma 2.26 implies that there exists $s \in S$ such that $p(sa) < p(a)$ or $p(as) < p(a)$. If $p(sa) < p(a)$, then by induction $l(a) \le l(sa) + 1 \le p(sa) + 1 \le p(a)$. The argument in the latter case is similar. This completes the proof. $\square$

Corollary 2.28 *Suppose that $a \in R$ and $s \in S$. If $l(sa) = l(a)$ then $sa = a$. If $l(as) = l(a)$ then $as = a$.*

In view of Proposition 2.27, the precise circumstances in which $l(sa) = l(a) + 1$ and $l(sa) = l(a) - 1$ are given by Lemma 2.22. Similarly, the circumstances in which $l(as) = l(a) + 1$ and $l(as) = l(a) - 1$ are given by Lemma 2.24. Our next aim is to prove the multiplication formula for the orbits $BaB, a \in R$, which exploits ormation.

If $(i, j) \in \Delta$, then let $x_{ij}(t) = 1 + tE_{ij}$ for $t \in K$. Then $X_{ij} = \{x_{ij}(t) | t \in K\}$ is the corresponding 'root subgroup'. We recall some basic facts about these subgroups, see [13]. A subset Γ of Δ is said to be closed if it has the following property

$$(i, j) \in \Gamma, (j, k) \in \Gamma, i \ne k \text{ imply that } (i, k) \in \Gamma \qquad (2.29)$$

If $\Gamma \subseteq \Delta^+$ let U_Γ be the subgroup of U generated by the X_{ij} with $(i, j) \in \Gamma$. If Γ is a closed subset of Δ^+, then every $u \in U_\Gamma$ may be uniquely written in the form

$$u = \prod_{(i,j) \in \Gamma} x_{ij}(t_{ij}) \qquad (2.30)$$

where $t_{ij} \in K$ and the product is taken in any fixed order. If $\Delta^+ = \Gamma' \sqcup \Gamma''$, where Γ', Γ'' are closed subsets of Δ^+, then

$$U = U_{\Gamma'} U_{\Gamma''} \text{ and } U_{\Gamma'} \cap U_{\Gamma''} = 1. \qquad (2.31)$$

Let $w \in W$. Then $\Phi'(w), \Phi''(w)$ are closed subsets of Δ^+. We define subgroups U_w', U_w'' of U by $U_w' = U_{\Phi'(w)}, U_w'' = U_{\Phi''(w)}$. Since $\Delta^+ = \Phi'(w) \sqcup \Phi''(w)$, we have

$$U_w' U_w'' = U = U_w'' U_w' \text{ and } U_w' \cap U_w'' = 1. \qquad (2.32)$$

Every element in BwB may be written in the form bwu'', where $b \in B$ and $u'' \in U_w''$ are uniquely determined. We use the partition (2.25) to define subgroups U_a', U_a'' for $a \in R$. Namely , let

$$\Theta'(a) = \Phi_{00}^+(a) \sqcup \Phi_{01}^+(a) \sqcup \Phi'(a), \quad \Theta''(a) = \Phi_{10}^+(a) \sqcup \Phi''(a). \qquad (2.33)$$

Note that for $a = w \in W$ we have $\Theta'(a) = \Phi'(w)$ and $\Theta''(a) = \Phi''(w)$.

Lemma 2.29 *If $a \in R$, then $\Theta'(a), \Theta''(a)$ are closed subsets of Δ^+ and $\Delta^+ = \Theta'(a) \sqcup \Theta''(a)$.*

Proof. Suppose that $(i, j) \in \Theta'(a)$ and $(j, k) \in \Theta'(a), i \neq k$. If $(i, j) \in \Phi_{00}^+(a) \sqcup \Phi_{01}^+(a)$, then $i \notin J(a)$ so $(i, k) \in \Phi_{00}^+(a) \sqcup \Phi_{01}^+(a) \subseteq \Theta'(a)$ because $i \neq k$. Suppose that $(i, j) \in \Phi'(a)$. Then $i \in J(a), j \in J(a)$ and $ai < aj$. Since $j \in J(a)$ and $(j, k) \notin \Theta'(a)$, we must have $(j, k) \in \Phi'(a)$. Thus $k \in J(a)$ and $aj < ak$. Thus $i \in J(a), k \in J(a)$ and $ai < ak$, so $(i, k) \in \Phi'(a) \subseteq \Theta'(a)$. Thus $\Theta'(a)$ is closed. Suppose that $(i, j) \in \Theta''(a)$ and $(j, k) \in \Theta''(a), i \neq k$. Then $i \in J(a)$ and $j \in J(a)$. Since $j \in J(a)$, we have $(i, j) \notin \Phi_{01}^+(a)$. Thus $(i, j) \in \Phi''(a)$. If $k \notin J(a)$ then, since $i \neq k$, we have $(i, k) \in \Phi_{10}^+(a) \subseteq \Theta'(a)$. If $k \in J(a)$, then $(j, k) \in \Phi''(a)$ so $aj > ak$. Thus $ai > ak$ so $(i, k) \in \Phi''(a) \subseteq \Theta''(a)$. Thus $\Theta''(a)$ is closed. The assertion $\Delta^+ = \Theta'(a) \sqcup \Theta''(a)$ follows from (2.25) with a^* in place of a. $\square$

Define subgroups U_a', U_a'' of U by

$$U_a' = U_{\Theta'(a)} \text{ and } U_a'' = U_{\Theta''(a)}. \qquad (2.34)$$

From (2.31) it follows that

$$U_a' U_a'' = U = U_a'' U_a' \text{ and } U_a' \cap U_a'' = 1. \qquad (2.35)$$

If $a = w \in W$, then U_a', U_a'' have the earlier meaning and (2.35) agrees with (2.32). We will need the fact that, if $i, j \in \{1, \dots, n\}$, then

$$E_{ij}a = \begin{cases} E_{i,ja} & \text{if } j \in I(a) \\ 0 & \text{otherwise} \end{cases} \qquad (2.36)$$

and

$$aE_{ij} = \begin{cases} E_{ai,j} & \text{if } i \in J(a) \\ 0 & \text{otherwise} \end{cases} \tag{2.37}$$

It follows that

$$x_{ij}(t)a = a \quad \text{if } j \notin I(a) \tag{2.38}$$
$$ax_{ij}(t) = a \quad \text{if } i \notin J(a)$$

and

$$x_{ij}(t)a = ax_{ia,ja}(t) \quad \text{if } i,j \in I(a) \tag{2.39}$$
$$ax_{ij}(t) = x_{ai,aj}(t)a \quad \text{if } i,j \in J(a).$$

We are ready to prove the formula for the product of $B \times B$-orbits $BaB, a \in R$. The obtained trichotomy replaces the duality of the classical case of $GL_n(K)$, [40], § 29. Recall that $BsB \cdot BwB$ is equal to $BswB$ if $l(sw) > l(w)$, or to $BswB \cup BwB$, if $l(sw) < l(w)$.

Theorem 2.30 *For every $a \in R$ and every simple reflection $s \in S$ we have*

$$BsB \cdot BaB = \begin{cases} BaB & \text{if } l(sa) = l(a) \\ BsaB & \text{if } l(sa) = l(a) + 1 \\ BsaB \cup BaB & \text{if } l(sa) = l(a) - 1 \end{cases}$$

Proof. We will show that

$$BsB \cdot BaB = \begin{cases} BaB & \text{if } \alpha_s \in \Psi_{00}(a) \\ BsaB & \text{if } \alpha_s \in \Psi_{10}(a) \sqcup \Psi'(a) \\ BsaB \sqcup BaB & \text{if } \alpha_s \in \Psi_{01}(a) \sqcup \Psi''(a) \end{cases}$$

Then the assertion will follow from the behaviour of the functions $m(a)$, $n(a)$ under left multiplication $a \mapsto sa$, determined in Lemma 2.22, and from the fact that $l(a) = m(a) + n(a)$, established in Proposition 2.27.

If $BsaB = BaB$ then $sa = a$ by Theorem 2.7. From Lemma 2.22 it follows that $\alpha_s \in \Psi_{00}(a)$. Thus $\alpha_s \in \Psi_{01}(a) \sqcup \Psi''(a)$ implies that $BsB \neq BsaB$.

The left-hand side may be replaced by sBa and equality may be replaced by inclusion provided that we show for $\alpha_s \in \Psi_{01}(a) \sqcup \Psi''(a)$ that sBa meets both $BsaB$ and BaB.

Write $U = U'_s U''_s$. We have $\Psi''(s) = \{\alpha_s\}$ and $\Psi'(s) = \Delta^+ \setminus \{\alpha_s\}$. From (2.9) it follows that $s\Psi'(s) = \Psi'(s)$, so that $sU'_s s = U'_s$. Let k be such that $\alpha_s = (k, k+1)$. Then $U''_s = X_{k,k+1}$. Hence $sB = sTU = TsU'_s U''_s = TsU'_s ssX_{k,k+1} \subseteq BsX_{k,k+1}$.

If $\alpha_s \in \Psi_{00}(a) \sqcup \Psi_{10}(a)$, then $k+1 \notin I(a)$, so from (2.38) it follows that $X_{k,k+1}a = a$. Then $sBa \subseteq Bsa \subseteq BsaB$. If $\alpha_s \in \Psi_{00}(a)$, then $sa = a$ by Lemma 2.21. Hence $sBa \subseteq BaB$. If $\alpha_s \in \Psi_{01}(a)$, then we must show that $sx_{k,k+1}(t)a \in BsB \cup BsaB$. This is clear for $t = 0$. If $t \neq 0$, let $h \in GL_n(K)$ be the diagonal matrix with entries $-t^{-1}, t$ in positions $k, k+1$, and the other diagonal entries equal to 1. It is easy to check that

$$sx_{k,k+1}(t) = hx_{k,k+1}(-t)x_{k+1,k}(t^{-1}).$$

Since $\alpha_s \in \Psi_{01}(a)$, we have $k \notin I(a)$, which in view of (2.38) implies that $x_{k+1,k}(t^{-1})a = a$. Therefore $sx_{k,k+1}(t)a = hx_{k,k+1}(-t)a \in BaB$, as desired. Suppose $\alpha_s \in \Psi_{11}(a)$. Then $k, k+1 \in I(a)$ and (2.39) implies that $x_{k,k+1}(t)a = ax_{ka,(k+1)a}(t)$. If $\alpha_s \in \Psi'(a)$, then $ka < (k+1)a$. Hence $sx_{k,k+1}(t)a \in BsaB$. If $\alpha_s \in \Psi''(a)$, then $\alpha_s \in \Psi'(sa)$ by Lemma 2.20. Thus, replacing a by sa we come to $sBsa \subseteq BaB$. Since $sBs \subseteq B \cup BsB$ by the remark preceding the theorem, it follows that $sBa = sBssa \subseteq (B \cup BsB)sa \subseteq BsaB \cup BaB$. This completes the proof of the theorem. $\square$

For the rest of this section we assume that $K = \mathbf{F}_q$ is a finite field with q elements. Our aim is to derive basic numerical information related to the presented structural approach.

Proposition 2.31 *For every* $n \geq 1$ *we have*

$$|GL_n(\mathbf{F}_q)| = (q-1)^n q^{n(n-1)/2} \prod_{i=1}^{n-1} (1 + q + \cdots + q^i).$$

Let $n_j = |GL_n(\mathbf{F}_q)| \, (q^{j(n-j)}|GL_j(\mathbf{F}_q)||GL_{n-j}(\mathbf{F}_q)|)^{-1}$, *for* $j = 1, \ldots, n$. *Then the semigroup* M_j/M_{j-1} *has* n_j *nonzero* $\mathcal{R}$-*classes,* n_j *nonzero* $\mathcal{L}$-*classes and*

$$|M_j \setminus M_{j-1}| = n_j^2 |GL_j(\mathbf{F}_q)|.$$

Proof. It is clear that the order of $GL_n(\mathbf{F}_q)$ is the number of ordered bases of the $\mathbf{F}_q$-space $\mathbf{F}_q^n$ (to a given $g \in GL_n(\mathbf{F}_q)$ associate the row vectors of g). The first vector can be chosen in $q^n - 1$ ways. Then there are $q^n - q$ choices for the second vector, $q^n - q^2$ for the third and finally $q^n - q^{n-1}$ choices for the last vector. Therefore

$$
\begin{aligned}
|GL_n(\mathbf{F}_q)| &= \prod_{i=0}^{n-1}(q^n - q^i) = \prod_{i=1}^{n-1} q^i \prod_{i=1}^{n}(q^i - 1) \\
&= (q-1)^n q^{n(n-1)/2} \prod_{i=1}^{n-1}(1 + q + \cdots + q^i).
\end{aligned}
$$

From Proposition 2.5 we know that the number of nonzero $\mathcal{R}$-classes of M_j/M_{j-1} is equal to the index of the subgroup $P_j = \{a \in GL_n(\mathbf{F}_q)| \; ae = eae\}$ in $GL_n(\mathbf{F}_q)$, where $e = e^2 = \begin{pmatrix} I & 0 \\ 0 & 0 \end{pmatrix}$ has rank j. Since the order of P_j is equal to $q^{j(n-j)}|GL_j(\mathbf{F}_q)||GL_{n-j}(\mathbf{F}_q)|$, the assertion on the number of $\mathcal{R}$-classes follows. A similar argument works for the $\mathcal{L}$-classes of M_j/M_{j-1}. In view of Theorem 2.3 this proves the formula for $|M_j \setminus M_{j-1}|$. □

In order to compute $|BaB|$ in terms of $l(a)$ we need another technical result.

Lemma 2.32 *If $a \in R$, then $BaB = BaU_a''$. Moreover, if $b_1 au_1 = b_2 au_2$ for some $b_1, b_2 \in B$ and $u_1, u_2 \in U_a''$, then $u_1 = u_2$ and $b_1 a = b_2 a$.*

Proof. First we show that if $a \in R$ and $u \in U$, then

$$ au \in Ua \text{ if and only if } u \in U_a'. \tag{2.40} $$

Suppose that $u \in U_a'$. Take $\Gamma = \Theta'(a)$ in (2.30) and write $u = \prod x_{ij}(t_{ij})$, where the order of the factors is chosen so that the terms with $(i,j) \in \Phi_{00}^+(a) \sqcup \Phi_{01}^+(a)$ appear on the left. From (2.38) we know that $ax_{ij}(t) = x_{ij}(t)$ for $(i,j) \in \Phi_{00}^+(a) \sqcup \Phi_{01}^+(a)$. Thus $au = a \prod x_{ij}(t_{ij})$, where the product is over $(i,j) \in \Phi'(a)$. If $(i,j) \in \Phi'(a)$, then $i \in J(a), j \in J(a)$ and $ai < aj$. Then $aj \in I(a)$ and $(aj)a = j$. It follows from (2.32) that if $t \in K$, then $ax_{ij}(t) = x_{ai,j}(t) = x_{ai,aj}(t)a \in Ua$. Therefore $au \in U_a'$, which proves the first implication.

Conversely, suppose that $u \in U$ and $au \in Ua$. We can write $u = u'u''$, where $u \in U_a', u'' \in U_a''$. Then $au' \in Ua$ by the first part of the argument. Thus $au'' \in Ua$. If $(i,j) \in \Phi_{10}^+(a)$, then $x_{ij}(t)a^* = a^*$ by (2.38). Write

$u'' = yz$, where $y \in U_{\Phi''(a)}$ and z is a product of factors $x_{ij}(t)$ with $(i,j) \in \Phi_{10}^+(a)$. Then $za^* = a^*$ so $aya^* = aua^* \in Uaa^*$. Since aa^* is a diagonal idempotent, it follows that aya^* is upper triangular. If Γ is a closed subset of Δ^+ and $v \in U_\Gamma$, then it follows by induction on the number of factors $x_{ij}(t)$ of v which are different from 1 that we may write

$$v = 1 + \sum_{(i,j)\in\Gamma} t_{ij} E_{ij} \tag{2.41}$$

for suitable $t_{ij} \in K$. Apply (2.41) with $\Gamma = \Phi''(a)$ and $v = y$. Since $ja^* = aj$ for $j \in J(a)$, we come to

$$aya^* = aa^* + \sum_{(i,j)\in\Phi''(a)} t_{ij} E_{ai,aj}.$$

Since $aya^* - aa^*$ is upper triangular and $ai > aj$ for $(i,j) \in \Phi''(a)$, it follows that $t_{ij} = 0$ for all $(i,j) \in \Phi''(a)$. Thus $y = 1$ and $z = u'' \in U_a''$. Now apply (2.41) with $\Gamma = \Phi_{10}^+(a)$ and $v = z$. Write

$$z = 1 + \sum_{(i,j)\in\Phi_{10}^+(a)} t_{ij} E_{ij}.$$

The indices j which occur here are not in $J(a)$. On the other hand, the elements of Ua are K-linear combinations of elements E_{ij} with $j \in J(a)$. Thus $t_{ij} = 0$ for all $(i,j) \in \Phi_{10}^+(a)$, so that $z = 1$. Therefore $u'' = yz = 1$ and $u = u' \in U_a'$. This completes the proof of (2.40).

Since $aT = Ta$, it follows from (2.35) and (2.40) that

$$BaB = BaTU = BaU_a'U_a'' \subseteq BaU_a'' \subseteq BaB.$$

Hence $BaB = BaU_a''$.

Suppose that $b_1au_1 = b_2au_2$ for some $b_1, b_2 \in B$ and $u_1, u_2 \in U_a''$. Then $au_2u_1^{-1} \in Ba$. From (2.40) it follows that $u_2u_1^{-1} \in U_a'$. Since $U_a' \cap U_a'' = 1$, this means that $u_1 = u_2$. Therefore $b_1a = b_2a$, as desired. $\square$

Corollary 2.33 *If $a \in R \subseteq M_n(\mathbf{F}_q)$ and $\mathrm{rank}(a) = r$, then $|BaB| = (q-1)^r q^{r(r-1)/2} q^{l(a)}$.*

Proof. Let $a = \sum_{k=1}^{r} E_{i_k j_k}$. If $b = \sum_{1 \leq i \leq j \leq n} t_{ij} E_{ij} \in B$, where $t_{ij} \in \mathbf{F}_q$ and $0 \neq t_{ii}$, then

$$ba = \sum_{k=1}^{r} \sum_{1 \leq i \leq i_k} t_{i,i_k} E_{i,j_k}.$$

Thus

$$|Ba| = (q-1)^r q^{\sum_{i \in I(a)} (i-1)}.$$

From Lemma 2.17 it follows that

$$|Ba| = (q-1)^r q^{r(r-1)/2} q^{m_{01}(a)}.$$

Choose $\Gamma = \Theta''(a) = \Phi_{10}^{+}(a) \sqcup \Phi''(a)$ in (2.30) From (2.22) we know that $|\Phi_{10}^{+}(a)| = m_{10}(a)$. In view of (2.18) and (2.20) we have $|\Phi''(a)| = |\Psi''(a^*)| = n(a^*) = n(a)$. Now the uniqueness in (2.30) implies that

$$|U_a''| = q^{m_{10}(a)+n(a)}.$$

From Lemma 2.32 it follows that

$$|BaB| = |Ba||U_a''| = (q-1)^r q^{r(r-1)/2} q^{m_{01}(a)+m_{10}(a)+n(a)}.$$

Since $l(a) = m(a) + n(a)$ by Proposition 2.27, the assertion follows. $\square$

2.4   Complete reducibility of $M_n(\mathbf{F}_q)$.

We turn to the complete reducibility problem for the monoid $M_n(\mathbf{F}_q)$ over a field K. For $K = \mathbf{C}$, the complex numbers, this problem was approached in a series of papers by Edigarjan and Faddeev, cf.[22], and independently by Munn (unpublished). Munn verified the assertion for $n \leq 4$. The solution was claimed by Faddeev in [23], but a complete proof has never been given. In [98], Putcha and the author proved much more generally that the complex semigroup algebra of a finite monoid of Lie type M is semisimple (see Chapter 8). Kovacs then obtained an elementary proof for the special case $M = M_n(\mathbf{F}_q)$, [66].

To present the proof of Kovacs we need an auxiliary notion. A matrix $c \in M_n(\mathbf{F}_q)$ is called a semiidempotent if $c^r = c^{r+1}$ for some $r \geq 1$. By the rank sequence $\sigma(c)$ of a semiidempotent c we mean the sequence $\mathrm{rank}(c), \mathrm{rank}(c^2), \dots, \mathrm{rank}(c^n)$. Consider the Jordan form of c. It is clear that $\mathrm{rank}(c^i) - \mathrm{rank}(c^{i+1})$ is the number of nilpotent Jordan blocks of degree greater than i and the nonnilpotent Jordan blocks are identity matrices. It follows that the rank sequence determines the Jordan form of a semiidempotent c. Since the Jordan form of a semiidempotent is a matrix with entries in $\{0,1\} \subseteq \mathbf{F}_q$, this implies that two semiidempotents have the same rank sequence if and only if they are conjugate in $M_n(\mathbf{F}_q)$.

If $a, b, c \in M_j \setminus M_{j-1}$ are such that $ca = b$, then $\bar{c}a = b$ for the unique isomorphism $\bar{c} : \mathrm{Im}(a) \longrightarrow \mathrm{Im}(b)$. We can treat $\bar{c}$ as a one-to-one partial map on $\mathbf{F}_q^n$ (that is, a map from a subspace of $\mathbf{F}_q^n$ into $\mathbf{F}_q^n$) and define partial maps $\bar{c}^2, \bar{c}^3, \ldots$ in the usual way. It is clear that $\bar{c}$ is a semiidempotent partial map (that is $\bar{c}^r = \bar{c}^{r+1}$ as partial maps for some $r \geq 1$) whenever so is c. By the rank sequence of $\bar{c}$ we then mean the sequence $\dim(\mathrm{Im}(\bar{c}^i)), i = 1, \ldots, n$. We are ready for the main result of this section. Here, Q_j and $n_j, j = 1, \ldots n$, have the same meaning as in Proposition 2.31.

Theorem 2.34 *Assume that K is a field such that $\mathrm{ch}(K)$ does not divide q. Then every sandwich matrix $Q_j, j = 1, \ldots, n$, is invertible in the matrix ring $M_{n_j}(K[GL_j(\mathbf{F}_q)])$ and*

$$K_0[M_n(\mathbf{F}_q)] \simeq \bigoplus_{i=1}^n M_{n_j}(K[GL_j(\mathbf{F}_q)]).$$

In particular, $K[M_n(\mathbf{F}_q)]$ is a semisimple algebra if and only if $\mathrm{ch}(K)$ does not divide the order of $GL_n(\mathbf{F}_q)$.

Proof. We will show that every contracted semigroup algebra $A_j = K_0[M_j/M_{j-1}], j = 1, \ldots, n-1$, has a left identity. Then the assertion will follow via Maschke's theorem. In fact, the existence of a left identity easily implies that $x \mapsto x \circ Q_j$ yields an isomorphism of A_j, viewed as the corresponding Munn algebra, onto $M_{n_j}(K[\mathbf{F}_q])$, see Proposition 4.13. (In particular, Q_j is also right invertible.)

Fix some $j < n$. Consider an element of A_j which is of the form $f = \sum_c k_{\sigma(c)} c$, where $k_{\sigma(c)} \in K$ and the summation runs over the set of semiidempotents c in $M_j \setminus M_{j-1}$. We claim that the coefficients $k_{\sigma(c)}$ can be chosen so that, whenever $a, b \in M_j \setminus M_{j-1}$, the sum of the $k_{\sigma(c)}$ over the c with $ca = b$ is 1 if $a = b$ and it is 0 otherwise. Then $fx = x$ in A_j for every $x \in M_j/M_{j-1}$, so that f is a left identity of A_j, as desired.

To prove the claim, note that the above condition leads to a system of linear equations in unknowns k_τ. Formally, this system consists of one equation for each pair a, b. Assume that the set $\{c \mid ca = b\}$ is nonempty (otherwise all coefficients in the respective equation vanish). We will show that

1. the coefficient of k_τ in the corresponding equation depends on a, b only via the rank sequence $\sigma(d(a, b))$, where $d(a, b) : \text{Im}(a) \longrightarrow \text{Im}(b)$ is the unique linear map such that $d(a, b)a = b$; so that equations corresponding to pairs a, b with a common $\sigma(d(a, b))$ are all the same. (Note that the coefficient of k_τ is the number of semi-idempotents of rank j with rank sequence τ that are extensions of $d(a, b)$ to endomorphisms of $\mathbf{F}_q^n$.)

2. this coefficient is 0 unless τ majorizes $\sigma(d(a, b))$ and it is a power of q if $\tau = \sigma(d(a, b))$.

This will imply that, when the equations and the unknowns are listed according to the lexicographic order on the set of the relevant sequences, the system is triangular with all diagonal coefficients being powers of q. Since the latter are units in K by the hypothesis, the system has a solution in K, which means that A_j has an identity, as desired.

To prove 1),2) fix some $a, b \in M_j \setminus M_{j-1}$ and assume that $ca = b$ for a semiidempotent c of rank j. Note that $\ker(a) = \ker(b)$. Denote by $\bar{c}$ the one-to-one map $\text{Im}(a) \longrightarrow \text{Im}(b)$ that is the restriction of c. Since $\bar{c}$ is uniquely determined by the pair a, b, we must have $\bar{c} = d(a, b)$. It follows that the rank sequence for c majorizes the rank sequence for $d(a, b)$ (because $\bar{c}^i$ is a restriction of c^i for $i = 1, 2, \dots$).

Put $d = d(a, b)$. Let $\hat{d} : \text{Im}(b) \longrightarrow \text{Im}(a)$ be the isomorphism such that $d\hat{d}, \hat{d}d$ are are identity maps on $\text{Im}(b), \text{Im}(a)$, respectively. It is clear that, for $i \geq 1$, the composition of partial maps $\hat{d}d^{i+1}$ is the restriction of d^i to the domain of d^{i+1}. Assume that $d^i(x) = \hat{d}d^i(y)$ for some x, y that lie in the domains of the respective partial maps. Then $d^i(y) = d\hat{d}d^i(y) = dd^i(x) \in \text{Im}(d^{i+1})$, so that $\hat{d}d^i(y) \in \text{Im}(\hat{d}d^{i+1})$. It follows that

$$\text{Im}(d^i) \cap \text{Im}(\hat{d}d^i) = \text{Im}(\hat{d}d^{i+1}) \qquad (2.42)$$

for $i \geq 1$, because the converse inclusion is clear.

Next, we show that

$$\text{Im}(d^i) = \text{Im}(\hat{d}d^{i+1}) \oplus (\ker(e) \cap \text{Im}(d^i)) \qquad (2.43)$$

whenever e is an extension of d to a semiidempotent with rank sequence equal to that of d.

The two summands on the right-hand side lie in $\text{Im}(a)$ and $\ker(e)$, respectively, so they form a direct sum (because $\text{Im}(a) \oplus \ker(e) = \mathbf{F}_q^n$). By (2.42) this sum is contained in $\text{Im}(d^i)$. Since $\text{Im}(d^i) \subseteq \text{Im}(e^i)$ and $\sigma(d) = \sigma(e)$, it follows that $\text{Im}(d^i) = \text{Im}(e^i)$. But $\dim(\ker(e) \cap \text{Im}(e^i)) = \dim(\text{Im}(e^i)) - \dim(\text{Im}(e^{i+1}))$ and $\dim(\text{Im}(\hat{d}d^{i+1})) = \dim(\text{Im}(d^{i+1}))$ imply then that the dimension of this sum is equal to that of $\text{Im}(d^i)$. This proves equality (2.43).

Conversely, assume that U is a complement of $\text{Im}(a)$ in $\mathbf{F}_q^n$ such that

$$\text{Im}(d^i) = \text{Im}(\hat{d}d^{i+1}) \oplus (U \cap \text{Im}(d^i)) \text{ for } i = 1, 2, \ldots \qquad (2.44)$$

Let e_U be the unique endomorphism of $\mathbf{F}_q^n$ that annihilates U and acts on $\text{Im}(a)$ as d. Then

$$e_U(\text{Im}(d^i)) = e_U(\text{Im}(\hat{d}d^{i+1})) = \text{Im}(e_U\hat{d}d^{i+1}) = \text{Im}(d\hat{d}d^{i+1}) = \text{Im}(d^{i+1}).$$

Since $\text{Im}(e_U) = \text{Im}(d)$, this implies by induction that $\text{Im}(e_U^i) = \text{Im}(d^i)$ for $i \geq 1$. Therefore, the rank sequences of e_U and d are equal. Now, d is the restriction of the semiidempotent c, so there exists $k \geq 1$ such that $\text{Im}(e_U^k) = \text{Im}(d^k) \subseteq \text{Im}(a)$ and $d^k = d^{k+1}$. Since e_U, d are equal on $\text{Im}(a)$, it follows that $e_U^{k+1} = e_U^k$. Therefore e_U is a semiidempotent.

We have shown that extensions of d to semiidempotents with the same rank sequence as d are in one-to-one correspondence with complements U of $\text{Im}(a)$ satisfying (2.44) (because we must have $e = e_{\ker(e)}$). Thus, to prove 2), it is enough to show that $\text{Im}(a)$ has a complement U satisfying (2.44) and the number of such complements is a power of q.

Let $k \geq 1$ be such that d is the identity map on $\text{Im}(d^k)$. If U is one of the desired complements of $\text{Im}(a)$, then it gives rise to a descending chain of subspaces

$$U_i = U \cap \text{Im}(d^i) \qquad (2.45)$$

that satisfy the following conditions:

$$U_k = U_{k+1} = \cdots = 0, \qquad (2.46)$$

$$U_i/U_{i+1} \text{ is a complement to } (U_{i+1} + \text{Im}(\hat{d}d^{i+1}))/U_{i+1}$$
$$\text{in } \text{Im}(d^i)/U_{i+1} \text{ for } i = 1, 2, \ldots, \qquad (2.47)$$

$$U/U_1 \text{ is a complement to } (U_1 + \text{Im}(a))/U_1 \text{ in } \mathbf{F}_q^n/U_1. \qquad (2.48)$$

Conversely, assume that $U, U_1, U_2, \ldots$ is a descending chain of subspaces of $\mathbf{F}_q^n$ that satisfies these three conditions. We will show that then U is a complement of $\text{Im}(a)$ and (2.44),(2.45) are satisfied. First, by induction we show that $U_i \cap \text{Im}(\hat{d}d^i) = 0$ for $i = k, k-1, \ldots, 1$. In fact, for $i = k$ the assertion is clear. If $k > i > 0$, then

$$
\begin{aligned}
U_i \cap \text{Im}(\hat{d}d^i) &= U_i \cap \text{Im}(\hat{d}d^i) \cap \text{Im}(d^i) \qquad (2.49)\\
&\quad \text{because by (2.47) } U_i \subseteq \text{Im}(d^i)\\
&= U_i \cap \text{Im}(\hat{d}d^{i+1}) \text{ by (2.43)}\\
&= U_{i+1} \cap \text{Im}(\hat{d}d^{i+1}) \text{ by (2.47)}\\
&= 0 \text{ by the inductive hypothesis.}
\end{aligned}
$$

In particular $U_i \cap \text{Im}(\hat{d}d^{i+1}) = 0$. Since (2.47) yields $\text{Im}(d^i) = U_i + \text{Im}(\hat{d}d^{i+1})$, it follows that this is a direct sum. But $\text{Im}(a) = \text{Im}(\hat{d}d)$, so in view of (2.48) we also get

$$U \cap \text{Im}(a) = U \cap (U_1 + \text{Im}(a)) \cap \text{Im}(a) = U_1 \cap \text{Im}(a) = 0.$$

Hence U is a complement of $\text{Im}(a)$. Finally, $U \cap (U_i + \text{Im}(\hat{d}d^{i+1})) = U_i + (U \cap \text{Im}(\hat{d}d^{i+1}))$ by the modular law, and the second component on the right-hand side is contained in $U \cap \text{Im}(a) = 0$, so that (2.44),(2.45) follow.

This shows that, if one considers (2.46) as a definition, next chooses $U_{k-1}, \ldots, U_1$ in that order subject only to (2.47) and chooses U subject to (2.48), then the U so obtained is a complement of $\text{Im}(a)$ that satisfies (2.44), and that each such complement is obtained this way. The number of complements of a subspace of an $\mathbf{F}_q$-space is a power of q. Hence, the total number of choices in the above construction also is a power of q. This completes the proof of 2).

It remains to prove that 1) holds. As above, suppose that $ca = b$ for some $a, b \in M_j \setminus M_{j-1}$ and a semiidempotent $c \in M_j$. Let $d = d(a, b)$. Constructing the chain of subspaces $U_{k-1}, \ldots, U_1, U$ we see from (2.47) that $U_{k-1} \oplus \text{Im}(\hat{d}d^k) = \text{Im}(d^{k-1})$. Since $\dim(\text{Im}(\hat{d}d^i)) = \dim(\text{Im}(d^i))$ for each i, it follows that the number of possible choices for U_{k-1} depends on $\dim(\text{Im}(d^k))$ and $\dim(\text{Im}(d^{k-1}))$ only. Similarly, by (2.47) we see that the number of possible choices for U_i (with $U_{k-1}, \ldots, U_{i+1}$ chosen already) depends on the dimensions of $\text{Im}(d^i)/U_{i+1}$ and of $(U_{i+1} +$

$\text{Im}(\hat{d}d^{i+1}))/U_{i+1}$. The latter subspace is isomorphic to $\text{Im}(\hat{d}d^{i+1})$ because, as seen in (2.49), we must have $U_j \cap \text{Im}(\hat{d}d^j) = 0$ for $j = i + 1$ once this has been verified for $j = k, k - 1, \ldots, i + 2$. Hence, by downward induction it follows that this number depends only on the rank sequence of d. The same applies to the number of possible choices for U (with $U_{k-1}, \ldots, U_1$ constructed already), which in view of (2.48) depends only on $\dim(U_1)$ and $\dim(\text{Im}(a)) = \dim(\text{Im}(d))$. Therefore, the number of extensions of d to semiidempotents with the same rank sequence as that of d depends only on this rank sequence. It follows that equations corresponding to pairs a, b with the same rank sequence are identical, as desired. This completes the proof of the theorem. $\square$

Once we know that $M_n(\mathbf{F}_q)$ is completely reducible over K, we can construct the complete set of nonequivalent irreducible representations. From Theorem 2.34 and its proof it follows that they are determined by the maps $x \to \phi(xf_j)$, where $f_j \in K_0[M_j]$ is the identity modulo $K_0[M_{j-1}]$ and ϕ is a homomorphism of $K_0[M_j/M_{j-1}] \simeq M_{n_j}(K[GL_j(\mathbf{F}_q)])$ onto a simple algebra (hence, ϕ is determined by an irreducible representation of $GL_j(\mathbf{F}_q)$). However, there is a direct way to construct all irreducible representations, avoiding the computation of f_j (which is Q_j^{-1} when M_j/M_{j-1} is identified with the completely 0-simple semigroup $\mathcal{M}(GL_j(\mathbf{F}_q), X_j, Y_j, Q_j)$). This follows from our last result in this section and from the description of complex irreducible representations of the group $GL_n(\mathbf{F}_q)$, cf.[30].

Theorem 2.35 *Let* $\phi : G_j \simeq GL_j(\mathbf{F}_q) \longrightarrow \text{End}(V), j \leq n$, *be an irreducible complex representation. Let* $W = V \otimes_\mathbf{C} X$, *where* X *is the* $\mathbf{C}$*-space with basis* X_j *and let* $e = \begin{pmatrix} I & 0 \\ 0 & 0 \end{pmatrix}$ *be the idempotent of rank* j. *Define* $\overline{\phi} : M_n(\mathbf{F}_q) \longrightarrow \text{End}(W)$, *for* $a \in M_n(\mathbf{F}_q), v \in V, x \in X_j$ *by*

$$\overline{\phi}(a)(v \otimes x) = \phi(g)(v) \otimes x' \text{ if } \text{rank}(ae) = j$$

where $ax = x'g$ *for some* $x' \in X_j, g \in G_j$, *and*

$$\overline{\phi}(a) = 0 \text{ if } \text{rank}(ae) < j.$$

Then $\overline{\phi}$ *is an irreducible representation of* $M_n(\mathbf{F}_q)$ *and it agrees with* ϕ *when restricted to* $V \otimes e\mathbf{C} \simeq V$. *Moreover, every irreducible complex representation of* $M_n(\mathbf{F}_q)$ *is equivalent to a representation of this type.*

Proof. We have seen in Theorem 2.3 that every ax of rank j can be uniquely written in the form $x'gy'$ for some $x' \in X_j, y' \in Y_j, g \in G_j$. Since $ax\mathcal{L}e$, it follows also that $y' = e$, so indeed $ax = x'g$.

Assume that $ax = yg, by = zh$ for some $a, b \in M_n(\mathbf{F}_q), x, y, z \in X_j$ and $g, h \in G_j$. Then $bax = byg = zhg$, and hence

$$
\begin{aligned}
\overline{\phi}(ba)(v \otimes x) &= \phi(hg)(v) \otimes z = \phi(h)(\phi(g)(v)) \otimes z \\
&= \overline{\phi}(b)(\phi(g)(v) \otimes y) = \overline{\phi}(b)(\overline{\phi}(a)(v \otimes x))
\end{aligned}
$$

which implies that $\overline{\phi}$ is a representation. It is clear that $\overline{\phi}(h)(v \otimes e) = h(v) \otimes e$ for $v \in V, h \in G_j$.

Note that $\mathbf{C}[M_n(\mathbf{F}_q)]$ acts on $\mathbf{C}[G_j] \otimes X$ by:

$$
a(g \otimes x) = \begin{cases} g'g \otimes x' & \text{where } ax = x'g' \text{ if } \operatorname{rank}(ax) = j \\ 0 & \text{if } \operatorname{rank}(ax) < j \end{cases}
$$

for a, g, x and x', g' as above. Since $\mathbf{C}_0[M_j/M_{j-1}]$ has an identity element f by Theorem 2.34, it is easy to see that $f(g \otimes x) = g \otimes x$. But $\overline{\phi}$ arises from the composition of this action with the natural homomorphism $\mathbf{C}[G_j] \otimes X \longrightarrow \operatorname{End}(V) \otimes X$. Therefore, $\overline{\phi}(f)$ is the identity map. It follows that for any nonzero submodule W' of W we have $\overline{\phi}(M_j)W' \neq 0$. Hence $\overline{\phi}(eM_j)W' \neq 0$ because $M_j e M_j = M_j$. Since $\overline{\phi}(ea)(v \otimes x) = \phi(eax)(v) \otimes e$ or it is 0, we have $\overline{\phi}(eM_j)W' \subseteq V \otimes e$. The fact that V is an irreducible G_j-module implies that $V \otimes e \subseteq W'$. Then $W' = W$ because $\overline{\phi}(x)(V \otimes e) = V \otimes x$ for every $x \in X_j$. This means that W is an irreducible $\mathbf{C}[M_n(\mathbf{F}_q)]$-module.

According to Theorem 2.34, irreducible representations of $M_n(\mathbf{F}_q)$ are in one-to-one correspondence with the irreducible representations of the groups $GL_j(\mathbf{F}_q), j = 1, \dots, n$. To complete the proof it is then enough to show that any two of the representations $\overline{\phi}, \overline{\phi'}$ constructed above are not equivalent. This is clear if ϕ, ϕ' are representations of $GL_j(\mathbf{F}_q), GL_k(\mathbf{F}_q)$, respectively, where $j \neq k$, because $\phi(M_i) = 0$ for $i < j$. If $j = k$, this is a consequence of the fact that the restriction of $\overline{\phi}$ to G_j agrees with ϕ. $\square$

Let $B \subseteq GL_n(\mathbf{F}_q)$ be the Borel subgroup considered in Section 2.1. Let $\epsilon = \epsilon_B = |B|^{-1} \sum_{b \in B} b \in \mathbf{C}[GL_n(\mathbf{F}_q)]$. It is known that the algebra $\epsilon \mathbf{C}[GL_n(\mathbf{F}_q)]\epsilon$, called the Hecke algebra of B, is isomorphic to the group ring $\mathbf{C}[W]$ of the symmetric group W, and it controls the decomposition of the permutation representation of G on G/B [18]. In a similar way, using Theorem 2.34 and the material of Section 2.3, it is shown in [129] that we have

$$\epsilon \mathbf{C}[M_n(\mathbf{F}_q)]\epsilon \simeq \mathbf{C}[R]$$

and the former has a strong flavour of the Hecke algebra, controlling representation theory for $GL_n(\mathbf{F}_q)$.

We note that modular representations of $M_n(\mathbf{F}_q)$ have also been studied. In particular, it was shown in [28] that $\mathbf{F}_p[M_n(F_p)]$ is not an algebra of finite representation type, that is, it has infinitely many non-equivalent indecomposable representations. Modular representations afforded by the space of homogeneous polynomials were further studied in [68]. Irreducible modular representations of $\mathbf{F}_p[M_n(\mathbf{F}_p)]$ have been described, and extensively used in a series of papers on algebraic topology, cf.[34]. Irreducible modular representations of finite monoids of Lie type (see Chapter 8) were considered in [111], [114].

Chapter 3

Structure of linear semigroups

In this chapter we present the general structure theorem for arbitrary subsemigroups S of $M_n(K)$. The starting idea is to study the ideal chain determined on S by the ideals $M_0 \subseteq M_1 \subseteq \cdots \subseteq M_n = M_n(K)$ and to see how S fills the egg-box pattern at every level $M_j \setminus M_{j-1}$. The theorem associates with S a collection of finitely many (at most 2^n) linear groups G_α and as many sandwich matrices P_α over these groups. Up to conjugation, G_α are all groups generated by the nonempty intersections $S \cap D$ with maximal subgroups D of $M_n(K)$. This description of S shows a strong analogy with Wedderburn's structure theorem for finite dimensional algebras. The philosophy is then to study S via the 'associated linear groups' G_α and the group action they determine on $\mathrm{cl}(S)$.

3.1 Uniform semigroups and associated groups

Our first aim is to introduce a class of subsemigroups of completely 0-simple semigroups that will be crucial for the approach presented in this chapter.

Proposition 3.1 *Let U be a subsemigroup of a completely 0-simple semigroup S with zero θ, which intersects nontrivially every nonzero $\mathcal{H}$-class of S. Fix a maximal subgroup D of S. Let G denote the subgroup* *of D generated by $U \cap D$. Then there exists a subsemigroup I of U and a Rees matrix presentation $S = \mathcal{M}(D, X, Y, Q)$ of S such that $I = \mathcal{M}(T, X, Y, Q)$ is a semigroup of matrix type over the subsemigroup $T = U^2 \cap D$ of D and $\tilde{I} = \mathcal{M}(G, X, Y, Q)$ is the smallest completely 0-simple subsemigroup of S containing U. Moreover, the construction of $\tilde{I}$ does not depend on the choice of D, in particular for any other maximal subgroup D' of S the subgroup of D' generated by $U \cap D'$ is equal to $\tilde{I} \cap D'$, so that it is isomorphic to G.*

Proof. Let X, Y be the sets of nonzero $\mathcal{R}, \mathcal{L}$-classes of S respectively, and let $x \in X, y \in Y$ be such that $D = S_{(x)}^{(y)} \setminus \{\theta\}$. Let $U_{(x)} = S_{(x)} \cap U, U^{(y)} = S^{(y)} \cap U$. Then $T = U_{(x)} U^{(y)} \setminus \{0\} \subseteq D$ satisfies $T = U^2 \cap D$. For every $i \in X, j \in Y$ choose nonzero elements $s_{iy} \in U_{(i)} U^{(y)}, u_{xj} \in U_{(x)} U^{(j)}$. Then $u_{xj} s_{iy} \in T^0$. Put $I = \bigcup_{i,j} s_{iy} T u_{xj} \cup \{\theta\}$. We claim that I is isomorphic to the semigroup of matrix type $\mathcal{M}(T, X, Y, Q)$ over T with sandwich matrix $Q = (q_{ji})$ defined by $q_{ji} = u_{xj} s_{iy}$. In fact, it is easy to see that the map $\phi : \mathcal{M}(T, X, Y, Q) \longrightarrow I$ defined by $\phi(t, i, j) = s_{iy} t u_{xj}$ is a semigroup homomorphism that maps $\mathcal{M}(T, X, Y, Q)$ onto I:

$$\begin{aligned} \phi((t,i,j)(t',i',j')) &= \phi((tq_{ji'}t', i, j')) = s_{iy} t q_{ji'} t' u_{xj'} \\ &= s_{iy} t u_{xj} s_{i'y} t' u_{xj'} = \phi((t,i,j)) \phi((t',i',j')) \end{aligned}$$

Suppose that $s_{iy} t u_{xj} = s_{ky} q u_{xl}$ for some $i, k \in X, j, l \in Y$, and $t, q \in T^0$. Since $S_{(x)}^{(y)}$ is a maximal subgroup of S and $s_{iy} \in S_{(i)}^{(y)}, t \in S_{(x)}^{(y)}, u_{xj} \in S_{(x)}^{(j)}$, it follows that $t = \theta$ whenever $s_{iy} t u_{xj}$ is the zero of S. Similarly, $q = \theta$ in this case. Otherwise, we must have $i = k, j = l$ because $s_{iy} t u_{xj}$ lies in $S_{(i)}^{(j)} \cap S_{(k)}^{(l)}$. Since S is completely 0-simple, there exist $a, b \in S$ such that $as_{iy}, u_{xj} b \in T$. The fact that $(as_{iy}) t (u_{xj} b) = a(s_{iy}) q(u_{xj} b)$ implies now that $t = q$. Therefore ϕ is an isomorphism, as claimed.

From Lemma 1.1 it is clear that $S = \bigcup_{i,j} s_{iy} D u_{xj} \cup \{\theta\}$. Therefore, the above shows that ϕ extends to an isomorphism $\mathcal{M}(D, X, Y, Q) \longrightarrow S$.

Put $\tilde{I} = \bigcup_{ij} s_{iy} G u_{xj} \cup \{\theta\} \subseteq S$. Since

$$s_{iy} G u_{xj} s_{ky} G u_{xl} \subseteq s_{iy} G u_{xl} \cup \{\theta\},$$

it follows that $\tilde{I}$ is a subsemigroup of S. But

$$\tilde{I} \cap S_{(x)}^{(y)} = s_{xy} G u_{xy} \cup \{\theta\} = G \cup \{\theta\}$$

because $s_{xy}, u_{xy} \in T$. It follows easily that $\tilde{I}$ is a completely 0-simple semigroup that can be identified with $\mathcal{M}(G, X, Y, Q)$. Let $z \in U$. Then there exist $c \in S_{(x)}, d \in S^{(y)}$ such that $czd \neq 0$. These c, d can be chosen so that $c, d \in I$. Therefore $czd \in T$. Now, if $c = (g, x, j), d = (h, k, y) \in \mathcal{M}(T, X, Y, Q)$ and $z = (f, i, l)$, then $czd = (g q_{ji} f q_{lk} h, x, y)$. Since $g, h, q_{ji}, q_{lk} \in T$, it follows that $f \in G$. Hence $z \in \tilde{I}$. Therefore $U \subseteq \tilde{I}$. Then $T \subseteq U \cap D \subseteq G$, so that G is the group generated by $U \cap D$, as well. It is clear that $\tilde{I}$ is the smallest completely 0-simple subsemigroup of S that contains I, so it is the smallest one that contains U. In particular, $\tilde{I}$ does not depend on the choice of the maximal subgroup D.

Now, if D' is another maximal subgroup of S, then by the above, the semigroups $\tilde{I'}, \tilde{I}$ coincide, so that $U \cap D'$ generates a maximal subgroup of $\tilde{I}$. This completes the proof of the proposition. $\square$

Let $S = \mathcal{M}(D, X, Y, P)$ be a completely 0-simple semigroup. A subsemigroup U of S will be called a uniform subsemigroup of S if U intersects each nonzero $\mathcal{H}$-class of a completely 0-simple subsemigroup $S' = \mathcal{M}(D, X', Y', P')$, where $X' \subseteq X, Y' \subseteq Y$ and P' is the corresponding $Y' \times X'$-submatrix of P. From the above result it follows that S has the smallest completely 0-simple subsemigroup $\hat{U}$ containing U. This $\hat{U}$ will be called the completely 0-simple closure of U in S. It can be viewed as a 'group approximation' of U. The common (up to isomorphism) group G generated by the intersections $U \cap D$ with maximal subgroups D of S' will be called the group associated to the uniform subsemigroup U of S. (Note that this definition does not imply that $\theta \in \hat{U}$. In fact, this is not so if $\theta \notin U$, or equivalently if all entries of P' are nonzero. However, including θ in $\hat{U}$ is sometimes convenient and produces no ambiguity.)

Uniform semigroups were introduced in [88]. A special class of such semigroups already appeared in the study of irreducible linear semigroups, [80], [106], (however, these were only certain subsemigroups of $M_n(K)$ and not of an arbitrary M_j/M_{j-1}.)

Remark We note that a semigroup U that is a uniform subsemigroup of two completely 0-simple semigroups S_1, S_2 with nonisomorphic completely 0-simple closures U_1, U_2 can be constructed. First, every almost polycyclic group G (a finite extension of a polycyclic group), which is not almost nilpotent, has a free noncommutative subsemigroup X, cf. Chapter 6. Therefore, the subgroup of G generated by X and the free group of rank equal to the rank of X both can be considered as completely 0-simple closures of X.

Here is another construction, in which the groups generated by the nonempty intersections of U with the maximal subgroups of U_1, U_2 are isomorphic. Let $S_1 = \mathcal{M}(G, 1, 2, P)$, where G is the free group on generators x, y and the sandwich matrix P is given by $p_{11} = 1, p_{21} = 1$. Let $U = V_1 \cup V_2$ with $V_1 = \{(g, 1, 1) | g \in \langle x, y \rangle^1 x\}$ and $V_2 = \{(g, 1, 2) | g \in \langle x, y \rangle^1 y\}$. It is easy to see that U is a cancellative subsemigroup of S_1 and S_1 is its completely 0-simple closure. On the other hand, U is isomorphic to the free semigroup $U' = \langle x, y \rangle \subseteq S_2 = G$ via the map $(g, 1, i) \mapsto g$. Clearly, S_2 is the completely 0-simple closure of U' in S_2.

However, such a situation cannot occur if every cancellative subsemigroup T of U satisfies the Ore condition: $xT \cap yT \neq \emptyset, Tx \cap Ty \neq \emptyset$ for every $x, y \in T$. Namely, assume that $a, b \in U$ are such that $a \mathcal{R} b$ in S_1. Since U is a uniform subsemigroup of S_1, there exist $x, y \in U$ such that $ax, by \in D$ for a maximal subgroup D of S_1. But $U \cap D$ is an Ore semigroup, so $axs = byt$ for some $s, t \in U \cap D$. It follows that

$$a \mathcal{R} b \text{ in } S_1 \text{ if and only if } au = bw \text{ for some } u, w \in U.$$

Since the same applies to S_2, the $\mathcal{R}$-class partition of U coming from S_1 is the same as that coming from S_2. A symmetric argument works for the $\mathcal{L}$-class partitions. Hence, if D_1 is a maximal subgroup of S_1 nontrivially intersected by U, then $U \cap D_1 = U \cap D_2$ for a maximal subgroup D_2 of S_2. The maximal subgroups G_1, G_2 of U_1, U_2 are the classical groups of quotients of the respective intersections, because they are generated by these intersections by Proposition 3.1. It follows that $G_1 \simeq G_2$ for an isomorphism which is the identity on $U \cap G_1$. We may write $U_1 = \mathcal{M}(G_1, X, Y, P_1), U_2 = \mathcal{M}(G_2, X, Y, P_2)$. Since U is a uniform subsemigroup of U_1, and of U_2, the sandwich matrices P_1, P_2 can be chosen as in the proof of Proposition 3.1. In particular, they are determined by U, so we can take $P_1 = P_2$. It follows that $U_1 \simeq U_2$.

It can be verified that, in the above case, the semigroup $\widehat{U} = U_1$ is the semigroup of quotients of U in the sense studied in [25] and [26]. Namely, every $s \in \widehat{U}$ can be written as $s = ux^{-1} = y^{-1}v$ for some $u, v \in U$ and some $x \in U \cap D, y \in U \cap D'$, where D, D' are maximal subgroups of

$\hat{U}$ and the inverses are taken in the respective groups. Moreover, semigroups U of this type admit an abstract characterization, not referring to $\hat{U}$. We note that, if a semigroup U has no free noncommutative subsemigroups, then every its cancellative subsemigroup satisfies the Ore condition.

We continue with some examples of uniform semigroups.

Examples. 1. A subsemigroup of a group; the multiplicative semigroup of a subring of a division algebra.

2. More generally, let S be the multiplicative semigroup of a prime right Goldie ring R, cf. [36]. Then S has an ideal I that is a uniform subsemigroup of some $\mathcal{M}(GL_j(F), X, Y, P)$, where F is a division algebra such that the classical quotient ring of R is isomorphic to $M_n(F)$. The same can be said about the image of a semigroup T under a homomorphism of its semigroup algebra $K[S]$ onto a prime right Goldie ring R. This is a consequence of the structure theorem presented in the next section, or more precisely, of its extension discussed in Chapter 9.

3. Let $S = \mathcal{M}(G, 2, 2, I)$, where G is the free nonabelian group on x, y and I is the 2×2-identity matrix. Then S can be considered as a subsemigroup of $M_2(K[G])$. Let U be the subsemigroup of S of the form

$$U = \begin{pmatrix} \langle x \rangle^1 & \langle x \rangle^1 y \\ y^{-1} \langle x \rangle^1 & y^{-1} \langle x \rangle^1 y \end{pmatrix}$$

(meaning that the only nonzero entry of every nonzero matrix of U lies in the corresponding indicated subset of G.) It is clear that U is a uniform subsemigroup of S and the associated group is the cyclic infinite group. Moreover, the proof of Proposition 3.1 shows that $U \simeq \mathcal{M}(\langle x \rangle, 2, 2, I)$. Note that the 'entries' of U do not generate a cyclic infinite group.

4. Let $S = M_n(\mathbf{Z}) \subseteq M_n(\mathbf{Q})$. Since for every $a \in M_n(\mathbf{Q})$ there exists $z \in \mathbf{Z}$ such that $za \in S$, it follows that S intersects all $\mathcal{H}$-classes of $M_n(\mathbf{Q})$. Therefore, for every j, $(S \cap M_j)/(S \cap M_{j-1})$ is a uniform subsemigroup of M_j/M_{j-1}.

5. Let X be a finitely generated free semigroup and $w \in X$ a primitive word in X (this means that w is not a power of its proper subword). Let $X_w = \{a \in X \,|\, a \text{ is not a subword of any } w^n, n \geq 1\}$. The Rees factor X/X_w has an ideal I_w with finite complement and such that I_w is a uniform subsemigroup of an inverse semigroup with fi-

nitely many $\mathcal{H}$-classes. In fact, there is a natural embedding of $I_w = \{a \in X \setminus X_w | a$ has w as a subword $\} \cup \{\theta\}$ into the semigroup $S = \mathcal{M}(\langle x \rangle, l(w), l(w), P)$, where $l(w)$ is the length of w, $\langle x \rangle^1$ is an infinite cyclic monoid and P is the matrix with diagonal $(1, 1, \ldots, 1, x)$ and zeros off the diagonal, cf. [87], Chapter 24.

We shall often use the fact that for certain classes of semigroups 'uniform' implies 'completely 0-simple'.

Corollary 3.2 *Let U be a uniform subsemigroup of a completely 0-simple semigroup S. If a power of every $a \in U$ is a von Neumann regular element of U, then U is completely 0-simple. In particular, this applies if U is π-regular.*

Proof. Let $\hat{U}$ be the completely 0-simple closure of U in S. Choose a maximal subgroup D of $\hat{U}$ and let $e = e^2 \in D$. If $a \in U \cap D$, then the hypothesis implies that a^{3k} is regular in U for some $k \geq 1$. Hence there exists $x \in U$ such that $a^{3k} = a^{3k}xa^{3k}$. Let $b = axa$. Then $a^{3k} = a^{3k-1}ba^{3k-1}$ and we must have $b \in U \cap D$. Cancellativity in D implies that $ba^{3k-2} = a^{3k-2}b = e$. Therefore, $e \in S$ and a is invertible in $S \cap D$. It follows that $S \cap D$ is a group. Since $S \cap D$ generates D (by Proposition 3.1), we have $D \subseteq S$. It follows easily that $U = \hat{U}$. $\square$

As we shall see below, more can be said in the case where U is a uniform subsemigroup of some principal factor M_j/M_{j-1} of the monoid $M_n(K)$.

Lemma 3.3 *Let $S \subseteq M_n(K)$ be a semigroup. Assume that for some maximal subgroups D_1, D_2 of $M_n(K)$, consisting of matrices of the same rank, there exist $a, b \in S$ such that $a(S \cap D_1)b \subseteq S \cap D_2$. Then the groups G_1, G_2 generated by $S \cap D_1, S \cap D_2$ in D_1, D_2, respectively, are conjugate in $M_n(K)$.*

Proof. Let j be the rank of matrices in D_1, D_2. Let e, f be the identities of D_1, D_2 respectively. We can replace a by any ax, where $x \in S \cap D_1$, and b by any yb, where $y \in S \cap D_1$. This allows to assume that $a, b \in M_j \setminus M_{j-1}$, and consequently $a \in fM_n(K)e, b \in eM_n(K)f$. Let $T \subseteq M_j/M_{j-1}$ be the union of the $\mathcal{H}$-classes of $M_j \setminus M_{j-1}$ intersected by the set $(S \cap D_1) \cup (S \cap D_2) \cup \{a, b\}$. Then $T \cup \{\theta\}$ is a completely 0-simple subsemigroup of M_j/M_{j-1} and $S\cap T$ determines modulo M_{j-1} a uniform subsemigroup U of $T \cup \{\theta\}$. Moreover, by Proposition 3.1, G_1, G_2 are maximal subgroups of the completely 0-simple closure of U in M_j/M_{j-1}. In particular, $bG_2a = G_1$. Since $ba \in S \cap G_1$, replacing a by some $az, z \in G_1$, we can assume that $ba = e$. Then $ab = aeb = abab \in G_2$, so that $ab = f$. Since the idempotents e, f have the same rank, there exists $g \in GL_n(K)$ such that $geg^{-1} = f$. Let $C_{GL_n(K)}(e)$ be the centralizer of e in $GL_n(K)$. Choose

$$h \in gC_{GL_n(K)}(e) \cap \{x \in GL_n(K)| \, x^{-1}a \in G_1\}.$$

(It is easy to see that this intersection is not empty. In fact, conjugating in $M_n(K)$ we can assume that $e = \begin{pmatrix} I & 0 \\ 0 & 0 \end{pmatrix}$. Then $g^{-1}a \in M_n(K)e$. Since $fg\mathcal{R}f\mathcal{R}a$ by Lemma 1.1 and g^{-1} acts on the set of $\mathcal{R}$-classes of $M_n(K)$ by left multiplication, we also have $g^{-1}a\mathcal{R}g^{-1}fg\mathcal{R}e$. Hence $g^{-1}a \in eM_n(K)$. Therefore $g^{-1}a \in D_1 = \begin{pmatrix} * & 0 \\ 0 & 0 \end{pmatrix}$. Hence, there exists $y \in C_{GL_n(K)}(e) = \begin{pmatrix} * & 0 \\ 0 & * \end{pmatrix}$ such that $y^{-1}g^{-1}a \in G_1$ and we can put $h = gy$.) Now $h^{-1}a \in G_1$ and

$$h^{-1}abh = h^{-1}fh = g^{-1}fg = e \in G_1.$$

But $bh\mathcal{L}fh\mathcal{L}e$ because $b\mathcal{L}f$ and h acts on the set of $\mathcal{L}$-classes of $M_n(K)$ by right multiplication. Since also $bh \in eM_n(K)$, it follows that $bh \in D_1$. The displayed formula implies that $bh \in G_1$. Therefore $h^{-1}G_2h = h^{-1}abG_2abh = G_1bG_2aG_1 = G_1$, as desired. $\square$

The following is an immediate consequence of Lemma 3.3 and of the definition of a uniform subsemigroup.

Corollary 3.4 *Let U be a uniform subsemigroup of some M_j/M_{j-1}. Assume that U intersects maximal subgroups D_1, D_2 of M_j/M_{j-1}. If G_1, G_2 are the groups generated by $U \cap D_1, U \cap D_2$, then G_1, G_2 are conjugate in $M_n(K)$.*

3.2 Structure theorem

Our aim in this section is to prove the main structure theorem for arbitrary linear semigroups. We also present the resulting philosophy of studying $S \subseteq M_n(K)$ via associated groups, the corresponding sandwich matrices and group actions. This is a natural extension of the structural approach to finite semigroups - via completely 0-simple principal factors, whose role is now played by the uniform components of S. The following theorem is due to the author, [88], and extends the result known before for π-regular linear semigroups, cf. [87], Theorem 3.10. We say that a subset A of a semigroup T is a 0-disjoint union of certain $A_\alpha, \alpha \in \mathcal{A}$, if $A = \bigcup_\alpha A_\alpha$ and $\bigcap_\alpha A_\alpha$ is either empty or it is the zero of T.

Theorem 3.5 *Let* $S \subseteq M_n(K)$ *be a semigroup. Define the sets*

$$S_j = \{a \in S |\ \mathrm{rank}(a) \leq j\}$$

$$T_j\ =\ \{a \in S_j |\ S^1 a S^1 \textit{ does not intersect maximal subgroups of } M_n(K)$$
$$\textit{contained in } M_j \setminus M_{j-1}\}.$$

Then

$$S_0 \subseteq T_1 \subseteq S_1 \subseteq T_2 \subseteq \cdots \subseteq S_{n-1} = T_n \subseteq S_n = S$$

are ideals of S *(if nonempty). Moreover*

1. *for every* j *we have* $T_j^{\binom{n}{j}} \subseteq S_{j-1}$, *so that* $N_j = T_j/S_{j-1}$ *is a nilpotent ideal of* S/S_{j-1},

2. $(S_j \setminus T_j) \cup \{\theta\} \subseteq M_j/M_{j-1}$ *is a 0-disjoint union of uniform subsemigroups* $U_1^{(j)}, \dots, U_{n_j}^{(j)}$, $n_j \leq \binom{n}{j}$, *of* M_j/M_{j-1} *that intersect different* $\mathcal{R}$- *and different* $\mathcal{L}$-*classes of* M_j/M_{j-1}; *moreover* N_j *does not intersect* $\mathcal{H}$-*classes of* M_j/M_{j-1} *intersected by* $S_j \setminus T_j$,

3. $U_i^{(j)} U_k^{(j)} \subseteq N_j$ *for every* $k \neq i$; *moreover* $U_i^{(j)} N_j, N_j U_i^{(j)} \subseteq N_j$ *and* $U_k^{(j)} N_j U_k^{(j)} = \theta$ *in* M_j/M_{j-1}. *In particular,* $U_i^{(j)}$ *can be considered as ideals of* S/T_j.

Proof. It is clear that every S_j, T_j is an ideal of S, if nonempty. Further, $T_j/S_{j-1} \subseteq M_j/M_{j-1}$ consists of nilpotents. Thus $T_j^{\binom{n}{j}} \subseteq S_{j-1}$ by Corollary 2.15. We then are left with proving assertions 2),3) on the Rees factors S_j/T_j.

Let $T = S_j/S_{j-1}$ and put $J = M_j/M_{j-1}, N = T_j/S_{j-1}$. By $\mathcal{H}, \mathcal{R}, \mathcal{L}$ we denote Green's relations on J and by θ the zero element of J. Suppose that $a \in T \setminus N$. Then $xay \in C$ for some $x, y \in (S/S_{j-1})^1$, where C is the union of all $T \cap D$, D - a nonzero maximal subgroup of J. The elements x, y can be chosen from T (replace x, y by xd, dy for any $d \in T \cap D$). Since xay is not nilpotent in T, it follows that $ayx \in C$. Therefore $az, za \in C$ for $z = yx \in T$.

Define a relation $\sim$ on C by

$$a \sim b \text{ if there exist } u, v \in T \text{ such that } uv\mathcal{H}a, vu\mathcal{H}b$$

It is clear that $\sim$ is reflexive and symmetric. Assume that $a \sim b, b \sim c$ for some $a, b, c \in C$. Then we also have $u'v'\mathcal{H}b, v'u'\mathcal{H}c$ for some $u', v' \in T$. Hence $u'\mathcal{R}v\mathcal{R}b$ and $u\mathcal{L}v'\mathcal{L}b$. Since $uv \neq \theta$, we have $uu' \neq \theta$ because $u'\mathcal{R}v$. Similarly, $v'u' \neq \theta$ implies $v'v \neq 0$. Then also $(v'v)(uu') \neq \theta$ and $(uu')(v'v) \neq \theta$ (note that $pqr = \theta$, $p, q, r \in J$, implies that $pq = \theta$ or $qr = \theta$). Since $v'vuu'\mathcal{H}v'u'\mathcal{H}c$ and $uu'v'v\mathcal{H}uv\mathcal{H}a$, this implies that $a \sim c$. This shows that $\sim$ is an equivalence relation on C.

Denote by $C_i, i \in I$, the equivalence classes of $\sim$ on C. It is clear that different classes do not intersect the same $\mathcal{L}$- or $\mathcal{R}$-classes of J. Let $D_i = \{a \in T \mid \text{there exists } t \in T \text{ such that } at, ta \in C_i\}$.

Assume that $a \in T \setminus N$. The first paragraph of the proof shows that $az, za \in C$ for some $z \in T$. Hence $az \sim za$, so there exists $i \in I$ such that $a \in D_i$. Therefore $T \setminus N = \bigcup_{i \in I} D_i$. It is clear that $D_i \cap D_k = \emptyset$ for $i \neq k$ because C_i, C_k intersect different $\mathcal{R}$- and $\mathcal{L}$-classes of J.

Let $z \in D_i, u \in D_k$. Then $xz, zx \in C_i, uy, yu \in C_k$ for some $x, y \in T$. In particular $zxz \neq \theta$ and $uyu \neq \theta$. If $xzuy \notin N$, then $xzuy \in D_m$ for some $m \in I$. Clearly $m = i$ because $xz\mathcal{R}xzuy$ and $m = k$ because $uy\mathcal{L}xzuy$. This contradicts the choice of $i \neq k$ and shows that $xzuy \in N$. Then $(zxz)(uyu) \in N$, and $zxz\mathcal{H}z, uyu\mathcal{H}u$. Since $zu\mathcal{H}zxzuyu$, from the definition of N it follows that $zu \in N$. This proves that $D_iD_k \subseteq N$.

Let $E_i = \{a \in J \mid a\mathcal{R}x, a\mathcal{L}y \text{ for some } x, y \in D_i\} \cup \{\theta\}$. First, we claim that D_i intersects all nonzero $\mathcal{H}$-classes of J contained in E_i. Choose $x, y \in D_i$. We know that there exist $u, z \in T$ such that $xu, ux, yz, zy \in C_i$. The fact that $ux \sim yz$ implies that there exist $v, v' \in T$ with $vv'\mathcal{H}yz$ and $v'v\mathcal{H}ux$. In particular $yz\mathcal{R}v, ux\mathcal{L}v$. Then clearly $y\mathcal{R}v, x\mathcal{L}v$. Similarly, since $xu \sim zy$, there exist $w, w' \in T$ with $ww'\mathcal{H}xu$ and $w'w\mathcal{H}zy$. Then $xu\mathcal{R}w, zy\mathcal{L}w$, so that $x\mathcal{R}w, y\mathcal{L}w$. It is easy to see that $v, w \in D_i$. (For example, $vv'\mathcal{H}yz$, $yz \in C_i$ imply that $vv' \in C_i$, so $v \notin N$ and hence $v \in D_k$ for some $k \in I$. But $v\mathcal{R}y$ implies then that $k = i$.) This establishes the claim.

Let $a \in E_i, a \neq \theta$. Then $a\mathcal{R}x$ for some $x \in D_i$. Hence there exists $t \in T$ such that $xt \in C_i$. Then $xt \in D_i \subseteq E_i$. Since $xt\mathcal{R}a$, it follows that the $\mathcal{R}$-class of a in J contains a maximal subgroup which is also contained in E_i. The same applies to the $\mathcal{L}$-class of a. Therefore, E_i is a completely 0-simple subsemigroup of J. We have also shown that $U_i^{(j)} = D_i \cup \{\theta\}$ is a uniform subsemigroup of E_i, and hence of J. Note also that $T \cap E_i = D_i$. In fact, we have seen already that every $a \in T \cap E_i$ cannot lie in D_k for $k \neq i$. Hence $a \in D_i \cup N$. Since D_i intersects all $\mathcal{H}$-classes of E_i, there exists $b \in D_i$ such that ab lies in a maximal subgroup of E_i. Therefore $a \notin N$, and so $a \in D_i$, as claimed.

The inclusions $NU_i^{(j)} \subseteq N, U_i^{(j)}N \subseteq N$ are clear because N is an ideal of T. Every nonzero element ab, for $a \in U_i^{(j)}, b \in U_k^{(j)}$, is in the $\mathcal{R}$-class of a and in the $\mathcal{L}$-class of b. Hence $U_i^{(j)}U_k^{(j)} \subseteq N$ if $i \neq k$.

If $su \notin N$ for some $s \in S/S_{j-1}, u \in U_i^{(j)}$, then the fact that $su\mathcal{L}u$ implies that $su \in U_i^{(j)}$. This and a symmetric argument show that $U_i^{(j)}$ is an ideal of S modulo T_j.

Suppose that $U_i^{(j)}NU_i^{(j)} \neq \theta$ for some $i \in I$. Then there exist $a, b \in D_i$ such that $axb \neq \theta$ for some $x \in N$. Now $axb \in N$ because N is an ideal in T. But $axb\mathcal{R}a, axb\mathcal{L}b$, so that $axb \in E_i$ and hence $axb \in D_i$. This contradicts the fact that $N \cap D_i = \emptyset$, completing the proof of 3).

Let $x \in C_i, y \in C_k$ for some $i \neq k$. If $z \in T$ is such that $(xzy)^2 \neq \theta$, then $xzy \in C$. Moreover $xzy\mathcal{R}x$ and $xzy\mathcal{L}y$. Therefore $xzy \in C_i \cap C_k$, a contradiction. This means that $(xzy)^2 = \theta$. It follows that $(egf)^2 = \theta$ for the idempotents $e, f \in J$ such that $e\mathcal{H}x, f\mathcal{H}y$ and every $g \in J$ which is $\mathcal{H}$-related in J to some element $z \in T$. From Corollary 2.13 we then see that there are at most $\binom{n}{j}$ $\sim$-classes on C. It follows that $|I| < \binom{n}{j}$. This completes the proof of the theorem. $\square$

Using the notation of the proof, define a relation ρ on C by: $a\rho b$ if there exist $a_1, \ldots, a_m \in C, m \geq 1$, such that $a_1 = a, a_m = b$, and for every $i = 1, \ldots, m - 1$, the elements a_i, a_{i+1} are $\mathcal{R}$-related or $\mathcal{L}$-related in J. Clearly, this is an equivalence relation on C. Moreover, if $a, b \in C$ and $a\mathcal{R}b$ (respectively $a\mathcal{L}b$), then $ab\mathcal{H}b, ba\mathcal{H}a$ (respectively $ab\mathcal{H}a, ba\mathcal{H}b$), so that $a \sim b$. This implies that the relation ρ is contained in $\sim$. The last paragraph of the proof shows in fact that there are at most $\binom{n}{j}$ ρ-classes on C.

The structure of S is therefore described in terms of the ideal chain $S = S_n \supseteq S_{n-1} \supseteq \cdots \supseteq S_0$, $S_j = S \cap M_j$ (note that we may have $S_i = \emptyset$ for some i). The Rees factors $S_j/S_{j-1} \subseteq M_j/M_{j-1}$ represent the 'layers' of S and are approached via the egg-box pattern on the completely 0-simple semigroup M_j/M_{j-1}. Recall that we adopt the convention $S_j/S_{j-1} = S_j$ if $S_{j-1} = \emptyset$.

Each nonempty $S_{j-1}, j \geq 1$, is an ideal in S_j and the Rees factor $S_{(j)} = S_j/S_{j-1}$ is a 0-disjoint union $\bigcup_\alpha U_\alpha \cup N$ of the maximal nil ideal N of $S_{(j)}$ and of certain semigroups $U_\alpha, \alpha \in A$, intersecting different $\mathcal{R}$- and different $\mathcal{L}$-classes of M_j/M_{j-1}. In particular, $U_\alpha U_\beta \subseteq N$ for $\alpha \neq \beta$. If the minimal rank of matrices in S is $j \geq 1$, so that $S_{j-1} = \emptyset$, then we must have $S_j = U_\alpha, A = \{\alpha\}$ and $N = \emptyset$. Every U_α is an 'order' in a unique completely 0-simple subsemigroup $\widehat{U}_\alpha$ of M_j/M_{j-1}. That is, U_α intersects all nonzero $\mathcal{H}$-classes of $\widehat{U}_\alpha$ and every maximal subgroup H of $\widehat{U}_\alpha$ is generated by $U_\alpha \cap H$. These U_α are uniquely determined by the above conditions. All U_α coming from all possible layers S_j/S_{j-1} are called the uniform components of S. With a slight abuse of language, the inverse image of $U_\alpha \setminus \{\theta\}$ in the corresponding $S_j \setminus S_{j-1}$ (θ denoting the zero of S_j/S_{j-1} here) will sometimes also be called a uniform component of S, depending on whether we view the matrices in $M_j \setminus M_{j-1}$ as elements of M_j/M_{j-1} or $M_n(K)$. If a uniform component U_α of S, treated as a subset of S, is an ideal of S or $U_\alpha \cup \{0\}$ is an ideal of S, then we call U_α an ideal uniform component of S. In this case $\widehat{U}_\alpha$, originally defined as a subset of M_j/M_{j-1}, will be identified with $(\widehat{U}_\alpha \setminus \{\theta\}) \cup \{0\} \subseteq M_n(K)$ (that is, the zero of $\widehat{U}_\alpha$ is replaced by the zero matrix). Also, if $0 \in S$, this allows to include the zero matrix into U_α, so that the latter is indeed an ideal of S.

By the nilpotent components of S we mean the maximal nil ideals N_j of $S_{(j)}$ (or, again, the inverse images in S of the sets of nonzero elements of N_j, if convenient), for all j.

The maximal subgroups of $\widehat{U}_\alpha$ are called the groups associated to S. In particular, the groups associated to S that come from the same U_α are conjugate by Corollary 3.4. Hence, $\widehat{U}_\alpha$ can be viewed as a 'group approximation' of U_α and it is called the completely 0-simple closure of U_α.

Thus, to every $S \subseteq M_n(K)$ there corresponds a collection of at most 2^n associated linear groups (these are linear groups generated by the nonempty intersections $S \cap D$ with the maximal subgroups D of $M_n(K)$) and sandwich matrices over these groups (in fact, over $S \cap D$) arising from Proposition 3.1.

Remark i) If $S \subseteq M_n(K)$ has a zero element θ, then Proposition 2.14 allows us to identify θ with the zero matrix by passing to $S \simeq eSe \subseteq eM_n(K)e \simeq M_t(K)$ for $e = 1 - \theta$ and $t = n - \text{rank}(\theta)$. If we first conjugate θ to a diagonal idempotent, then from the description of the egg-box patterns on $M_n(K), eM_n(K)e$ it is clear that this identification does not affect the structure of S coming from Theorem 3.5.

ii) We note that uniform components can be described in terms of the rank of certain matrices. Namely, $a \in S$ is in a uniform component of S, if $\text{rank}(a) = \text{rank}(xay) = \text{rank}((xay)^2)$ for some $x, y \in S$. From the above proof it follows easily that the uniform components of S consisting of matrices of rank j are the equivalence classes of the relation: $a \leftrightarrow b$ if $a\mathcal{H}xby$ for some $x, y \in S_j \setminus S_{j-1}$, defined on every nonempty $T = S_j \setminus T_j$. As a consequence of Proposition 3.6 we shall see that this relation coincides with the one given by the following condition: there exist $x, y \in S_j \setminus S_{j-1}$ such that $x\mathcal{R}a, x\mathcal{L}b, y\mathcal{R}b, y\mathcal{L}a$.

iii) The proof also shows that any subsemigroup T of a completely 0-simple semigroup J is a 0-disjoint union $\bigcup_{\alpha \in A} U_\alpha \cup N$ for a nil of index two (but not necessarily nilpotent) ideal N of T and uniform subsemigroups U_α of T (A can be an infinite set, though) that satisfy the remaining assertions of the theorem. These U_α are called uniform components of T.

iv) There is a striking analogy of the assertion of Theorem 3.5 with that of Wedderburn's theorem on the structure of finite dimensional algebras. Namely, N_j plays the role of the nilpotent radical, while the uniform components $U_i^{(j)}$ behave like the simple blocks of the algebra modulo the radical. (More exactly, they correspond to 'orders' in simple algebras.) The connection with irreducible representations will be discussed in Chapter 4.

v) It is easy to see that any homomorphic image and any ideal of a π-regular semigroup is also π-regular. Therefore, if $S \subseteq M_n(K)$ is π-regular (for example periodic, Zariski closed or regular), then from Corollary 3.2 it follows that every uniform component of S is completely 0-simple.

vi) If S is finitely generated, then $S \subseteq M_n(R)$ for a finitely generated domain $R \subseteq K$. Therefore, S is a subdirect product of finite subsemigroups $S_\beta, \beta \in B$, which are the images of S in the algebras $M_n(K_\beta)$ for certain finite fields K_β such that R is a subdirect product of $K_\beta, \beta \in B$. The important point here is that the numerical invariants coming from Theorem 3.5 are for all S_β bounded by a function of n.

More can be said on the egg-box pattern of the nilpotent radicals N_j of the factors $S_j/T_j \subseteq M_j/M_{j-1}$ arising from Theorem 3.5. A part of the available information is given below.

Proposition 3.6 *Let S be a subsemigroup of a completely 0-simple semigroup M. Assume that $S = U_1 \cup \cdots \cup U_r \cup N$ where U_i are the uniform components of S and N is the maximal nil ideal of S. Then there exist partitions $X = X_1 \cup \cdots \cup X_{r+1}, Y = Y_0 \cup \cdots \cup Y_r$ of the sets X, Y of $\mathcal{R}$- and $\mathcal{L}$-classes of M respectively, such that for*

$$M_{ij} = \{a \in M \mid a\mathcal{R}x, a\mathcal{L}y \text{ for some } x \in X_i, y \in Y_j\}$$

and

$$N_{ij} = N \cap M_{ij},$$

we have

1. *$N = N_1 \cup N_2 \cup N_3 \cup N_4 \cup \{\theta\}$, where $N_1 = \bigcup_{1 \leq j < i \leq r} N_{ij}$, $N_2 = \bigcup_{i=1}^r N_{i,0}$, $N_3 = \bigcup_{j=1}^r N_{r+1,j}$, $N_4 = N_{r+1,0}$.*

2. *for every $i, j, k \in \{1, \ldots, r\}$, U_i intersects all $\mathcal{H}$-classes of M contained in M_{ii}, and N intersects all $\mathcal{H}$-classes of M contained in M_{ij} or none of them; if $N_{ij} \neq \emptyset$ and $N_{jk} \neq \emptyset$, then $N_{ik} \neq \emptyset$,*

3. *$U_i U_l = \theta$ whenever $1 \leq i < l \leq r$; moreover if $k \leq r, j \geq 1$, then $N_{ij} N_{kl} = \theta$ if $i \leq k$ or $j \leq l$.*

In particular, after a suitable permutation of the sets X, Y, the above partitions of X, Y give the following egg-box pattern on S

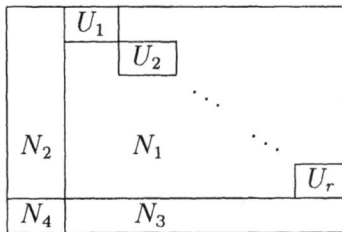

Proof. Let X_i, Y_i be the sets of nonzero $\mathcal{R}, \mathcal{L}$-classes respectively, of M that intersect U_i. Put $X_{r+1} = X \setminus \bigcup_{i=1}^r X_i, Y_0 = Y \setminus \bigcup_{i=1}^r Y_i$.

Let $i, j \in \{1, \ldots, r\}$. Assume that $a \in U_i, b \in U_j, i \neq j$, are such that there exists $z \in S$ with $a\mathcal{R}z$ and $b\mathcal{L}z$. Since U_i, U_j are uniform subsemigroups of M, there exist $c \in U_i, d \in U_j$ such that $c\mathcal{R}a, d\mathcal{L}b$ and $c^2 \neq \theta \neq d^2$. Then $cz \neq \theta$ and $zd \neq \theta$, so that $czd \neq \theta$. If $x \in U_i$ is such that $x\mathcal{L}c$ and $y \in U_j$ satisfies $y\mathcal{R}d$, then $xzy \in U_i z U_j$ satisfies $xzy\mathcal{R}x, xzy\mathcal{L}y$. Since from Theorem 3.5 we know that U_i, U_j intersect all $\mathcal{H}$-classes of the completely 0-simple semigroups M_{ii}, M_{jj} respectively, it follows that $U_i z U_j \subseteq N_{ij}$ intersects all $\mathcal{H}$-classes of M contained in M_{ij}. Suppose additionally that $a\mathcal{L}w, b\mathcal{R}w$ for some $w \in S$. Then a symmetric argument shows that N intersects all $\mathcal{H}$-classes of M contained in M_{ji}. Hence there exists $v \in S$ such that $v\mathcal{R}d$ and $v\mathcal{L}c$. Then $vz\mathcal{H}d, zv\mathcal{H}c$, which implies that $c \sim d$, where $\sim$ is the relation used in the proof of Theorem 3.5. This contradicts the fact that $i \neq j$. We have shown that $N_{ji} = \emptyset$ whenever $N_{ij} \neq \emptyset$.

Assume that $N_{ij} \neq \emptyset$ and $N_{jk} \neq \emptyset$. We have seen that $xzy \neq \theta$ for some $z \in N_{ij}, x \in U_i, y \in U_j$. Also, we can choose $w \in N_{jk}$ with $w\mathcal{R}y$. Then $xzw \neq \theta$. Hence $xzw \in N_{ik}$, so $N_{ik} \neq \emptyset$. This completes the proof of 2).

Suppose that for some $i_1, \ldots, i_m \in \{1, \ldots, r\}, m \geq 2$, we have $N_{i_m i_1} \neq \emptyset$ and $N_{i_j i_{j+1}} \neq \emptyset$ for $j = 1, \ldots, m - 1$. The above shows that $N_{i_m i_{m-1}} \neq \emptyset$. Since we also have $N_{i_{m-1} i_m} \neq \emptyset$, this is a contradiction. Consider the graph $\Gamma = (V, E)$, where $V = \{1, \ldots, r\}$ and $(i, j) \in E$ if $N_{ij} = \emptyset$. From the above it follows that there exists i with $(i, j) \in E$ for every j. Therefore, an induction applied to the graph $\Gamma' = (V', E')$, where $V' = V \setminus \{i\}$ and $E' = E \cap (V' \times V')$, allows to show that there exists a reordering of the set $\{1, \ldots, r\}$ such that $(i, j) \in E$ for $i < j$. This means that $N_{ij} = \emptyset$ whenever $i < j$.

If $N_{ij} N_{kl} \neq \theta$ for some $j \geq 1, k \leq r$, then $U_j U_k \neq \theta$, so that $j \geq k$. But $j < i, l < k$ imply then that $k < i$ and $l < j$. This completes the proof. $\square$

The structural approach resulting from Theorem 3.5 raises the question of the status of arbitrary cancellative subsemigroups of the given $S \subseteq M_n(K)$ as well as of the independence of the structural description of S of the embedding of S into a full matrix monoid $M_m(F)$ for a field F and $m \geq 1$.

First, as seen in Remark after Proposition 3.1, two isomorphic free noncommutative semigroups $X, X' \subseteq GL_n(K)$ can have nonisomorphic completely 0-simple closures. If G is a free nonabelian subgroup of $GL_j(K), j > 1$, (see Chapter 6) then the semigroup M_j/M_{j-1}, defined for $M_n(K)$, contains an isomorphic copy of the semigroup $S_1 = \mathcal{M}(G, 1, 2, P)$ described in this remark. Therefore, the free semigroup $\langle x, y \rangle$ has two faithful representations into M_j/M_{j-1} with nonisomorphic completely 0-simple closures. There is also a further reason for the dependence of the decomposition of S on the matrix representation. For example, if J is a completely prime ideal of $S \subseteq GL_n(K)$ (that is, $S \setminus J$ is a subsemigroup of S), consider the embedding $\phi : S \longrightarrow M_{2n}(K)$ given by $s \mapsto \begin{pmatrix} s & 0 \\ 0 & I \end{pmatrix}$ for $s \in S \setminus J$ and $s \mapsto \begin{pmatrix} s & 0 \\ 0 & 0 \end{pmatrix}$ for $s \in J$. Clearly, S and $\phi(S)$ intersect different number of layers $M_j \setminus M_{j-1}$ of $M_n(K), M_{2n}(K)$ respectively, so their decompositions coming from Theorem 3.5 are different.

So, speaking about uniform components of S we shall always have in mind the given embedding $S \subseteq M_n(K)$. However, the set of the groups associated to a linear semigroup S has in many cases 'maximal members' which are determined uniquely up to isomorphism - they do not depend on the matrix presentation of S.

Proposition 3.7 *Let $S \subseteq M_n(K)$ be a semigroup with no free noncommutative subsemigroups contained in maximal subgroups of $M_n(K)$. Assume that T is a cancellative subsemigroup of S and j the minimal rank of matrices in T. Then T has no noncommutative free subsemigroups and T has a group of quotients H which is isomorphic to the group $\mathrm{gp}(T \cap D)$ for any maximal subgroup D of $M_n(K)$ consisting of matrices of rank j and intersecting T. In particular, if $\phi : S \longrightarrow M_m(K')$ is an embedding, for a field K' and some $m \geq 1$, and G is a group associated to $\phi(S)$, then G is isomorphic to a subgroup of a group associated to S.*

Proof. For $e = e^2 \in D$ we put $I = T \cap eM_n(K)$. Then I is a right ideal of T and $T \cap D$ is a left ideal of I. Since a one-sided ideal of a free noncommutative semigroup contains a free noncommutative semigroup, it follows that T has no such subsemigroups. Therefore T satisfies the Ore condition. Therefore, within the group H of right quotients of T

(which is also its group of left quotients) we have

$$H = TT^{-1} \supseteq II^{-1} \supseteq IT(IT)^{-1} = ITT^{-1}I^{-1} = IHI^{-1} = H.$$

Further

$$
\begin{aligned}
\mathrm{gp}(T \cap D) &= (T \cap D)(T \cap D)^{-1} \supseteq I(T \cap D)(I(T \cap D))^{-1} \\
&= I\,\mathrm{gp}(T \cap D)I^{-1} = II^{-1} = H.
\end{aligned}
$$

Finally, assume that E is a maximal subgroup of $M_m(K')$ such that $U = \phi(S) \cap E \neq \emptyset$. Let $G = \mathrm{gp}(U) \subseteq M_m(K')$. Then $T = \phi^{-1}(U)$ is a cancellative subsemigroup of S and therefore $TT^{-1} \simeq \mathrm{gp}(T \cap D)$ for a maximal subgroup D of $M_n(K)$. Since $U \simeq T$, G embeds into $\mathrm{gp}(T \cap D)$. Now $\mathrm{gp}(T \cap D) \subseteq \mathrm{gp}(S \cap D)$ and the result follows. $\square$

The conditions of Theorem 3.5 imply that the ideal chain $S_0 \subseteq T_1 \subseteq S_1 \subseteq \cdots \subseteq S_n = S$ can be refined in the following way.

Corollary 3.8 *Let $S \subseteq M_n(K)$. Then there exists an ideal chain*

$$I_1 \subset I_2 \subset \cdots \subset I_r = S$$

such that $r \leq 2^n + n - 1$ and I_1 and each Rees factor I_k/I_{k-1} is either nilpotent of index at most $\binom{n}{j}$ for some $j \leq n$ or it is a uniform subsemigroup of some M_j/M_{j-1}.

A chain of this type will be called a structural chain of S. (Note that it may be not unique because of the possible permutation of the order of uniform components coming from the same rank in $M_n(K)$.)

The following corollary generalizes the result on chains of principal ideals of π-regular semigroups [87], Theorem 3.13, improving also the bounds given there.

Corollary 3.9 *Let $S \subseteq M_n(K)$ be a semigroup.*

1. *Assume that $v_1 \cdots v_q \in S_j \setminus S_{j-1}$ for some $j \in \{1, \dots, n\}$ and some $v_i \in S_j$, where $q = 2\binom{n}{j}$. Then there exists a uniform component U of S contained in $S_j \setminus S_{j-1}$ such that $v_k, v_{k+1} \in U$ for some k. In particular $v_k v_{k+1} \in U$.*

2. Let $t = 2^n \prod_{i=1}^{n} \binom{n}{i}$. Assume that $a_1 S^1 \supset a_2 S^1 \supset \cdots \supset a_t S^1 \neq 0$ for some $a_i \in S$. Put $a_0 = 1$. Then there exists a uniform component U of S such that $a_k = a_i u, a_m = a_i u w$ for some $i < k < m$ and $u, w \in U$ such that $u \mathcal{R} u w$ in $M_n(K)$.

3. If S is π-regular, then the length of every chain of nonzero principal right (two-sided) ideals of S is less than t (t^2 respectively).

Proof. Let $S_{(j)} = S_j / S_{j-1}$. Let W be a uniform component of $S_{(j)}$. To prove 1) we can assume that $W \neq S_{(j)}$. Then from Proposition 3.6 it follows that $W(S_{(j)})^1 (S_{(j)} \setminus W)(S_{(j)})^1 W = \theta$ in $S_{(j)}$. Suppose 1) does not hold. Then each W contains at most one element of the set $V = \{v_1, \ldots, v_q\}$. But N_j is nilpotent of index $\leq q/2$ and $WW' \subseteq N_j$ for every uniform component $W' \neq W$ of S_j. Therefore $v_1 \cdots v_q$ must be zero in $S_{(j)}$, a contradiction. Thus, 1) follows.

2) By induction on $n - j$ we show that:

if $a_1 S^1 \supset \cdots \supset a_p S^1$ for some $a_i \in S \setminus S_{j-1}$, and $p \geq r_{j-1} = 2^{n-j+1} \prod_{i=j}^{n} \binom{n}{i}$, then the assertion of 2) is valid for this chain.

For $j = 1$ this will prove 2). If $j = n$, then $S \setminus S_{j-1} \subseteq GL_n(K)$ is a uniform component of S. Thus, the assertion is trivial in this case. Assume that $j < n$. As before, let $q = 2\binom{n}{j}$. Then $r_{j-1} = q r_j$. Suppose that $p \geq r_{j-1}$. Let $v_r \in S$ be such that $a_{r+1} = a_r v_{r+1}, r = 1, \ldots, p - 1$. Put $v_1 = a_1$. If $w_r = v_r \cdots v_{r+r_j-1} \notin S_j$, for some $r \in \{1, \ldots, (q-1)r_j + 1\}$, then the chain $v_r S^1 \supset \cdots \supset v_r \cdots v_{r+r_j-1} S^1$ satisfies the induction hypothesis. Hence, we can find elements u, w in a uniform component U of S such that $u \mathcal{R} u w$ in $M_n(K)$ and either $v_r \cdots v_k = v_r \cdots v_i u, v_r \cdots v_m = v_r \cdots v_i u w$ for some $i < k < m$ with $i \geq r$, or $v_r \cdots v_k = u$ and $v_r \cdots v_m = u w$ for some $k < m, k \geq r$. Then $a_k = a_i u, a_m = a_i u w$, or $a_k = a_{r-1} u, a_m = a_{r-1} u w$, and we are done. Thus, we can assume that every $w_r, r = 1, \ldots, (q-1)r_j + 1$, lies in S_j. But and $a_{q r_j} = w_1 w_{r_j+1} \cdots w_{(q-1)r_j+1} \in S_j \setminus S_{j-1}$. Hence, 1) applied to this product shows that $u = w_{l r_j + 1}, w = w_{(l+1)r_j+1}, u w$ all belong to a uniform component U of S for some $l \in \{0, \ldots, q-2\}$. Then $u \mathcal{R} u w$ in $M_n(K)$. Since $a_{(l+1)r_j} = a_{l r_j} u, a_{(l+2)r_j} = a_{l r_j} u w$, this completes the inductive proof of 2).

3) If S is π-regular, then every uniform component U of S is completely 0-simple by Corollary 3.2. Suppose that $a_1 S^1 \supset \cdots \supset a_t S^1 \neq 0$

for some $a_i \in S$. The elements u, w found in 2) must satisfy $uU^1 = uwU^1$. Hence $uS^1 = uwS^1$ and so $a_k S^1 = a_m S^1$, a contradiction. Therefore, the assertion on chains of principal right ideals follows. Let $r = t^2$. Suppose that there exist $b_1, \ldots, b_r \in S$ such that $S^1 b_1 S^1 \supset \cdots \supset S^1 b_r S^1 \neq 0$. Then $x_i b_i y_i = b_{i+1}$ for some $x_i, y_i \in S^1$. Put $x_0, y_0 = 1$ and define

$$t_{i+1} = b_1 y_0 \cdots y_i, \quad z_{i+1} = x_i \cdots x_0 b_1$$

for $i = 0, \ldots, r-1$. Then $t_1 S^1 \supseteq \cdots \supseteq t_r S^1$ and $S^1 z_1 \supseteq \cdots \supseteq S^1 z_r$. Let $T = \{t_1 S^1, \ldots, t_r S^1\}, Z = \{S^1 z_1, \ldots, S^1 z_r\}$. We know that $|T| < t$ and $|Z| < t$. Let $W = \{1, \ldots, r\}$. Define a function $\phi : W \longrightarrow T \times Z$ by $\phi(i) = (t_i S^1, S^1 z_i)$. Since $r = t^2$, there exist $\alpha, \beta \in W, \alpha \neq \beta$, such that $\phi(\alpha) = \phi(\beta)$. Hence $t_\alpha S^1 = t_\beta S^1, S^1 z_\alpha = S^1 z_\beta$, and so

$$
\begin{aligned}
S^1 b_\alpha S^1 &= S^1 x_{\alpha-1} \cdots x_0 b_1 y_0 \cdots y_{\alpha-1} S^1 = S^1 z_\alpha y_0 \cdots y_{\alpha-1} S^1 \\
&= S^1 z_\beta y_0 \cdots y_{\alpha-1} S^1 = S^1 x_{\beta-1} \cdots x_0 b_1 y_0 \cdots y_{\alpha-1} S^1 \\
&= S^1 x_{\beta-1} \cdots x_0 t_\alpha S^1 = S^1 x_{\beta-1} \cdots x_0 t_\beta S^1 \\
&= S^1 x_{\beta-1} \cdots x_0 b_1 y_0 \cdots y_{\beta-1} S^1 = S^1 b_\beta S^1.
\end{aligned}
$$

This contradicts the supposition and proves 3). $\square$

Our basic approach, resulting from the structure theorem, is to study $S \subseteq M_n(K)$ via its uniform components. The aim here is to reduce problems on S to questions on the associated linear groups and the properties of the corresponding sandwich matrices. This approoach can be successful once we are able to transfer the properties from the cancellative subsemigroups of type $S \cap D$, D a maximal subgroup of $M_n(K)$, to those of the associated groups $\mathrm{gp}(S \cap D)$. However, to understand the global structure of S one often needs also to handle the relationship between different uniform components. It seems that the action of the associated groups on the uniform components of S that lie below is crucial here. This approach may be described as follows.

Lemma 3.10 *Let D be a maximal subgroup of $M_n(K)$ and $e = e^2 \in D$. Consider the monoid $M(e) = eM_n(K)e \simeq M_{\mathrm{rank}(e)}(K)$. If U is a uniform component of $S \subseteq M_n(K)$ intersecting $M(e)$, then $U \cap M(e)$ is a uniform component of $S \cap M(e)$. If additionally S is π-regular, then $S \cap D$ is a group and it acts by conjugation on $U \cap M(e)$.*

Proof. Let j be the rank of matrices in U. Since $M(e) = eM_n(K) \cap M_n(K)e$, from Lemma 1.1 it follows that $M(e)$ cuts a 'rectangle' from $M_j \backslash M_{j-1}$ (viewed via its egg-box pattern). Fix some $f = f^2 \in M_j \backslash M_{j-1}$ such that $N = fM_j \cap U \cap M(e) \neq \emptyset$. We view $V = U \cup \{\theta\}$ as a subsemigroup of M_j/M_{j-1}. Then $N \cup \{\theta\}$ can be considered as a right ideal in $(U \cap M(e)) \cup \{\theta\} \subseteq M_j/M_{j-1}$. Suppose that $N^2 = \theta$ in V. Let $g \in D$. Then $NUg \backslash \{\theta\}$ is contained in a single $\mathcal{R}$-class of M_j/M_{j-1} and so $NVg \subseteq N \cup N_j$, where N_j is the nilradical of S_j/S_{j-1}. Hence $(NVg)N = \theta$ and consequently $(gNV)^2 = \theta$. Since NV is a right ideal of V, it intersects all $\mathcal{H}$-classes of M_j/M_{j-1} that come from the elements of $U \cap fM_j$. Hence gNV and gV intersect the same $\mathcal{H}$-classes. Therefore $(gV)^2 = \theta$.

If $VgV \neq \theta$, then it is a nonzero ideal of V, so it intersects all $\mathcal{H}$-classes intersected by V. Therefore $g(VgV) \neq \theta$ (because g does not annihilate N), which is impossible. Hence $VgV = \theta$. Since $g \in S \cap D$ is arbitrary, Lemma 1.17 implies that $VeV = \theta$. But then $VN = VeN \subseteq VeV = \theta$. Since V is a uniform subsemigroup of M_j/M_{j-1} and $N \subseteq V$, it follows that $N = \theta$, a contradiction. This means that we cannot have $N^2 = \theta$. Therefore, $(U \cap M(e)) \cup \{\theta\}$ has no 'rows' with zero multiplication. A similar argument works for the 'columns', hence $(U \cap M(e)) \cup \{\theta\}$ is a uniform subsemigroup of M_j/M_{j-1}. Every element $a \in (S \cap M(e)) \backslash U$ which is $\mathcal{R}$- or $\mathcal{L}$-related to an element in $U \cap M(e)$ must be in the nilpotent radical N_j of S_j/S_{j-1}. Hence, a lies in the nilpotent radical of $(S \cap M(e))_j/(S \cap M(e))_{j-1}$. This implies that $U \cap M(e)$ is a uniform component of $S \cap M(e)$. The remaining assertion is clear. $\square$

We conclude with some simple examples. In particular they show that, in general, there is no obvious connection between the sandwich matrices and the associated groups of S and those of its closure $\mathrm{cl}(S)$.

Examples. 1. Let $S \subseteq M_n(K)$ be the semigroup of all diagonal matrices. Then S is a semilattice of groups (each of type $(K^*)^j, j \leq n$). S is regular, and at every rank j it has exactly $\binom{n}{j}$ groups (=uniform components of S.) This shows that the bound on the number of uniform components in Theorem 3.5 is sharp.
2. Let S be the semigroup of all upper triangular matrices in $M_n(K)$. S is π-regular by Corollary 1.5. We use the notation of Theorem 3.5.

It is easy to check that $N_j \setminus \{\theta\}$ consists of rank j matrices that have $< j$ nonzero diagonal entries (applying the Jordan form of any matrix a of this type we see that a power of a has rank $< j$, so N_j indeed is a nil ideal of S_j/S_{j-1}.) Also, $S_j/T_j = U_1^{(j)} \cup \cdots \cup U_{\binom{j}{j}}^{(j)}$ for completely 0-simple semigroups U_i with maximal subgroups isomorphic to the group of upper triangular matrices in $GL_j(K)$. Each $U_i^{(j)} \setminus \{\theta\}$ consists of matrices with a given pattern of zero entries on the diagonal ($U_i^{(j)} \setminus \{\theta\}$ is a subsemigroup and $U_i^{(j)} U_k^{(j)} = \theta$ for $k \neq i$, so these indeed are the uniform components of S.) It is easily seen that this also leads to a decomposition of S as a semilattice of nilpotent extensions of completely 0-simple semigroups $U_j^{(i)}$.

3. Let $S \subseteq M_n(K)$ be the set of all monomial matrices. That is, $a \in S$ if each row (and each column) of a has at most one nonzero entry. It is easy to check that S is a regular semigroup whose only ideals are $S_j = S \cap M_j, j = 0, 1, \ldots, n$. Moreover, S_j/S_{j-1} is isomorphic to a completely 0-simple semigroup $\mathcal{M}(G_j, \binom{n}{j}, \binom{n}{j}, I_j)$, where G_j is the group of all monomial matrices in $GL_j(K)$ and I_j is the identity matrix. This can be compared to the monoid R used in Section 2.1.

4. Let $S \subseteq M_n(K)$ be a Zariski closed connected monoid with group of units G. Then S is π-regular and the uniform components of S are of the form GaG, where $\text{rank}(a) = \text{rank}(a^2)$, cf.[108].

5. Let $S = \langle s, e \rangle \subseteq M_2(\mathbf{Q})$, where

$$ s = \begin{pmatrix} a & 1 \\ 0 & a \end{pmatrix}, \ e = \begin{pmatrix} 1 & 0 \\ 1 & 0 \end{pmatrix} $$

for some $a > 0$. Let also $S' = \langle s^{-1}, e \rangle$. First, we claim that the set U of rank one matrices of S is a uniform component of S and $S = U \cup \langle s \rangle$ (so it is a union of its two uniform components.) Clearly, $U = \langle e, es^n, s^n e; n = 1, 2, \ldots \rangle$ and U has no nilpotents. Since $s^n es^k \in U$ for $n, k \geq 0$, U intersects each $\mathcal{H}$-class of $M_2(\mathbf{Q})$ which is $\mathcal{R}$-related to some $s^n e$ and $\mathcal{L}$-related to some es^k. Therefore U is a uniform component of S and the associated group is isomorphic to the subgroup of $\mathbf{Q}^*$ generated by the nonzero entries of the matrices in $\langle es^n e; n = 0, 1, \ldots \rangle$. It is clear that $\text{cl}(S) = \text{cl}(S')$. However, the entries of the first row of each matrix es^{-n} have opposite signs, while the entries of every $t \in S$ are nonnegative. It follows that S' intersects some $\mathcal{L}$-classes of $M_2(\mathbf{Q})$ that do not intersect S. Consequently, $\text{cl}(S)$ intersects more $\mathcal{L}$-classes of $M_2(\mathbf{Q})$ than S.

Let D be the maximal subgroup of $M_2(\mathbf{Q})$ containing e. The $(1,1)$-entry of the matrix $es^{-n}e$ is equal to $(a-n)a^{-n-1}$, so that it is negative for some $n \geq 1$. Therefore the groups G, G' generated by $S \cap D, S' \cap D$, respectively, do not satisfy $G' \subseteq G$. It follows that the maximal subgroup of $\mathrm{cl}(S)$ containing e is bigger than G.

3.3 Closure

Our main objective in this section is to study the closure $\mathrm{cl}(S)$ of a semigroup $S \subseteq M_n(K)$ and to discuss connections between the structure of S and of $\mathrm{cl}(S)$.

When studying the structure of S it is often convenient to replace S by its linear homomorphic image S' in such a way that the given uniform component U of S becomes an ideal of S'.

The following lemma, though rather technical, is essential for our approach. It allows us, in particular, to consider a linear semigroup with an ideal uniform component I as a subsemigroup of a linear semigroup in which $\hat{I}$ is an ideal.

Lemma 3.11 *Assume that U is a uniform component of a semigroup $S \subseteq M_n(K)$. Let $\hat{U}$ be the completely 0-simple closure of U in the corresponding M_j/M_{j-1} and G a maximal subgroup of $\hat{U}$. Then*

1. *S and the semigroup S_G generated by $S \cup G$ intersect the same $\mathcal{H}$-classes of $M_n(K)$ contained in $M_n(K) \setminus M_{j-1}$,*

2. *$\hat{U}$ is a uniform component of S_G. Therefore $\hat{U}$ is an ideal of the Rees factor $\hat{S} = S_G/J$, where*

$$J = \{a \in S_G G S_G | \mathrm{rank}(a) < j \text{ or } \mathrm{rank}((bac)^2) < j \\ \text{for all } b, c \in S_G\}.$$

In particular, if U is an ideal uniform component of S, then $\hat{S} = S \cup \hat{U} = S_G$.

Proof. Since $G \cup I$ generates $\hat{U}$, it is clear that $\hat{U}$ is contained in a uniform component V of S_G. Let $a \in \hat{U}$ be nonzero (the set of nonzero elements of $\hat{U}$ is identified with a subset of S_G) and let $s \in S$. Then, since U is a uniform subsemigroup of $\hat{U}$, there exists $a' \in \hat{U}$ such that

$a'\mathcal{H}a$ in $M_n(K)$. If rank$(as) = j$, then we have $aM_n(K) = asM_n(K)$ and $M_n(K)a' = M_n(K)a$. Hence rank$(a's) = j$ and $as\mathcal{H}a's$. This means that the $\mathcal{H}$-class of as in $M_n(K)$ contains an element of S. The same holds for sa. A similar argument shows that, if $z \in S_G$ is of rank j and is in the $\mathcal{H}$-class of an element of S, then, for any $t \in S \cup G$, each of the elements tz, zt also is in the $\mathcal{H}$-class of $M_n(K)$ intersected by S, whenever it is of rank j. Therefore, an induction on the length of the elements of S_G as words in $S \cup G$ allows to show that S and S_G intersect the same $\mathcal{H}$-classes of $M_n(K)$ consisting of matrices of rank j. In view of the structure theorem (see also Proposition 3.6) this implies that $\widehat{U}$ intersects the same $\mathcal{H}$-classes of $M_n(K)$ as V.

Assume that as is in the $\mathcal{H}$-class of an element of U for some $a \in \widehat{U}$ and $s \in S$. Choose $e \in \widehat{U}$ such that $as = ase$. Since $U\widehat{U} = \widehat{U}$, we can write $e = xy$ for some $x \in U$, $y \in \widehat{U}$. Then $asx\mathcal{H}a'sx$ for an element $a' \in U$ such that $a'\mathcal{H}a$. The structure theorem applied to S implies that $a'sx \in U$ and, since sx and $a'sx$ are in the same $\mathcal{L}$-class of $M_n(K)$, we must also have $sx \in U$. Hence $as = asxy \in \widehat{U}U\widehat{U} \subseteq \widehat{U}$. This shows that $\widehat{U}S \cap V \subseteq \widehat{U}$ because V, U intersect the same $\mathcal{H}$-classes of $M_n(K)$. Similarly one gets $S\widehat{U} \cap V \subseteq \widehat{U}$. Again by an induction argument, this implies that $V = \widehat{U}$ is a uniform component of S_G. Since the matrices of rank j in $S_G G S_G$ are contained in $\widehat{U} \cup N$, where N is the nilpotent component of S_G consisting of matrices of rank j (see Proposition 3.6), $\widehat{U}$ is an ideal of $\widehat{S}$.

Finally, assume that U is an ideal uniform component of S. Suppose that $a \in \widehat{U}$ and $s \in S$ are such that $as \in J$. As above, there exist $v \in U, u \in \widehat{U}$ such that $uv \in \widehat{U}$ satisfies $uva = a$. Choose If $u' \in U$ with $u'\mathcal{H}u$. Then $uvas \in J$ implies that $u'vas \in J \cap U$. Therefore $u'vas = 0$. It follows that $as = 0$ because $u'va\mathcal{L}a$. Similarly, $sa \in J$ implies $as = 0$. This proves that $S \cup \widehat{U}$ is a semigroup and $\widehat{U}$ is its ideal. Consequently $S_G = S \cup \widehat{U}$. Moreover $J = \{0\}$ or $J = \emptyset$, so that $S_G = \widehat{S}$. This completes the proof. $\square$

If $S_G \subseteq M_n(K)$ is the semigroup constructed above, then $\widehat{I}$ can be considered as a subsemigroup of $S_G/(S_G \cap M_{j-1}) \subseteq M_n(K)/M_{j-1}$. In particular, Lemma 1.8 often allows to study $\widehat{I} \subseteq S_G/(S_G \cap M_{j-1})$ as linear semigroups. A further step in this direction will be made in Lemma 3.26.

We are now ready to get some insight into the construction of cl(S).

Remark Let $M_{(0)}, \dots, M_{(n)}$ be the sets of matrices in $M_n(K)$ of ranks $0, \dots, n$ respectively. Put $S_{(i)} = S \cap M_{(i)}$. Let j be maximal with $S_{(j)} \neq \emptyset$. Write $S^{(0)} = S$. Consider the nonempty intersections $S \cap D$ with maximal subgroups D of $M_n(K)$ whose elements are of rank j. We define $S^{(1)} = \langle S, \bigcup(S \cap D)^{-1} \rangle$, where $(S \cap D)^{-1}$ denotes the set of inverses in D of elements of $S \cap D$ and the summation runs over all such D. Next, construct $S^{(2)} \supseteq S^{(1)}$ proceeding in the same way with respect to $S^{(1)}$ and the set $S^{(1)}_{(j-1)}$ of matrices of rank $j-1$ in $S^{(1)}$. After j steps we reach a subsemigroup $S^{(j)}$ of $M_n(K)$. It is clear that, in each step $r \geq 1$, $S^{(r)}_{(k)} = S^{(r-1)}_{(k)}$ for $k \geq j - r + 2$. Moreover, from Lemma 3.11 it follows that

$$S^{(r)}_{(j-r+1)} = \langle S^{(r-1)}_{(j-r+1)}, (A^{(r-1)}_{(j-r+1)})^{-1} \rangle \cap M_{(j-r+1)},$$

where $A^{(m)}_{(k)}$ denotes the set of 'group elements' in $S^{(m)} \cap M_{(k)}$. (Note also that, in each step r, it is enough to invert the elements of $S \cap D$ for only finitely many maximal subgroups D in $M_{(j-r+1)}$ - one for each uniform component of $S^{(r)}$ coming from this level.) This easily implies that $S^{(j)} = \mathrm{cl}(S)$. Moreover, each element of $S^{(r)}$ is a word $w(s_1, \dots, s_q)$ in some $s_1, \dots, s_q \in S^{(r-1)}$ that allows local inverses of those s_i that lie in $A^{(r-1)}_{(j-r+1)}$. Therefore, each $z \in \mathrm{cl}(S)$ is an iterated word $z = w_j(w^{(1)}_{j-1}(\dots), \dots, w^{(n_j-1)}_{j-1}(\dots))$ of this type.

Next we show that $\mathrm{cl}(S)$ is determined by the closures of finitely generated subsemigroups of S.

Lemma 3.12 *Let $S \subseteq M_n(K)$ be a semigroup. Then*

1. *$\mathrm{cl}(S) = \bigcup_T \mathrm{cl}(T)$, where the union runs over all finitely generated subsemigroups T of S,*

2. *for every finitely generated subsemigroup R of $\mathrm{cl}(S)$ there exists a finitely generated subsemigroup T of S such that $R \subseteq \mathrm{cl}(T)$.*

Proof. Let $A = \bigcup_T \mathrm{cl}(T)$. If $s_1, s_2 \in A$, then the definition of A implies that there exist finitely generated subsemigroups T_1, T_2 of S such that $s_i \in \mathrm{cl}(T_i)$ for $i = 1, 2$. Clearly, $s_1 s_2 \in \mathrm{cl}(\langle T_1, T_2 \rangle)$. Since $\langle T_1, T_2 \rangle \subseteq S$ is finitely generated, it follows that $s_1 s_2 \in A$. Hence A is a semigroup.

Moreover, $\mathrm{cl}(T_1)$ is π-regular, so $s_1^m = s_1^m t s_1^m$ for some $t \in \mathrm{cl}(T)$ and $m \geq 1$. Therefore, A is π-regular. Since $S \subseteq A \subseteq \mathrm{cl}(S)$, it follows that $A = \mathrm{cl}(S)$.

Assume that $R = \langle v_1, \dots, v_k \rangle$ for some $v_1, \dots, v_k \in \mathrm{cl}(S)$. From 1) it follows that $v_i \in \mathrm{cl}(V_i)$ for a finitely generated subsemigroup V_i of S, $i = 1, \dots, k$. Then $R \subseteq \mathrm{cl}(\langle V_1, \dots, V_k \rangle)$ and $\langle V_1, \dots, V_k \rangle \subseteq S$ is finitely generated. The assertion follows. $\square$

The following result explains the basic relation between the components of S and the components of its closure.

Theorem 3.13 *Let U be a uniform component of $S \subseteq M_n(K)$ and N a nilpotent component of S. Then*

1. *$N = N' \cap S = N'' \cap S$ for some nilpotent components N', N'' of the Zariski closure $\overline{S}$, and of $\mathrm{cl}(S)$ respectively,*

2. *$U = U' \cap S = U'' \cap S$ for some uniform components U', U'' of $\overline{S}, \mathrm{cl}(S)$ respectively.*

Proof. We view N as a subset of the appropriate $M_j \setminus M_{j-1}$. If $z \in N$, then $(zx)^2 \in M_{j-1}$ for every $x \in S$. Since M_{j-1} is a closed subset of $M_n(K)$, it follows that $(zx)^2 \in M_{j-1}$ for every $x \in \overline{S}$. Thus, $z\overline{S}$ is a right ideal of $\overline{S}$ which is nil modulo M_{j-1}. Therefore $z \in N'$, where N' is the nilpotent component of $\overline{S}$ consisting of matrices of rank j. Hence $N \subseteq N'$. Clearly, $N' \cap S$ is a nil ideal of S modulo M_{j-1}. Hence $N' \cap S \subseteq N$, and so $N = N' \cap S$. Since $S \subseteq \mathrm{cl}(S)$, we have $\overline{\mathrm{cl}(S)} = \overline{S}$. The above applied to $\mathrm{cl}(S)$ yields $N' \cap \mathrm{cl}(S) = N''$, where N'' is the nilpotent component of $\mathrm{cl}(S)$ consisting of matrices of rank j. Hence $N'' \cap S = N' \cap S = N$.

Assume that U_1, U_2 are uniform components of S consisting of matrices of rank j. Let $N \subseteq M_j \setminus M_{j-1}$ be the nilpotent component of S. From Theorem 3.5 it follows that $U_1 \subseteq U_1', U_2 \subseteq U_2'$ for some uniform components U_i' of $\overline{S}$. Also, if $U_1 \neq U_2$ and $a \in U_1, b \in U_2$, then $axb \in N$, so that $(axby)^2 \in M_{j-1}$, for all $x, y \in S$. As above, this implies that $(axby)^2 \in M_{j-1}$ for all $x, y \in \overline{S}$. Hence a, b are not in the same uniform component of $\overline{S}$ (and hence of $\mathrm{cl}(S)$) so that $U_1' \neq U_2'$. Now $U_1 \subseteq U_1' \cap S$ and it follows that the latter does not intersect uniform components of S other than U_1. Moreover $U_1' \cap S$ does not intersect $N = N' \cap S$ because

$U_1' \cap N' = \emptyset$. Hence $U_1 = U_1' \cap S$. Similarly, since $U_1 \subseteq U_1''$ for a uniform component U_1'' of $\mathrm{cl}(S)$ and $\overline{\mathrm{cl}(S)} = \overline{S}$, we get $U_1 \subseteq U_1'' = \mathrm{cl}(S) \cap V$ for a uniform component V of $\overline{S}$. Hence $U_1 \subseteq V, U_1 \subseteq U_1'$ imply that $V = U_1'$. This completes the proof. $\square$

The following example shows that $\mathrm{cl}(S)$ may have more uniform components than S.

Example Let $S = \langle x, e \rangle \subseteq M_3(\mathbf{Q})$, where

$$x = \begin{pmatrix} 1 & 1 & 0 \\ 1 & 0 & 0 \\ 0 & 0 & 1 \end{pmatrix}, \quad e = \begin{pmatrix} 1 & 0 & 0 \\ 0 & 0 & 0 \\ 0 & 0 & 1 \end{pmatrix}.$$

Then $ex^{-1}e = \begin{pmatrix} 0 & 0 & 0 \\ 0 & 0 & 0 \\ 0 & 0 & 1 \end{pmatrix} \in \mathrm{cl}(S)$. The $(1,1)$ entry of x^n is positive, so that $ex^n e$ is a matrix of rank 2 for every $n \geq 1$. The same is true of every $a \in \langle ex^n, x^n e, ex^n e \,|\, n \geq 1 \rangle$. This easily implies that the set of elements of rank 2 in S forms a uniform component U of S and $S = U \cup \langle x \rangle$. Therefore $\mathrm{cl}(S)$ has more components than S because $ex^{-1}e$ lies in a uniform component of $\mathrm{cl}(S)$.

As noted before, the uniform components of $\mathrm{cl}(S)$ can intersect more $\mathcal{R}$- (and also more $\mathcal{L}$-) classes of $M_n(K)$ than those of S. However, this does not happen in the following special case.

Proposition 3.14 *Let U be a uniform component of a semigroup $S \subseteq M_n(K)$ such that U intersects finitely many $\mathcal{R}$-classes of $M_n(K)$. Then the uniform components U', U'' of the Zariski closure $\overline{S}$, and of $\mathrm{cl}(S)$ respectively, containing U intersect the same $\mathcal{R}$-classes of $M_n(K)$ as U.*

Proof. Let R be the union of the $\mathcal{R}$-classes of $M_n(K)$ intersecting U. Then $R = \bigcup_{i=1}^{r} e_i M_n(K) \setminus M_{j-1}$ for some $e_i = e_i^2 \in M_n(K), r \geq 1$, where j is the rank of matrices in U. Each $e_i M_n(K)$ is a closed subset of $M_n(K)$. Hence $Z = R \cup M_{j-1}$ is closed. Choose $u \in U$. If $s \in S$, then either $su \in M_{j-1}$ or $su\mathcal{L}u$. In the latter case $su \in U \cup N$, where N is the nilpotent component of S consisting of matrices of rank j. Hence $(sut)^2 \in Z$ for every $s, t \in S$, so it follows that we also have $(sut)^2 \in Z$ for every $s, t \in \overline{S}$.

Let $W = \{w \in U' | w\mathcal{L}u \text{ for some } u \in U\} \setminus R$. Suppose that $W \neq \emptyset$. Choose $w \in W$. Since U is a uniform component of S, we can find $u \in U$ such that $w\mathcal{L}u$ and $u^2 \in U$. Hence $wu\mathcal{H}w$ in $M_n(K)$. The first paragraph of the proof shows that $(wut)^2 \in Z$ for every $t \in \overline{S}$. But $wu \in U'$ and U' is a uniform component of $\overline{S}$. Hence t can be chosen so that $wut \in D$ for a maximal subgroup D of $M_n(K) \setminus M_{j-1}$. So $(wut)^2 \in D$. Now $(wut)^2 \in Z = R \cup M_{j-1}$ implies that $wut \in R$. On the other hand, since $wut\mathcal{R}w$ and $w \notin R$, it follows that $wut \notin R$, a contradiction. This shows that $W = \emptyset$, proving the assertion on U'. Since $U'' \subseteq U'$, this completes the proof. $\square$

Uniform components of the type considered above have also the following important property.

Proposition 3.15 *Assume that a uniform component U of a semigroup $S \subseteq M_n(K)$ intersects finitely many $\mathcal{R}$-classes of $M_n(K)$. Then for every maximal subgroup G of $\mathrm{cl}(S)$ intersecting U we have $G = \mathrm{gp}(S \cap G)$.*

Proof. We will show that, for every extension $S' \subseteq S'_D = \langle S', (S' \cap D)^{-1} \rangle$, for a maximal subgroup D of $M_n(K)$, used in the construction of $\mathrm{cl}(S)$ (see the remark preceding Lemma 3.12), we have $\mathrm{gp}(S' \cap G) = \mathrm{gp}(S'_D \cap G)$. Since $\mathrm{cl}(S)$ is obtained from S in finitely many steps of this type, the assertion will follow.

On the other hand, if V is the uniform component of some $S'_D \subseteq \mathrm{cl}(S)$ and $U \subseteq V$, then the pair V, S'_D inherits the hypothesis on U, S by Proposition 3.14. Therefore, it is enough to prove this for $S' = S$.

Let V be the uniform component of $\mathrm{cl}(S)$ containing $S \cap G$. Let W be the union of $\mathcal{H}$-classes of $M_n(K)$ intersecting U. From Proposition 3.14 we know that V intersects the same $\mathcal{R}$-classes of $M_n(K)$ as U. It is enough to show that $S_D \cap W \subseteq \widehat{U}$ because $\mathrm{gp}(S \cap G)$ is a maximal subgroup of $\widehat{U}$ and $S_D \cap G \subseteq S_D \cap W$. Consider an arbitrary element $x \in S_D \cap W$. Then $x = s_1 b_1 s_2 \ldots b_{k-1} s_k$ for the inverses b_i of some elements of $S \cap D$ and some $s_i \in S \cup \{1\}$. Clearly, we may assume that $k \geq 2$. There exist $u, w \in U$ such that $uxw \in W$. If we show that $uxw \in \widehat{U}$, then it is easy to see that also $x \in \widehat{U}$, as desired. Therefore, replacing s_1 by us_1 and s_k by $s_k w$, we can assume that $s_1, s_k \in U$ and $x_1 = s_1 b_1 \ldots s_{k-1}, x_2 = b_k s_k$ are contained in V. Hence $x_2 \in W$ (use the definition of W and the fact that $x_2\mathcal{L}s_k$ and $s_k \in U$).

First, consider the element x_2. Write $b = b_k, s = s_k$. Let b is the inverse in D of an element $a \in S \cap D$. It is clear that the group $H = gp(S \cap D)$ acts transitively on the set of $\mathcal{R}$-classes of $M_n(K)$ intersected by Hs by left multiplication. In view of Theorem 3.5 this implies that $Hs \subseteq V$. Therefore, $as \in V$. Since $as \in S$, we have $as \in S \cap V \subseteq U$, see Theorem 3.13. Since $bs \in W$, we must have $bs = fbs$ for an idempotent $f \in \hat{U}$. Now

$$as = (a^2 f)(bs) \in U \subseteq V,$$

which implies in particular that $a^2 f \in V$. Hence $a^2 f \in V \cap S_G = \hat{U}$ by Lemma 3.11. Therefore $as \in U, bs \in W$ imply in view of the displayed equality that $bs \in \hat{U}$. In particular, this shows that $x_2 \in \hat{U}$.

Let F be the maximal subgroup of $\hat{U}$ containing f. Since $fx_2 = x_2$, we have $x = (x_1 v)(z x_2) \in W$, where $v \in U \cap F$ and $z \in F$ is the inverse of v in F. We know that $z x_2 \in \hat{U}$ because $x_2 \in \hat{U}$. Moreover $x_1 v = s_1 b_1 \cdots b_{k-2}(s_{k-1} v) \in W$ has a shorter presentation than x. Therefore, an induction allows us to assume that $x_1 v \in \hat{U}$. Then $x = (x_1 v)(z x_2) \in \hat{U}$, as desired. This proves our claim on S_D, completing the proof of the proposition. □

Even more can be said in case S is finitely generated.

Proposition 3.16 *Assume that U is a uniform component of a semi-group $S \subseteq M_n(K)$ such that U intersects finitely many $\mathcal{R}$-classes of $M_n(K)$. If S is finitely generated, then the group associated to S that comes from U also is finitely generated.*

Proof. Choose $e = e^2 \in D$ for a maximal subgroup D of $M_n(K)$ that intersects U. If $\text{rank}(e) = j$, then the image I of $U \cup \{0\}$ in M_j/M_{j-1} is a uniform subsemigroup. Put $I' = \hat{I}e$, where $\hat{I}$ is the completely 0-simple closure of I in M_j/M_{j-1} and let G denote the subgroup of D generated by $U \cap D = S \cap D$. Here $K_0[\hat{I}]$ is an ideal of $K_0[S_G/T]$ for an ideal T of the semigroup S_G constructed in Lemma 3.11 (see Corollary 3.8). $G = e\hat{I}e \setminus \{\theta\}$ acts on I' by right multiplication. Hence $K_0[I']$ is a right $K[G]$-module. Note that it is a free $K[G]$-module of finite rank, say r, by the hypothesis. Indeed, picking one element from every nonzero $\mathcal{H}$-class of $\hat{I}$ intersected by I', we get a basis $e_1, \ldots, e_r$ of this module. We can choose $e_1 = e$. Now, each $s \in S$ acts by left multiplication on $K_0[I']$, so it determines an endomorphism of $K_0[I']$. Hence, we get a homomorphism

$\phi : K_0[S] \longrightarrow M_r(K[G])$, which maps S into column monomial matrices $M_r(G \cup \{0\})$ over G^0. By the hypothesis $\phi(S) = \langle \phi(s_1), \ldots , \phi(s_t) \rangle$ for some $s_i \in S$. Let $H \subseteq G$ be the group generated by the nonzero entries of all matrices $\phi(s_i)$. It is clear that $S \cap D \subseteq H$. Since G is generated by $S \cap D$, it follows that $G = H$ is a finitely generated group. $\square$

3.4   0-simple semigroups

From the point of view of the ideal structure of a semigroup S the knowledge of 0-simple semigroups that can arise as principal factors of S is crucial. The aim of this section is to discuss this problem for linear semigroups S. We focus on the nature of 0-simple linear semigroups, showing in particular that they are uniform semigroups of certain special type, the study of which reduces in some sense to simple subsemigroups of linear groups.

We start with a very useful, but not widely known, result of Jones, [50]. Recall that a 0-simple semigroup with a nonprimitive idempotent contains the bicyclic semigroup, that is the semigroup given by the presentation $B = \langle p, q \rangle, pq = 1$. Moreover, a 0-simple π-regular semigroup must be completely 0-simple, [15], Theorem 2.54 and Theorem 2.55.

Proposition 3.17 *Assume that S is a 0-simple semigroup with no nonzero idempotents. Then*

1. *The bicyclic semigroup is a homomorphic image of a subsemigroup of S.*

2. *If the relation $\mathcal{R}$ is trivial on S, then S contains a free noncommutative subsemigroup.*

Proof. If S has a zero element, then it is denoted by θ. Suppose that $x = xy$ for some nonzero $x, y \in S$. 0-simplicity of S implies that there exist $s, t \in S$ such that $y = sxt$. Hence $x = xy = xsxt$ and $sx = (sx)^2 t$, where $sx \neq \theta$. Therefore $(sx)\mathcal{R}(sx)^2$ in S.

First, consider the case where $\mathcal{R}$ is trivial on S. The above shows that $x \neq xy$ for every $\theta \neq x, y \in S$. There exists $b \in S$ such that $b^2 \neq \theta$ (otherwise S is nil, so it is π-regular and hence completely 0-simple, which is not possible because S is nil). Now $b = ab^2c$ for some nonzero $a, c \in S$. We claim that the elements ab, a^2b generate a free subsemigroup of S.

Suppose that u, v are distinct (nonempty) words in ab, a^2b that are equal in S. Let w be the longest common initial segment of u, v as words in ab, a^2b (w can be the empty word). Then $u = wu_1, v = wv_1$ for some words u_1, v_1. Note that u_1, v_1 are not empty. (Otherwise $w = wv_1$ or $w = wu_1$, respectively, in S, which is not possible because $\mathcal{R}$ is trivial on S.) Let for example $u_1 = (a^2b)u_2$ and $v_1 = (ab)v_2$, where u_2, v_2 may now be empty. For any $i \geq 1$ inductively we show that

$$b = ab^2c = (ab)(ab^2c)c = (ab)^2bc^2 = \cdots = (ab)^ibc^i.$$

Next, $b = ab^2c = a(ab^2c)bc = (a^2b)b(cbc)$, so that by induction

$$ab = a(ab^2c) = (a^2b)bc = (a^2b)^2b(cbc)c = \cdots = (a^2b)^ib(cbc)^{i-1}c.$$

Multiplying $u_1 = (a^2b)u_2$ on the right by elements of the tye $bc^i, b(cbc)^ic$ several times (depending on whether the resulting element ends with a power of ab or a power of a^2b) we can find, in view of the two displayed identities, an element $x \in \langle a, b, c \rangle$ such that $u_1x = ab$. Then

$$wab = wu_1x = wv_1x = (wab)v_2x.$$

This contradicts the hypothesis on S, and establishes 2).

To prove 1) it is now enough to consider the case where $\mathcal{R}$ is nontrivial on S. Then $xy = x \neq 0$ for some $x, y \in S$, so by the first paragraph of the proof there exist $a, b \in S \setminus \{0\}$ such that $a = a^2b$. We will show that $T = \langle a, b \rangle$ maps onto the bicyclic semigroup $B = \langle p, q \mid pq = 1 \rangle$. Using the relation $a^2b = a$ we can write each $t \in T$ in the form $t = v(ab)^ma^n$, where $m \geq 0, n \geq 0$ and $v \in V = \langle b, (ab)b, (ab)^2b, \ldots \rangle^1$. Put $s = ab$. Let $l(v)$ denote the length of v in V. Then $as = a$ implies that $a^{l(v)}v = s$ for all $v \in V \setminus \{1\}$ and also

$$(*) \qquad a^{l(v)+1}v = a, \quad a^{l(v)+1}(vs^ma^n) = a^{n+1} \text{ for all } v \in V.$$

We claim that, for any two elements $vs^ma^n, v's^ia^j \in T$ as above

$$vs^ma^n = v's^ia^j \text{ implies that } l(v) = l(v'), n = j.$$

Assume for example that $l(v) \geq l(v')$. By $(*)$ applied twice

$$a^{n+1} = a^{l(v)+1}(vs^ma^n) = a^{l(v)+1}(v's^ia^j) = a^{l(v)-l(v')}a^{j+1}.$$

Since a is not nilpotent (because $\theta \neq a = a^2 b = \cdots = a^k a b^k$ for all $k \geq 1$) and S has no nonzero idempotents, it follows that $n - j = l(v) - l(v')$. Suppose that $l(v) > l(v')$. Since $a^{l(v)} v = s$ and $a^k b^k = s$ for $k \geq 1$, we get

$$a^{l(v)}(vs^m a^n)b^n = (a^{l(v)}v)s^{m+1} = s^{m+2}$$

and

$$
\begin{aligned}
a^{l(v)}(v's^i a^j)b^n &= a^{l(v)-l(v')}(a^{l(v')}v')s^i(a^j b^j)b^{n-j} \\
&= a^{l(v)-l(v')}s^{i+2}b^{n-j} = a^{n-j}b^{n-j} = s.
\end{aligned}
$$

Then $s^{m+2} = s$, hence a power of s is a nonzero idempotent, a contradiction. Thus $l(v) \leq l(v')$, so that $l(v) = l(v')$ and $n = j$, as claimed.

Now, the formula $\phi(vs^m a^n) = q^{l(v)}p^n$ defines a surjection $\phi : T \longrightarrow B$. We shall repeatedly use the equalities $as = a, a^{l(v)}v = s$ for $v \neq 1$. It is easy to see that $l(ws^m v) = l(w) + l(v)$ for $v, w \in V, v \neq 1$.

Let $v' \in V$ and $i, j \geq 0$. Write $z = vs^m a^n v's^i a^j, x = vs^m a^n, y = v's^i a^j$. Assume that $v' = 1$. Then $z = vs^{m+i}a^{n+j}$ if $n = 0$ and $z = vs^m a^{n+j}$ if $n \neq 0$. Hence, in both cases $\phi(z) = p^{l(v)}q^{n+j} = p^{l(v)}q^n p^0 q^j = \phi(x)\phi(y)$.

Hence, assume that $v' \neq 1$. First consider the case where $l(v') \geq n$. Then $z = vs^m a^n v's^i a^j = vs^{m+1}ws^i a^j$ for some $w \in V$ with $l(w) = l(v') - n$ if $n \neq 0$, and $z = vs^m v's^i a^j$ if $n = 0$. Hence, in both cases $\phi(z) = q^{l(v)+l(v')-n}p^j = q^{l(v)}p^n q^{l(v')}p^j = \phi(x)\phi(y)$.

Finally, if $l(v') < n$, then $z = vs^m a^n v's^i a^j = vs^m a^{n-l(v')}s^{i+1}a^j = vs^m a^{n-l(v')+j}$. Hence, $\phi(z) = q^{l(v)}p^{n-l(v')+j} = \phi(x)\phi(y)$ in this case, too. Thus, ϕ is the desired homomorphism. $\square$

Corollary 3.18 *Let J be a 0-simple principal factor of a semigroup $S \subseteq M_n(K)$. If S has no free noncommutative subsemigroups or J contains a nonzero idempotent, then J is a completely 0-simple semigroup.*

Proof. Suppose that the relation $\mathcal{R}$ is nontrivial on J. Then $x = xy \in J$ for some nonzero $x, y \in J$. Since J is 0-simple and it is of the form A/B for some ideals $B \subset A$ of J, it follows that the set $A \setminus B$ of nonzero elements of J can be considered as a subset of M_j/M_{j-1} for some j. Then $x = xy \neq \theta$ in M_j/M_{j-1} leads to $y = y^2$ by Lemma 1.1. Thus, in view of Proposition 3.17, either of the hypotheses implies that J has a nonzero idempotent. But S cannot have infinite chains of idempotents.

Therefore J has a primitive idempotent, so it is completely 0-simple. $\square$

We now concentrate on 0-simple linear semigroups. Several ideas and examples presented below come from [126], where simple semigroups of matrices were considered.

Lemma 3.19 *Let $S \subseteq M_n(K)$ be a nontrivial 0-simple semigroup. Then there exists j such that the nonzero elements of S have rank j and the image $\tilde{S}$ of S in M_j/M_{j-1} is a uniform subsemigroup of M_j/M_{j-1} isomorphic to S. Moreover, for every maximal subgroup D of $M_n(K)$, $S \cap D$ is a simple subsemigroup of D whenever it is nonempty.*

Proof. The first assertion is an immediate consequence of the structure theorem. We only have to identify the zero of S (if it has one) with the zero of the corresponding M_j/M_{j-1}.

Let $a \in S \cap D$. If $b \in S \cap D$, then $b = xa^3y$ for some $x, y \in S$ because $Sa^3S = S$. Thus, $xa\mathcal{R}b$ and $ay\mathcal{L}b$ in $M_n(K)$, which implies that $xa, ay \in S \cap D$. This means that $(S \cap D)a(S \cap D) = S \cap D$, as desired. $\square$

If $|S| \neq 1$, then define $\tilde{S}$ as the image of $S \subseteq M_n(K)$ in $M_n(K)/M_{j-1}$, where j is the minimal rank of a nonzero element of S. (Note that, if S has a zero, it is not necessarily the zero matrix.)

Theorem 3.20 *The following conditions are equivalent for any semigroup $S \subseteq M_n(K), |S| \neq 1$,*

1. *S is 0-simple,*

2. *$S = S^2$, $\tilde{S}$ is a uniform subsemigroup of some M_j/M_{j-1} and $S \cap D$ is simple for every maximal subgroup D of $M_n(K)$ intersecting S,*

3. *$\tilde{S}$ is a uniform subsemigroup of some M_j/M_{j-1}, $S \cap D$ is simple for every maximal subgroup D of $M_n(K)$ intersecting S and $S \cap H_1 H_2 = (S \cap H_1)(S \cap H_2)$ for every $\mathcal{H}$-classes H_1, H_2 of $M_n(K)$ intersecting S,*

4. *$S \setminus \{\theta\} = \bigcup y(S \cap D)x$ for a maximal subgroup D of $M_n(K)$ such that $S \cap D$ is simple, where the summation runs over $x \in X \cap S, y \in Y \cap S$ for some subsets X, Y of the $\mathcal{R}$-, respectively $\mathcal{L}$-class of the idempotent $e \in D$ in $M_n(K)$.*

Proof. Assume that S is 0-simple. By Lemma 3.19 $\tilde{S}$ is a uniform subsemigroup of M_j/M_{j-1} for some j. Let $s \in H_1, t \in H_2$ for $\mathcal{H}$-classes H_1, H_2 of $M_n(K)$ not contained in M_{j-1}. Let $S^{(s)}, S_{(t)}$ be the intersections of $\tilde{S}$ with the $\mathcal{L}$-class of s, respectively $\mathcal{R}$-class of t in M_j/M_{j-1}. Then $S^{(s)}S_{(t)}$ is an ideal of $\tilde{S}$. If it is nonzero, we must have $\tilde{S} = S^{(s)}S_{(t)}$. What contributes to the $\mathcal{H}$-class of st is $(S \cap H_1)(S \cap H_2)$. Hence 3) is a consequence of 1).

If $\tilde{S}$ is uniform, then for every $a \in S$ there exist $b, c \in S$ such that $bc\mathcal{H}a$ in $M_n(K)$. Therefore, 2) follows from 3).

If 2) holds, then it is easy to see that every nonzero ideal I of S contains the set $\bigcup S \cap D$ where the summation runs over all maximal subgroups of $M_n(K)$. From Proposition 2.14 it follows that S/I it must be nilpotent. Since $S = S^2$, this means that $I = S$, which establishes 1).

Assume that 4) holds. The definition of a uniform semigroup is then satisfied for $\tilde{S}$, viewed as a subsemigroup of $M_j/M_{j-1}, j = \text{rank}(e)$. This and the fact that $S \cap D$ is simple imply that every nonzero ideal of S contains $S \cap D$. Therefore it contains S, hence S is 0-simple.

If S is 0-simple and $s \in S \cap D$, then, using the above notation, we come to $\tilde{S} = (S^{(s)}S_{(s)})(S^{(s)}S_{(s)}) = S^{(s)}(S \cap D)S_{(s)}$. Hence, 4) is a consequence of 1). This completes the proof. $\square$

Corollary 3.21 *Assume that $S \subseteq M_n(K)$ is a 0-simple semigroup and j is the common rank of nonzero elements of S. If M_j/M_{j-1} is identified with $\mathcal{M}(GL_j(K), X, Y, P)$, then for every subsets X', Y' of X, Y respectively, the corresponding subsemigroup $\tilde{S} \cap \mathcal{M}(GL_j(K), X', Y', P') \subseteq M_j/M_{j-1}$ also is 0-simple whenever it is a uniform subsemigroup of M_j/M_{j-1}.*

Proof. This is a direct consequence of condition 3) in Theorem 3.20. $\square$

Assume that condition 4) of the theorem is satisfied. Then, conjugating $e = e^2 \in D$ to the appropriate diagonal form, we may define $T = \{t \in GL_j(K)| \begin{pmatrix} t & 0 \\ 0 & 0 \end{pmatrix} \in S \cap D\}$. Then the nonzero elements of S are of the form

$$\begin{pmatrix} u & 0 \\ c & 0 \end{pmatrix}\begin{pmatrix} h & 0 \\ 0 & 0 \end{pmatrix}\begin{pmatrix} v & d \\ 0 & 0 \end{pmatrix} = \begin{pmatrix} uhv & uhd \\ chv & chd \end{pmatrix} = \begin{pmatrix} g & gb \\ ag & agb \end{pmatrix},$$

where $h, u, v \in T \simeq S \cap D$, $g = uhv$, $a = cu^{-1}$, $b = v^{-1}d$, and a, b are rectangular matrices of the appropriate sizes. Denote the displayed matrix by $s(g, a, b)$. For fixed a, b let $T_{ab} = \{g \in T| \; s(g, a, b) \in S\}$. Let $j = \text{rank}(e)$. Without loss of generality, assume that S has a zero θ. So, there exist sets A, B of rectangular $(n - j) \times j$, $j \times (n - j)$ matrices respectively, and a collection of subsets $T_{ab} \subseteq GL_j(K)$, $a \in A, b \in B$, such that

$$S = \{s(g, a, b)| \; a \in A, b \in B, g \in T_{ab}\} \cup \{\theta\}$$

where

$$T_{ab}(I + ba')T_{a'b'} \subseteq T_{ab'} \quad \text{if} \quad \text{rank}(I + ba') = j$$

with $e = \begin{pmatrix} I & 0 \\ 0 & 0 \end{pmatrix}$. Moreover $T_{(ab)} = \bigcup T_{ab}$, where the summation runs over all a, b with $s(g, a, b)$ in a given $\mathcal{H}$-class of $M_n(K)$, is a simple subsemigroup of $GL_j(K)$ if $\text{rank}(I + ba) = j$.

Using the notation of the proof we see also that, if $S \subseteq M_n(K)$ is 0-simple, then $S_{(t)}S^{(s)} = S \cap D$ for every $s, t \in \tilde{S}$ such that $s\mathcal{R}e, t\mathcal{L}e$, because $(S \cap D)^2 \subseteq S^{(s)}S_{(t)}$. The proof of Proposition 3.1 shows, in view of condition 4) that $S = \bigcup_\alpha S_\alpha$ for a collection of (0-simple) semigroups of matrix type $I_\alpha \simeq \mathcal{M}(S \cap D, X, Y, P_\alpha)$. However, as shown in Example 2 below, S is not necessarily isomorphic to a semigroup of the latter form.

We continue with some examples of 0-simple linear semigroups that are not completely 0-simple.

Examples. 1. It is well known and easy to check that

$$S_1 = \left\{ \begin{pmatrix} a & b \\ 0 & 1 \end{pmatrix} \Big| \; a, b \in \mathbf{R}^+ \right\}, S_2 = \left\{ \begin{pmatrix} a & b \\ 0 & c \end{pmatrix} \Big| \; a, b, c \in \mathbf{R}^+ \right\}$$

are simple subsemigroups of $GL_2(\mathbf{R})$.
2. Let

$$S = \left\{ \begin{pmatrix} a & b & 0 \\ 0 & c & 0 \\ a & b & 0 \end{pmatrix}, \begin{pmatrix} a & b+c & 0 \\ 0 & c & 0 \\ 0 & 0 & 0 \end{pmatrix} \Big| \; a, b, c \in \mathbf{R}^+ \right\}.$$

It is easy to see that $S = (S \cap D_1) \cup (S \cap D_2)$ for two maximal subgroups

D_1, D_2 of $M_3(\mathbf{R})$. These groups are determined by the idempotents

$$e_1 = \begin{pmatrix} 1 & 0 & 0 \\ 0 & 1 & 0 \\ 1 & 0 & 0 \end{pmatrix}, e_2 = \begin{pmatrix} 1 & 0 & 0 \\ 0 & 1 & 0 \\ 0 & 0 & 0 \end{pmatrix}.$$

Theorem 3.20 implies that S is 0-simple (use condition 2)). The completely 0-simple closure of S in M_2/M_1 is isomorphic to $\mathcal{M}(G, 2, 1, P)$, where G is the group of 2×2 upper triangular matrices with positive diagonal entries and $p_{11} = p_{12} = 1$. However, the simple subsemigroups

$$S \cap D_1 \simeq \left\{ \begin{pmatrix} a & b+c \\ 0 & c \end{pmatrix} \right\}, S \cap D_2 \simeq \left\{ \begin{pmatrix} a & b \\ 0 & c \end{pmatrix} \right\}$$

are not isomorphic.

3. Let H, G be subgroups of a group D. Assume that GxH is a subsemigroup of D for some $x \in D$. Then $xHGx \subseteq GxH$. Hence, for any $g \in G, h \in H$ there exist $g_1, g_2 \in G, h_1, h_2 \in H$ such that

$$(GxH)gxh(GxH) = G(xHgx)hGxH \supseteq G(g_1xh_1)hGxH$$
$$= Gg_1(xh_1hGx)H \supseteq Gg_1(g_2xh_2)H = GxH.$$

Therefore $S = GxH$ is a simple semigroup. This allows to construct simple subsemigroups in $GL_2(K)$ that are not groups. For example, if $G = H$ is the group of diagonal matrices in $SL_2(\mathbf{R})$ and x is a transvection, it is easy to check that S is of this type.

The theorem above reduces the description of 0-simple subsemigroups of $M_n(K)$ to simple subsemigroups of $GL_j(K)$ for $j \leq n$. However, not many examples of the latter type have been constructed. An example of a simple irreducible subsemigroup of $GL_2(K)$ that is not a group will be presented in Section 4.1. It is a subsemigroup of a finitely generated group. However, if S itself is finitely generated, examples of this type cannot be constructed.

Proposition 3.22 *A finitely generated 0-simple semigroup $S \subseteq M_n(K)$ is completely 0-simple.*

Proof. We know that $S = \langle s_1, \ldots, s_m \rangle$ embeds into M_j/M_{j-1} for some $j \leq n$. It is clear that the completely 0-simple closure of S in M_j/M_{j-1}

intersects finitely many $\mathcal{H}$-classes of the latter semigroup. Choose $t_0 \in A = \{s_1, \ldots, s_m\}, t_0 \neq \theta$. Since $S^2 = S$, we have $t_0 = t_1 u_1$ for some $u_1 \in S$ and $t_1 \in A$. Next $t_1 = t_2 u_2$ and similarly $t_i = t_{i+1} u_{i+1}$ for all $i \geq 1$, where $u_i \in S, t_i \in A$. There exist $j < k$ such that $t_j = t_k$. Therefore $t_k = t_j = t_k u_k \cdots u_{j+2} u_{j+1}$. This implies that $e = u_k \cdots u_{j+1}$ is a nonzero idempotent, cf. Lemma 1.1. The assertion follows because $e \in S$ must be a primitive idempotent. $\square$

3.5 Some reduction techniques

The aim of this section is to establish some technical results that provide useful reductions when the structure theorem is applied. They often allow to simplify the structural chains of S and the structure of the groups associated to S. One can then approach many problems by induction on the length of a structural chain or on certain numerical invariants of the associated groups.

We start with a result saying that every uniform component of S is, in a rather strong sense, controlled by any of its intersections with maximal subgroups of $M_n(K)$.

Lemma 3.23 *Let U be a uniform component of a semigroup $S \subseteq M_n(K)$. Fix any maximal subgroup D of $M_n(K)$ that intersects U. Then there exists a finite set $Z \subseteq U$ such that for every $a \in U$ there exist $x, y \in Z$ with $xay \in S \cap D$.*

Proof. We view U as a subset of M_j/M_{j-1} for some $j \leq n$. From Lemma 1.6 we know that $V = \Lambda^j(U) \cup \{0\}$ is a subsemigroup of the full linear monoid of dimension $t = \binom{n}{j}$ over K. Moreover, every nonzero matrix in V has rank 1 and V is a uniform subsemigroup of the completely 0-simple subsemigroup M_1 of $M_t(K)$.

Let $W = \bigcap W_u \subseteq K^t$ where $W_u = \ker(u)$ and the intersection runs over all $u \in V, u \neq 0$. There exist $w_1, \ldots, w_r \in V$, $r \leq t$, such that $W = W_{w_1} \cap \cdots \cap W_{w_r}$. Suppose that $w_i v = 0$ for some $v \in V, v \neq 0$, and every $i = 1, \ldots, r$. Then $\text{Im}(v) \subseteq \ker(w_i)$, so that $\text{Im}(v) \subseteq W_{w_1} \cap \cdots \cap W_{w_r}$. Thus $\text{Im}(v) \subseteq W_u$ for every $0 \neq u \in V$. This means that $Vv = 0$, contradicting the fact that V is a uniform subsemigroup of M_1. Hence, for each $0 \neq v \in V$ there exists k such that $w_k v \in V \setminus \{0\}$. A dual argument allows to find elements $z_1, \ldots, z_q, q \leq t$, such that for every $0 \neq v \in V$ we have $vz_i \in V \setminus \{0\}$ for some i. Therefore $w_k v z_i \neq 0$. Let $B \subseteq U$ be a finite set that contains an inverse image of $\{w_1, \dots, w_r, z_1, \dots, z_q\}$. From Lemma 1.6 it follows that, if $a \in U$, then there exist $x, y \in B$ such that $xay \in U$.

Let T be the union of all $\mathcal{H}$-classes H of $M_n(K)$ such that for every $h \in H$ there exist $b \in B, c \in B$ with $c\mathcal{R}h$ and $b\mathcal{L}h$. For every $\mathcal{H}$-class H of $M_n(K)$ contained in T, there exist $x', y' \in U$ such that $x'(U \cap H)y' \subseteq U \cap D$ (because $U \cup \{\theta\}$ is a uniform subsemigroup of M_j/M_{j-1}). Since there are finitely many such $\mathcal{H}$-classes and $xay \in U$, the result follows. $\square$

If U is an ideal uniform component of S, then the sandwich matrix of U actually comes from its finite submatrix in a very simple way.

Lemma 3.24 *Assume that $U \subseteq M_j \setminus M_{j-1}$ is a uniform component of a semigroup $S \subseteq M_n(K)$ such that $U \cup \{0\}$ is a subsemigroup of $S \cup \{0\}$. Let D be a maximal subgroup of $M_n(K)$ intersecting U and $e = e^2 \in D$, $\mathrm{rank}(e) = j$. Let $X \subseteq U \cap M_n(K)e$ be a set of representatives of the nonzero $\mathcal{R}$-classes of $M_n(K)$ intersected by U and $Y \subseteq U \cap eM_n(K)$ a set of representatives of the nonzero $\mathcal{L}$-classes of $M_n(K)$ intersected by U. Let $X' \subseteq X, Y' \subseteq Y$ be bases of the subspaces of $M_n(K)$ spanned by the sets X, Y, respectively. Denote by C the set of $\mathcal{H}$-classes of $M_n(K)$ that are $\mathcal{R}$-related to some $x \in X'$ and $\mathcal{L}$-related to some $y \in Y'$. Then $T = \bigcup_{H \in C}(S \cap H) \cup \{\theta\} \subseteq M_j/M_{j-1}$ is a uniform subsemigroup of M_j/M_{j-1} contained in $U \cup \{\theta\}$. Moreover $|X| \leq jn, |Y| \leq jn$, and the sandwich matrix $P = (p_{yx})_{x \in X, y \in Y}$ of U can be chosen so that for every $x \in X$ and $y \in Y$ we have*

$$p_{yx} = \begin{cases} \overline{y}^t P' \overline{x} & \text{if it is in } D \\ \theta & \text{otherwise} \end{cases}$$

where $P' = (p_{yx})_{x \in X', y \in Y'}$ and $\overline{u}, \overline{v}$ are the coordinate vectors of u, v in the bases X', Y', respectively.

Proof. Suppose that $aT = 0$ in $M_n(K)$ for some $a \in T \setminus \{\theta\}$. For every $u \in U$ there exists $q \in X$ such that $q\mathcal{R}u$. Hence $q = \sum_i \lambda_i x_i$, where $\lambda_i \in K$ and $x_i \in X' \subseteq T$. It follows that $aq = 0$, so that $au = 0$. Therefore $aU = 0$. This contradicts the fact that $U \cup \{\theta\}$ is a uniform subsemigroup of M_j/M_{j-1}. A dual argument shows that $Ta \neq 0$ in $M_n(K)$ for $\theta \neq a \in$

T. This implies that each 'row' (and each 'column') of T intersects a maximal subgroup of M_j/M_{j-1}. Therefore, T is a uniform subsemigroup of M_j/M_{j-1}.

Let $x \in X, y \in Y$. Then $x = \sum_i \alpha_i x_i, y = \sum_k \beta_k y_k$ for some $\alpha_i, \beta_k \in K$ and $x_i \in X', y_k \in Y'$. Hence $yx = \sum_{i,k} \alpha_i \beta_k y_k x_i$. Since yx, and every $y_k x_i$, is either the zero matrix or it is in D, it is enough to define $p_{yx} = yx$ for $x \in X, y \in Y$, (see the proof of Proposition 3.1 for the way a sandwich matrix of U can be determined). It is clear that the cardinality of X, and of Y, does not exceed jn. $\square$

From Lemma 1.8 we know that every $S/(S \cap M_j)$ is an 'almost linear' semigroup and it is linear whenever S is finitely generated. This allows us often to reduce the study of U to the case where the matrices in U have the least rank among all nonzero matrices in S. In particular, Lemma 3.24 implies that the sandwich matrix of every uniform component of S is determined by its finite submatrix. The next reduction step is given below. It allows to 'cut off' the nilpotent component from the given layer of S and also to 'separate' the uniform components of this layer.

Lemma 3.25 *Let K be a finitely generated field. Assume that $S \subseteq M_n(K)$ is a semigroup and J is the ideal of matrices in S of the least nonzero rank, and zero if it is in S. Let $U_1, \ldots, U_r$ be the uniform components of J and N the maximal nilpotent ideal of J. Then there exist congruences $\rho_1, \ldots, \rho_r$ on S such that the natural homomorphism $\pi : S \longrightarrow S/\rho_1 \times \cdots \times S/\rho_r \times S/J$ satisfies the following conditions:*

1. *the kernel of the induced homomorphism $K_0[S] \longrightarrow K_0[\pi(S)]$ of the contracted semigroup algebras is nilpotent,*

2. *$\pi(S)$ is a linear semigroup such that the ideal $\pi(J)$ consists of all matrices in $\pi(S)$ of minimal nonzero rank, and zero if it is in $\pi(S)$, and $\pi(J)$ has no nonzero nilpotent ideals,*

3. *each $S/\rho_j \times S/J$ is a linear semigroup such that U_j/ρ_j (identified with $U_j/\rho_j \times \{\theta\}$) is its ideal uniform component containing all matrices of the least nonzero rank,*

4. *the homomorphism $S \longrightarrow S/\rho_j$ is one-to-one when restricted to every intersection $U_j \cap D$ with a maximal subgroup D of $M_n(K)$. In particular, the groups associated to U_j and U_j/ρ_j are isomorphic.*

Proof. We may assume that $0 \in S$, adjoining it to S, if necessary. We shall also assume that $J \cap GL_n(K) = \emptyset$, because otherwise the assertions are clear with π the identity map. Let $\widehat{U}_i$ be the smallest completely 0-simple subsemigroup of $M_n(K)$ containing U_i. From Lemma 3.11 it follows that $S \subseteq S' \subseteq M_n(K)$ for $S' = (S \setminus J) \cup J'$ with $J' = \widehat{U}_1 \cup \cdots \cup \widehat{U}_r \cup N'$ and N' the nil radical of J'. Clearly, $\widehat{U}_i$ are the uniform components of J'. Since, under any homomorphism of S', the image of $\widehat{U}_i$ is a completely 0-simple closure of the image of U_i, it follows that it is enough to prove all assertions for S'. Therefore, we shall assume that $S = S'$.

Let $I_j = \{x \in K\{S\} | U_j S^1 x S^1 U_j = 0\}$, where $K\{S\}$ denotes the subalgebra spanned in $M_n(K)$ by S. Define congruences $\rho_j, j = 1, \ldots, r$, on S by

$$a\rho_j b \text{ if } a - b \in I_j \text{ for } a, b \in S.$$

Since S/ρ_j embeds into the finite dimensional algebra $K\{S\}/I_j$, it is a linear semigroup. From Theorem 3.5 we know that $U_j N U_j = 0$ and $U_j U_k U_j = 0$ for $j \neq k$. Hence the homomorphism $\pi_j : S \longrightarrow S/\rho_j$ maps $J \setminus U_j$ to the zero of S/ρ_j. Suppose that $s, t \in J$ are such that $\pi_j(s) = \pi_j(t)$ is nonzero. Then $s, t \in U_j$ and $K_0[U_j S^1](s - t)K_0[S^1 U_j] = 0$ in $K_0[S]$.

By Lemma 1.8 $S/\rho_1 \times \cdots \times S/\rho_r \times S/J$ is a linear semigroup because $J = S \cap M_{k-1}$ for some $k < n$. Moreover, choosing for S/J in Lemma 1.8 r big enough, we can ensure that every two matrices of different ranks are mapped to matrices of different ranks under the homomorphisms $S \longrightarrow S/\rho_j \times S/J$ and under π. In particular, $\pi(J), \pi_j(J) \times \{\theta\}$ contain all matrices of minimal nonzero rank in $\pi(S)$ and $\pi_j(S) \times S/J$, respectively. But $\pi_j(J) = U_j/\rho_j$. Therefore, conditions 2) and 3) are satisfied.

Put $\rho = \bigcap_{j=1}^r \rho_j$. It is clear that $\pi(S)$ can be identified with S/ρ', where $s\rho't$ if and only if $s = t$ or $s, t \in J$ and $s\rho t$. Moreover, the kernel L of $K_0[S] \longrightarrow K_0[\pi(S)]$ is the subspace of $K_0[S]$ spanned by the set $\{s - t | s\rho't\}$. Assume that $s_i\rho't_i$ for some $s_i, t_i \in J, i = 1, 2, 3$. Let $z = (s_1 - t_1)(s_2 - t_2)(s_3 - t_3)$. If all s_i, t_i are in U_j for some j, then we have seen that $z = 0$ in $K_0[S]$. Otherwise $z \in K_0[N]$ because $U_j U_k \subseteq N$ for $j \neq k$. Hence $L^3 \subseteq K_0[N]$. Since $N^n = 0$, L is nilpotent. This shows that 1) holds.

It is clear that ρ_j is trivial on cancellative subsemigroups of U_j, so

4) follows. $\square$

Using the notation of the lemma above, let $J_j = (J \backslash U_j) \cup \{0\}$. Clearly π factors through the composition of the natural homomorphisms

$$S \longrightarrow S/N \longrightarrow S/J_1 \times \cdots \times S/J_r.$$

U_j can be considered as an ideal of S/J_j. In order to make the latter a linear semigroup we had to factor out the congruence coming from the middle annihilator of U_j. The point is that the resulting kernel is rather small.

Our next reduction procedure is concerned with homomorphic images of the groups associated to S. It extends a classical result on linear groups. The proof heavily depends on the techniques developed in this case [83], [134]. It is based on a refinement of the proof of [134], Theorem 6.4, see Theorem 6.1. Recall that an element g of a subgroup G of $M_n(K)$ is called unipotent if all its eigenvalues $\lambda \in \overline{K}$ lie in the set $\{0, 1\}$ and g is semisimple if it is conjugate in $M_n(\overline{K})$ to a diagonal matrix.

Proposition 3.26 *Let $T = S \cap D \neq \emptyset$ for a semigroup $S \subseteq M_n(K)$ and a maximal subgroup D of $M_n(K)$. Assume that $G \subseteq D$ is the group generated by T and H is a normal subgroup of G which is Zariski closed in G. Let U be the uniform component of S containing T and $S_G = \langle S, G \rangle \subseteq M_n(K)$. Then there exists a homomorphism $\phi : S_G \longrightarrow M_r(K), r \geq 1$, such that*

1. $\phi(T) \subseteq D'$ for a maximal subgroup D' of $M_r(K)$,

2. the subgroup G' of D' generated by $\phi(T)$ contains $D' \cap \phi(S)$ and is isomorphic to G/H' for some $H' \subseteq H$,

3. $f = f^2 \in \phi(G)$ is a diagonal matrix and $\phi(H)$ consists of scalar matrices in $f M_r(K) f$,

4. $J = \phi(U)$ is an ideal uniform component of $\phi(S)$ and G' is a maximal subgroup of the completely 0-simple closure $\hat{J}$ of J in $M_r(K)$.

Moreover, the image of any unipotent (respectively, semisimple) element of G is a unipotent (semisimple) element of $\phi(G)$. In particular, if H is unipotent, then $\phi(H) = f$ and $\phi(G) \simeq G/H$.

Proof. Lemma 3.11 implies that it is enough to find a homomorphism ϕ such the corresponding assertions hold for $S_G, G, \hat{U}$ in place of S, T, U. In fact, if $\phi : S_G \longrightarrow M_r(K)$ is such a homomorphism, then $D' \cap \phi(S) \subseteq D' \cap \phi(S_G) = D' \cap \phi(e)\phi(S_G)\phi(e) = D' \cap \phi(G) = G'$, where $e = e^2 \in G$. Since $\phi(U)$ intersects all $\mathcal{H}$-classes of the completely 0-simple ideal $\phi(\hat{U})$ of $\phi(S_G)$, all assertions will follow. Therefore, we consider the case $S = S_G$ only. In particular $T = G$.

From Lemma 3.25 it follows that, passing to a homomorphic image of S, we can assume that U is an ideal uniform component of S. We will find a homomorphism $\phi : S \longrightarrow M_r(K), r \geq 1$, such that $\phi(G) \simeq G/H'$ for a normal subgroup H' of G contained in H and 3) is satisfied. Then $\phi(U)$ (with zero, if it is in $\phi(S)$) is a completely 0-simple ideal of $\phi(S)$ and $\phi(e)\phi(S)\phi(e) = \phi(eUe) = \phi(G)$ is a maximal subgroup of $\phi(S)$ (or $\phi(G) \cup \{0\}$ if $0 \in \phi(S)$). Let D' be the maximal subgroup of $M_r(K)$ containing $\phi(e)$. Then, again $D' \cap \phi(S) = D' \cap \phi(e)\phi(S)\phi(e) = \phi(G)$. Hence, assertions 1) - 4) will follow.

To construct ϕ we shall proceed along the lines of the proof of Theorem 6.4 in [134]. We may assume that $e = \begin{pmatrix} I & 0 \\ 0 & 0 \end{pmatrix}$. Let $\text{rank}(e) = k$. Let KY, KX denote the polynomial rings in indeterminates $X_{ij}, i, j = 1, \ldots, n$; $X_{ij}, i, j = 1, \ldots, k$, respectively, that are the coordinate rings of $M_n(K)$ and $eM_n(K)e$. Let J be the annihilator ideal of H in KX, m the maximal total degree of any polynomial from a fixed finite generating set for J, A_m the subspace of polynomials of degree $\leq m$ in $KY, J_m = J \cap A_m$. Then each $s \in S$ acts on A_m via the map $g \longrightarrow g^s$, defined by $g^s(Y) = g(sY)$. Here Y denotes the collection of the appropriate indeterminates written in the form of an $n \times n$ matrix and sY corresponds to the product of s and Y, hence the substitution of $\sum_k s_{ik}X_{kj}$ in place of X_{ij}. This yields a homomorphism of S into $\text{End}_K A_m$, inducing on A_m a structure of a $K[S]$-submodule of KY. We may identify S with its image in $\text{End}_K(A_m)$.

Let $E(A_m)$ be the exterior algebra of A_m and let $\sigma : \text{End}_K A_m \longrightarrow \text{End}_K E(A_m)$ be the homomorphism induced by the exterior power, see Section 1.3. Now $T_m = A_m \cap KX$ is a $K[G]$-submodule of A_m and $\text{End}_K T_m$ is mapped by σ into $\text{End}_K E(T_m) \subseteq \text{End}_K E(A_m)$.

Let $P_1, \ldots, P_t$ be a basis of J_m over K and $w = P_1 \wedge \cdots \wedge P_t \in E(A_m)$. Let $W = \sum_{s \in S} \sigma(s)(w)K \subseteq E(A_m)$. It is a $K[S]$-module with a $K[G]$-submodule $W_G = \sum_{g \in G} \sigma(g)(w)K$. This determines a representation

$\sigma' : S \longrightarrow \text{End}_K W$ such that $\sigma'_{|G} : G \longrightarrow \text{End}_K W_G$.

As in the proof of Theorem 6.4 in [134] we see that H is the normalizer of J_m in G. From the proof of Theorem 6.3 in [134] we know that $\sigma'(H)$ coincides with the set of scalar matrices in $\sigma'(G) \subseteq fM_r(K)f$, where $\text{End}_K W$ is identified with $M_r(K), r = \dim_K W$, and $f = \sigma'(e) = \begin{pmatrix} I & 0 \\ 0 & 0 \end{pmatrix}$. Indeed, if $h \in H, g \in G$, then by Lemma 1.6

$$\sigma'(h)\sigma'(g)(w) = \sigma'(g)\sigma'(g^{-1}hg)(w) =$$
$$\sigma'(g)(\det(g^{-1}hg)w) = \det(h)\sigma'(g)(w),$$

where det stands for the appropriate 'local determinant' of the elements of H as matrices in $\text{End}_K(A_m)$. So, $\sigma'(h)$ acts as the scalar $\det(h)$ on W. Moreover, the kernel of this representation restricted to G is contained in H.

Finally, in each of the steps in the proof, unipotent (semisimple) matrices of G are mapped to unipotent (semisimple) matrices of the respective image of G. Therefore, the result follows. $\square$

The following example shows that, in general, one cannot accomplish the isomorphism $\phi(G) = G/H$ in the above reduction step.

Example Let $S \subseteq M_2(\mathbf{Q})$ be the multiplicative semigroup of all matrices of rank at most one. S is completely 0-simple with a Rees presentation $S = \mathcal{M}(Q^*, X, Y, P)$, where Y is the set of all reduced row echelon forms of nonzero matrices of S, X is the transpose of Y, and the sandwich matrix $P = (p_{yx})_{y \in Y, x \in X}$ is such that p_{yx} is equal to the $(1,1)$-entry of the matrix yx. Consider the sequence y_n of elements of Y with the first row $(1, n), n = 1, 2, \ldots$, and the sequence of elements x_n of X with the first column $(1, -1/n)^t, n = 1, 2, \ldots$. It is clear that the submatrix $Q = (q_{ij})_{i,j \geq 1}$ of P determined by these sequences satisfies $q_{ii} = 0$ and $q_{ij} \neq 0$ for $i \neq j$. Let $\phi : Q^* \longrightarrow \{1\}$ be the trivial homomorphism. Let $\phi(P) = (\hat{p}_{yx})$, where $\hat{p} = 0$ if $p = 0$ and $\hat{p} = 1$ otherwise. It is easy to see that the determinant of every submatrix $\hat{Q}_n = (\hat{q}_{ij})_{i,j=1,\ldots,n}$, $n > 1$, is nonzero. Let $S' = \mathcal{M}(\{1\}, X, Y, \phi(P))$. The above implies that the contracted semigroup ring $\mathbf{Q}_0[S']$ contains subalgebras isomorphic to the matrix algebras $M_n(\mathbf{Q})$ for all $n \geq 1$ (see Proposition 4.13).

Let $e \in X \cap Y$ be the unique idempotent. Then $G = eSe \setminus \{0\}$ is a maximal subgroup of S, $G \simeq \mathbf{Q}^*$. Suppose that $\pi : S \longrightarrow M_t(\mathbf{Q})$

is a homomorphism such that $\pi(G)$ is trivial. Then π factors through S'. This is a contradiction because π determines a homomorphism of algebras $\mathbf{Q}_0[S] \longrightarrow M_t(\mathbf{Q})$.

We conclude with a result which shows that the completely 0-simple closure $\widehat{U}$ of a uniform component U of S can be viewed, in some cases, as a localization. We say that the sandwich matrix P of $\widehat{U}$ is invertible over the group ring $K[G]$ if $\widehat{U} \simeq \mathcal{M}(G, r, r, P), r < \infty$, and P has an inverse in the matrix ring $M_r(K[G])$. Clearly, a Rees matrix presentation of $\widehat{U}$ is chosen here, but it is well known that, if we also have $\widehat{U} \simeq \mathcal{M}(G, r, r, P')$, then P' is invertible over $K[G]$ (see Section 1.1). Note also that $K_0[\widehat{U}] \simeq \mathcal{M}(K[G], r, r, P)$ (the Munn algebra over $K[G]$) and the latter is isomorphic to $M_r(K[G])$ via the map $a \mapsto a \circ P$, where $\circ$ denotes the usual matrix product, as explained in Section 4.2.

Lemma 3.27 *Assume that U is a uniform subsemigroup of its completely 0-simple closure $\widehat{U}$ such that the sandwich matrix is invertible over $K[G]$ and $U \cap G$ satisfies the right Ore condition for every maximal subgroup G of $\widehat{U}$. Then $K_0[\widehat{U}]$ is a right localization of $K_0[U]$.*

Proof. Write $\widehat{U} = \mathcal{M}(G, r, r, P)$, with P invertible in $M_r(K[G])$. Recall that the element of $\widehat{U}$ with $g \in G$ at the (i, j)-entry, and zeroes elsewhere, is denoted by (g, i, j). Let $\widehat{U}_{ij}$ be the set of nonzero elements of $\widehat{U}$ that are in row i and column j. Put $U_{ij} = \widehat{U}_{ij} \cap U$. We use the fact that $\widehat{U}_{j1} = U_{j1} U_{i1}^{-1}$, where $\widehat{U}_{i1}$ is a group from the first column. Indeed, $p_{1i} \neq 0$, and $U_{i1} U_{i1}^{-1} = \widehat{U}_{i1}$ by the hypothesis, so that

$$(*) \qquad U_{j1} U_{i1}^{-1} \supseteq U_{j1} U_{i1} U_{i1}^{-1} \supseteq U_{j1} \widehat{U}_{i1} = \widehat{U}_{j1}$$

In particular, for every element $y \in \widehat{U}_{j1}$ there exists $u \in U_{i1}$ such that $yu \in U_{j1}$, because $e = e^2 \in \widehat{U}_{i1}$ is a right identity for U_{j1}.

View $K_0[U]$ as a subset of $M_r(K[G])$, see Section 1.1, (but they have different multiplication of course). Define the subset $C = \{P^{-1} \circ s \mid s \in A\}$ of $M_r(K[G])$, where A is the set of diagonal matrices with entries in G. Clearly C consists of invertible elements in $K_0[\widehat{U}]$. Let C' consist of those elements of C that lie in $K_0[U]$ (when treated as elements of $K_0[\widehat{U}]$). It is enough to show that for every matrix $z \in K_0[\widehat{U}]$ there exists $c = P^{-1} \circ s \in C'$ such that $zc \in K_0[U]$. But $zc = z \circ P \circ c = z \circ s$. So we need to find s such that $z \circ s$ and $P^{-1} \circ s$ are in $K_0[U]$. Hence, for each $1 \leq q \leq r$ and for the finitely many elements from the support of the q-th columns of P^{-1} and z (we treat them as elements of $K_0[\widehat{U}]$) it is enough to find $t = s_q = (g, q, q) \in U$ such that t multiplies these elements (on the right) into U.

For simplicity assume that $q = 1$. So, given a finite collection of elements $x_k = (h_k, i_k, 1)$ in $\widehat{U}$, we need an element $t = (g, 1, 1) \in U_{11}$ such that $x_k \circ t = (h_k g, i_k, 1) \in U$ for every k. But $U_{11} U_{i1} \subseteq U_{11}$ implies that replacing x_k by $x'_k = x_k \circ a$ for any fixed $a \in U_{11}$ it is enough to find an element $u \in U_{i1}$ such that $x'_k u \in U_{i_k 1}$ for all k.

Existence of such u follows from $(*)$ above ($(*)$ allows first to find u for the first of the elements x'_k, and then adjust it step by step by right multiplication by elements of U_{i1} so that it works for all x'_k - like in the process of finding a common denominator for finitely many fractions). This proves the lemma. □

Chapter 4

Irreducible semigroups

A semigroup $S \subseteq M_n(K)$ is irreducible if $S \neq \{0\}$ and the column vector space K^n has no proper S-invariant subspaces, where S acts on K^n by left multiplication. In other words, K^n is an irreducible $K\{S\}$-module. As in the case of linear groups, this class has certain special properties and it provides a general technique for dealing with arbitrary semigroups of matrices. In this chapter we first discuss general results on irreducible semigroups. Then, the theory of irreducible representations of a semigroup $S \subseteq M_n(K)$ is set up in terms of the structure of S. As an intermediate step, we describe all irreducible representations of a completely 0-simple semigroup. Finally, as an application, results on triangularizability of linear semigroups are obtained. The basic information on the properties of irreducible linear groups can be found in [83], [134].

4.1 Structure

If K is algebraically closed then, by Burnside's theorem, S is irreducible if and only if it is absolutely irreducible, that is $K\{S\} = M_n(K)$. For an arbitrary field K, the following result is an easy consequence of the standard facts on the structure of finite dimensional algebras.

Lemma 4.1 *Assume that $S \subseteq M_n(K)$ is an irreducible semigroup. Then $K\{S\}$ is a simple algebra and $1 \in K\{S\}$. Moreover, if S has a zero element, then either it is the zero matrix or $S = \{1\}$ and $n = 1$.*

Proof. $K\{S\}$ is a semisimple algebra because otherwise the Jacobson radical J of $K\{S\}$, being nilpotent, yields a proper S-invariant subspace JK^n of K^n. If $K\{S\}$ has a central idempotent $e \neq 0$, then eK^n is an invariant subspace. Therefore $e = 1$, so that $K\{S\}$ must be simple. If S has a zero element θ, then, in view of Proposition 2.14, we must have $\theta = 0$ or $\theta = 1$. The assertion follows. □

For any subset $A \subseteq M_n(K)$ by $R(A)$ we denote the row space of A, that is, the space spanned by the rows of all matrices in A. $C(A)$ stands for the column space of A.

Lemma 4.2 *If $S \subseteq M_n(K)$ is irreducible, then S satisfies any of the following equivalent conditions*

   1. $\dim_K R(S) = \dim_K C(S) = n$,

   2. $K\{S\}K^n = K^n$ and if $Sv = 0$ for some $v \in K^n$, then $v = 0$,

   3. if $Sx = 0$ or $xS = 0$ for some $x \in M_n(K)$, then $x = 0$.

Proof. Assume that 1) holds. Let $e_1, \ldots, e_n$ be the standard basis of K^n. Then $K\{S\}K^n = K\{S\}(\sum_i e_i K) = \sum_i K\{S\}e_i = C(S) = K^n$. Moreover, $Sv = 0$ implies that, for every $w \in R(S)$, the scalar product of w and v is zero, so we must have $v = 0$. Hence 2) is a consequence of 1). If 2) holds and $xS = 0$ for some $x \in M_n(K)$, then $0 = xK\{S\}K^n = xK^n$, so that $x = 0$. Assume that 3) holds. If $C(S) \neq K^n$, then it is easily seen that there exists $x \in M_n(K), x \neq 0$, such that $xS = 0$, a contradiction. This and a symmetric argument show that 1) holds.

Finally, if S is irreducible, then 2) is satisfied because $K\{S\}K^n$ and $\{v \in K^n \mid Sv = 0\}$ are S-invariant subspaces. □

Irreducibility of S carries heavy consequences for the structure of the 'bottom layer' of S (that is, of the ideal consisting of matrices of the least nonzero rank and the zero matrix, if it is in S). Recall that $S \subseteq M_n(K)$ is completely reducible if K^n is a direct sum of S-invariant subspaces V_i on which S acts irreducibly. That is, $0 \neq SV_i \subseteq V_i$ and V_i has no proper S-invariant subspaces. As before, M_j stands for the ideal of matrices of rank at most j in $M_n(K)$.

Proposition 4.3 *Assume that $S \subseteq M_n(K)$ is a semigroup. Let j be the minimal rank of nonzero matrices in S. Then*

1. *if S is completely reducible, then $S_j = S \cap M_j$ is a 0-disjoint union of ideal uniform components of S. Moreover, for every maximal subgroup D of $M_n(K)$ intersecting $S_j \setminus \{0\}$, $D \cap S_j$ is completely reducible as a subsemigroup of $eM_n(K)e \simeq M_j(K)$, where $e = e^2 \in D$.*

2. *if S is irreducible, then S_j is the only ideal uniform component of S and $D \cap S_j$ is an irreducible subsemigroup of $eM_n(K)e$.*

Proof. 1) Let $K^n = V_1 \oplus \cdots \oplus V_r$ for irreducible $K\{S\}$-submodules V_i. If N is a nilpotent ideal of S, then each NV_i is S-invariant, so it must be trivial. Hence, S has no nonzero nilpotent ideals. Therefore, the first assertion follows from the structure theorem in Section 3.2. We have $eK^n = eV_1 + \cdots + eV_r$. Choose $a \in D \cap S_j$. Then there exists $b \in D$ such that $ba = e = ba$. If $ev_1 + \cdots + ev_r = 0$ for some $v_i \in V_i$, then $av_i + \cdots + av_r = 0$. Since $aV_i \subseteq V_i$, the subspaces aV_i form a direct sum and we have $av_i = 0$ for every i. Then $ev_i = bav_i = 0$. Hence $eK^n = eV_1 \oplus \cdots \oplus eV_r$. Since $aV_i \subseteq eV_i$ and $\mathrm{rank}(a) = \mathrm{rank}(e)$, we have $aV_1 \oplus \cdots \oplus aV_r = eV_1 \oplus \cdots \oplus eV_r$. Consequently $eV_i = aV_i \subseteq V_i$ for every i. Choose $0 \neq w \in eV_i$. From Corollary 1.5 we know that $\mathrm{gp}(S_j \cap D) = \mathrm{cl}(S_j \cap D) \subseteq K\{S_j \cap D\}$. Moreover, Lemma 3.11 implies that $eSe \subseteq \mathrm{cl}(S_j \cap D)$. Now $K\{S_j \cap D\}w \supseteq eK\{S\}ew = eK\{S\}w = eV_i$ because $w \in V_i$ and the latter is an irreducible $K\{S\}$-module. This proves that each eV_i is an irreducible $K\{S_j \cap D\}$-module.

2) Suppose U, V are two different ideal uniform components of S. Then $UV \subseteq U$ and $UV \subseteq V$. Since U, V do not intersect nontrivially, we must have $UV = 0$. But VK^n is an S-invariant subspace of K^n. This contradicts Lemma 4.2. Therefore, the hypothesis and 1) imply that $U = S_j$ indeed is the unique ideal uniform component of S. An argument as in 1) shows that K^n is an irreducible $K\{S \cap D\}$-module. $\square$

The key reduction step reads as follows.

Proposition 4.4 *Let $S \subseteq M_n(K)$ be a semigroup. Then there exist natural numbers $n_0 = 0, n_1, \ldots, n_r$ and diagonal idempotents $e_1, \ldots, e_r$ such that the nonzero entries of e_i are in columns $n_{i-1} + 1, \ldots, n_i$, respectively, $n = n_1 + \cdots + n_r$, and S is conjugate to a semigroup $T \subseteq \sum_{i \leq j} e_i M_n(K) e_j$ such that for every i either $e_i T e_i = 0$ or $e_i T e_i$*

is an irreducible subsemigroup of $e_i M_n(K) e_i \simeq M_{n_i}(K)$. That is, T is block upper triangular

$$
T \subseteq \begin{pmatrix}
T^{(1)} & * & * & * \\
0 & T^{(2)} & * & * \\
& & \ddots & \\
0 & & 0 & T^{(r)}
\end{pmatrix}
$$

and such that every nonzero projection $T^{(i)} = e_i T e_i$ of T onto a diagonal block is irreducible.

Proof. If S is irreducible, then $r = 1$ and $S = T = T^{(1)}$. Otherwise, there exists a proper S-invariant subspace $V \subseteq K^n$. Choose a chain of S-invariant subspaces $0 \neq V_1 \subseteq V_2 \subseteq V_r = V$ of maximal length. Then there exists a basis $v_1, \ldots, v_n$ of K^n such that $v_1, \ldots, v_{k_i}$ is a basis of V_i for $i = 1, \ldots, r$ and some k_i. Conjugating in $M_n(K)$ we can bring S to a block upper triangular form with the sizes of diagonal blocks $n_1 = k_1$ and $n_i = k_i - k_{i-1}$ for $i = 2, \ldots, r$. The choice of the chain V_i easily implies that the block diagonal projections are irreducible whenever nonzero. $\square$

Corollary 4.5 *We have $K\{S\} = K\{U_1\} + \cdots + K\{U_m\} + \mathcal{J}(K\{S\})$, where $\mathcal{J}(K\{S\})$ denotes the Jacobson radical of the algebra $K\{S\}$ and $U_1, \ldots, U_m$ are uniform components of S. Moreover, if S is in the block triangular form of Proposition 4.4 and π is the corresponding block diagonal projection, then $\mathcal{J}(K\{S\}) = \ker(\pi) \cap K\{S\}$.*

Proof. From Corollary 3.8 we know that $R = K\{S\}$ has an ideal chain $0 = R_0 \subset R_1 \subset \cdots \subset R_t = R$ such that every R_i / R_{i-1} is either nilpotent or it is the image in R/R_{i-1} of a subalgebra $A_i = K\{U_i\}$ for a uniform component U_i of S. We put $A_i = 0$ for factors of the former type. By induction on t we show that $R = A_1 + \cdots + A_t + \mathcal{J}(R)$ for any algebra R of this type. If $t = 1$ the assertion is clear. It is easy to see that $R' = R/R_1$ satisfies the induction hypothesis with respect to the chain $R'_i = R_i/R_1, i = 1, \ldots t$, and the subalgebras A'_i defined as the images in R' of A_i. Therefore $R' = A'_1 + \cdots + A'_t + \mathcal{J}(R)$. Since $R_1 \subseteq \mathcal{J}(R)$ (if it is nilpotent) or $R_1 = A_1$, we come to $R = A_1 + \cdots + A_t + \mathcal{J}(R)$.

Assume that S is in the block triangular form of Proposition 4.4. Let $T^{(i)} \subseteq e_i M_n(K) e_i$ be an irreducible block diagonal projection of S. Then

$K\{T^{(i)}\}$ is a simple algebra, so the algebra $\pi(K\{S\})$ is a semisimple. Since $\ker(\pi)$ is nilpotent we get $\mathcal{J}(K\{S\}) = \ker(\pi) \cap K\{S\}$. $\square$

It is clear that every completely reducible semigroup $S \subseteq M_n(K)$ is conjugate to a block diagonal semigroup with irreducible diagonal blocks. Moreover, the algebra $K\{S\}$ is semisimple.

The above form of S fits well into our structural approach.

Lemma 4.6 *Let $S \subseteq M_n(K)$ be in the block upper triangular form of Proposition 4.4. Let $\pi : S \longrightarrow M_n(K)$ be the corresponding block diagonal projection. Then*

1. $\mathrm{rank}(\pi(a)) = \mathrm{rank}(a)$ *for every element $a \in S$ which belongs to a uniform component of S,*

2. $\mathrm{cl}(\pi(S)) = \pi(\mathrm{cl}(S))$ *and for every $e = e^2 \in \mathrm{cl}(S)$ there exists $g \in GL_n(K) \cap \left(\sum_{i \le j} e_i M_n(K) e_j \right)$ such that $\pi(e) = g^{-1} eg$.*

Proof. Let $R = \sum_{i \le j} e_i M_n(K) e_j$, $Z = \sum_{i=1}^{r} e_i M_n(K) e_i$ and let $\phi : R \longrightarrow Z$ be the natural projection. Clearly, $\pi = \phi_{|S}$. R is π-regular by Proposition 1.3, so that $\mathrm{cl}(S) \subseteq R$.

Let $a \in R$. Recall that $\mathrm{rank}(a)$ is the dimension of the K-column space of a. Since ϕ is a linear map, it is easy to see that columns $i_1, \ldots, i_k$ of a are independent whenever columns $i_1, \ldots, i_k$ of $\phi(a)$ are independent. Therefore $\mathrm{rank}(a) \ge \mathrm{rank}(\phi(a))$.

Assume that $f = f^2 \in R$. It is well known that every two decompositions of 1 into a sum of minimal orthogonal idempotents of R are conjugate, cf.[24], 18.23. Therefore there exists $x \in R \cap GL_n(K)$ such that $x^{-1} f x$ is diagonal. Then $\phi(x^{-1} \phi(f) x) = \phi(x^{-1}) \phi(f) \phi(x) = \phi(x^{-1} f x) = x^{-1} f x$, so the foregoing implies that $\mathrm{rank}(x^{-1} \phi(f) x) \ge \mathrm{rank}(x^{-1} f x)$. Hence $\mathrm{rank}(\phi(f)) \ge \mathrm{rank}(f)$ and the equality follows. Since $\mathrm{rank}(\phi(f)) = \mathrm{rank}(x^{-1} f x)$, these two idempotents are conjugate in the semisimple algebra $Z \subseteq R$. Hence $f, \phi(f)$ are conjugate in R, which proves the second part of 2).

If a is an element of a uniform component U of S, then there exists $x \in S$ such that $ax \in U$ and $ax \mathcal{H} f$ for an idempotent $f \in M_n(K)$. Then $f \in \mathrm{cl}(S) \subseteq R$. Clearly, $\phi(ax) \mathcal{H} \phi(f)$, so these matrices have the same rank. Since we know that $\mathrm{rank}(f) = \mathrm{rank}(\phi(f))$, it follows that $\mathrm{rank}(\phi(ax)) = \mathrm{rank}(\phi(f)) = \mathrm{rank}(f) = \mathrm{rank}(ax)$. Hence $\mathrm{rank}(a) \ge$

$\mathrm{rank}(\phi(a)) \geq \mathrm{rank}(\phi(ax)) = \mathrm{rank}(ax) = \mathrm{rank}(a)$. Therefore $\mathrm{rank}(a) = \mathrm{rank}(\phi(a))$, which proves 1).

Next, observe that $\phi(\mathrm{cl}(S))$ is π-regular as a homomorphic image of $\mathrm{cl}(S)$. Since it contains $T = \phi(S)$, we have $\phi(\mathrm{cl}(S)) \supseteq \mathrm{cl}(T)$. On the other hand, $V = \{a \in R \mid \phi(a) \in \mathrm{cl}(T)\}$ is a π-regular subsemigroup containing S. In fact, let $v \in V$. We know that v^n lies in a subgroup H of R, so there exists $u \in H$ such that $v^n u v^n = v^n$. Then $\phi(H)$ is a subgroup of Z and $\phi(u)$ is the inverse of $\phi(v^n)$ in $\phi(H)$. Since $\phi(V) = \mathrm{cl}(T)$ is π-regular, and $\phi(v^n) \in \phi(V)$, it follows that $\phi(u)$ is the inverse of $\phi(v)^n$ in a maximal subgroup of $\phi(V)$. Therefore $u \in V$, so that V is indeed π-regular. Hence $V = \mathrm{cl}(V) \supseteq \mathrm{cl}(S)$ and consequently $\mathrm{cl}(T) = \phi(V) \supseteq \phi(\mathrm{cl}(S))$. This implies that $\mathrm{cl}(\phi(S)) = \phi(\mathrm{cl}(S))$, which completes the proof. $\square$

The following lemma is crucial for applications of the structure theorem in the context of reducibility and triangularizability of semigroups of matrices.

Lemma 4.7 *Assume that $S \subseteq M_n(K)$ is in the block upper triangular form of Proposition 4.4. As above, write $R = \sum_{i \leq j} e_i M_n(K) e_j$ and $Z = \sum_i e_i M_n(K) e_i$. If $\phi_i : R \longrightarrow Z_i = e_i Z e_i$ is the natural projection on an irreducible diagonal block, then*

1. *there exists a uniform component T of S such that $\phi_i(T)$, or $\phi_i(T) \cup \{0\}$ if $0 \in \phi_i(S)$, is an ideal of $\phi_i(S)$ and it intersects the same $\mathcal{H}$-classes of $M_n(K)$ as a uniform component V of $\phi_i(S)$,*

2. *if D is a maximal subgroup of $M_n(K)$ intersecting $\phi_i(T)$ and $e = e^2 \in D$, then $\mathrm{gp}(\phi_i(T) \cap D) = \mathrm{gp}(V \cap D)$ and $\phi_i(T) \cap D$ is irreducible as a subsemigroup of $e M_n(K) e \simeq M_{\mathrm{rank}(e)}(K)$,*

3. *there exists a maximal subgroup D' of $M_n(K)$ such that $\phi_i(T \cap D') \subseteq D$ and $\mathrm{gp}(\phi_i(T \cap D')) = \mathrm{gp}(\phi_i(T) \cap D)$.*

Proof. We may assume that $0 \in S$, adjoining the zero matrix to S, if necessary. Suppose that the image in Z_i of every uniform component of S is zero. Then $\phi_i(S)$ is a nil semigroup, so it must be nilpotent, which contradicts its irreducibility. Hence we can choose a uniform component T of S consisting of matrices of minimal possible rank j such that $\phi_i(T) \neq 0$. Then $\phi_i(S_{j-1}) = 0$ because the choice of T implies that $\phi_i(S_{j-1})$ is a nil ideal of $\phi_i(S)$. We know that the j-th layer of S is of the form $S(j) = U_1 \cup \cdots \cup U_q \cup N$, where U_i are uniform components of S and N is the nilradical of $S(j)$. Say $T = U_1$. Clearly $\phi_i(N) = 0$. It follows that $\phi_i(T) \cup \{0\} = \phi_i(S_j)$ is an ideal of $\phi_i(S)$. Consequently, it is irreducible (as a subsemigroup of $Z_i = e_i M_n(K) e_i \simeq M_j(K)$). Since ϕ_i preserves the relations $\mathcal{R}$ and $\mathcal{L}$, $\phi_i(T)$ intersects all nonzero $\mathcal{H}$-classes of a completely 0-simple subsemigroup of the corresponding Rees factor $(R \cap M_k)/(R \cap M_{k-1}) \subseteq M_k/M_{k-1}$ (k = the rank of matrices in $\phi_i(T)$). Hence, the structure theorem applied to $\phi_i(S)$ easily implies that $\phi_i(T)$ is contained in a uniform component V of $\phi_i(S)$. Since $\phi_i(T) \cup \{0\}$ is an ideal of $V \cup \{0\}$, $V\phi_i(T)V \subseteq \phi_i(T) \cup \{0\}$. But $V\phi_i(T)V$ meets all $\mathcal{H}$-classes of $M_n(K)$ that intersect V (by the definition of a uniform semigroup applied to V). This proves 1).

We know that $\phi_i(T)$ is irreducible in Z_i. From Proposition 4.3 it follows that $\phi_i(T \cap D)$ is irreducible as a subsemigroup of $e M_n(K) e$.

Since $\phi_i(T) \cap D$ is an ideal in $V \cap D$, it is easy to see that $\mathrm{gp}(\phi_i(T) \cap D) = \mathrm{gp}(V \cap D)$. (In fact, $gh \in \mathrm{gp}(\phi_i(T) \cap D)$ for $g \in V \cap D, h \in \phi_i(T) \cap D$, implies that $g \in \mathrm{gp}(\phi_i(T) \cap D)$, so that the nontrivial inclusion follows). This completes the proof of 2).

If $a \in T$ is such that $\phi_i(a) \in D$, then $\phi_i(a^2) \in D$. Hence $\mathrm{rank}(a) = \mathrm{rank}(a^2)$ because $\phi_i(S_{j-1}) = 0$. Therefore $a \in D'$ for a maximal subgroup D' of $M_n(K)$. Clearly $\phi_i(R \cap D') \subseteq D$. Since $(T \cap D')T(T \cap D') \subseteq (T \cap D') \cup S_{j-1}$, it follows that

$$\phi_i(T \cap D')(\phi_i(T) \cap D)\phi_i(T \cap D') \subseteq \phi_i((T \cap D')T(T \cap D')) \cap D \subseteq \phi_i(T \cap D').$$

Hence $\phi_i(T) \cap D \subseteq \mathrm{gp}(\phi_i(T \cap D'))$. On the other hand, it is clear that we also have $\phi_i(T \cap D') \subseteq \phi_i(T) \cap D$. Therefore 3) follows. $\square$

Clearly, the simplest class of uniform semigroups consists of subsemigroups of $GL_n(K)$. We continue with an example of a simple subsemigroup of $GL_n(K)$ which is not a group (and so it is not completely 0-simple), but it is irreducible. It is due to Kelarev, [60], and answers the question asked in [106].

Example Let F be a free nonabelian group with free generators x, y. It is well known that F contains a free subgroup G of infinite rank with free generators $x_1, x_2, \ldots$. We define a chain of subsemigroups of

G inductively by

$$S^{(1)} = \langle x_1 \rangle, \quad S^{(n+1)} = \langle x_{n+1}, S^{(n)} x_{n+1}^{-1} G^{(n)} \rangle$$

for $n \geq 1$, where $G^{(n)}$ denotes the subgroup generated by $S^{(n)}$. Clearly $S^{(n)} = (S^{(n)} x_{n+1}^{-1}) x_{n+1} \subseteq S^{(n+1)}$ and $G^{(n)}$ is the free group generated by $x_1, \ldots, x_n$. Put $S = \bigcup_{n \geq 1} S^{(n)}$. If $s, t \in S$, then $s x_{n+1}^{-1} t^{-1} \in S$, where $n \geq 1$ is such that $s, t \in S_n$. Therefore $s = (s x_{n+1}^{-1} t^{-1}) t x_{n+1} \in StS$. This shows that S is a simple semigroup.

By induction on n we prove that $S^{(n)}$ does not contain the identity e of G. This is clear for $S^{(1)} = \langle x_1 \rangle$. Assume that $e \notin S^{(n)}$ for some $n \geq 1$. Every element $t \in S^{(n+1)}$ has a unique presentation of shortest length $t = t_1 t_2 \cdots t_m$ with $t_i \in G^{(n)} \cup \langle x_{n+1}, x_{n+1}^{-1} \rangle$. Let $t = y_1 \cdots y_r$ where $r \geq 1$ and all y_i are in $\langle x_{n+1} \rangle \cup S^{(n)} x_{n+1}^{-1} G^{(n)}$. By induction on r we first show that:

  i) $t_1 \in \langle x_{n+1} \rangle \cup S^{(n)}$,
  ii) if some $t_i \in \langle x_{n+1} \rangle$, then $i = m$ or $t_{i+1} \in S^{(n)}$.

If $r = 1$, then $t = x_{n+1}^k$ for some $k \geq 1$ or $t \in S^{(n)} x_{n+1}^{-1} G^{(n)}$. Since $e \notin S^{(n)}$, in the latter case we have $t_1 \in S^{(n)}, t_2 = x_{n+1}^{-1}$, and $m = 2$ or $m = 3$. Hence, i) and ii) are satisfied.

Let $r > 1$. Let $y_2 \cdots y_r = t_1' \cdots t_p'$ be the shortest presentation with $t_i' \in G^{(n)} \cup \langle x_{n+1}, x_{n+1}^{-1} \rangle$. By the induction hypothesis (condition i) applied to $t_1' \cdots t_p'$) we have $t_1' \in \langle x_{n+1} \rangle \cup S^{(n)}$.

Assume first that $y_1 = x_{n+1}^k, k \geq 1$. If $t_1' \in S^{(n)}$, then $t_1 = y_1, m = p + 1$, and $t_{i+1} = t_i'$ for $i = 2, \ldots, p$. The induction hypothesis applied to $y_2 \cdots y_p$ implies that ii) holds. Hence, consider the case $t_i' \in \langle x_{n+1} \rangle$. It follows that $t_1 = y_1 t_1', m = p$, and $t_i = t_i'$ for $i = 2, \ldots, p$. Therefore, ii) holds in this case, too. It is clear that 1) is satisfied.

Hence, assume that $y_1 = s x_{n+1}^{-1} g$ for some $s \in S^{(n)}, g \in G^{(n)}$. Suppose that $t_1 \notin \langle x_{n+1} \rangle \cup S^{(n)}$. Since $s \in S^{(n)}$, we must have $t_1' = g^{-1}$ and $t_2' = x_{n+1}$, or $g = 1$ and $t_1' = x_{n+1}$. The induction hypothesis (condition ii) applied to $y_2 \cdots y_r$) implies that either $t_3' \in S^{(n)}$ and $t = s t_3' \cdots t_p'$, or $t_2' \in S^{(n)}$ and $t = s t_2' \cdots t_p'$, respectively. Hence $t_1 = s t_3'$ or $t_1 = s t_2'$ is in $S^{(n)}$. This contradicts our supposition and proves that i) holds for t. Since $t = s x_{n+1}^{-1} g t_1' \cdots t_p'$, we must have $t = s t_3' \cdots t_p'$ or $t = s t_2' \cdots t_p'$, so it is easy to see that ii) also is satisfied. This completes the proof of the inductive claim.

From i) applied to $t = e$ it now follows that S_{n+1} does not contain the identity of G. In fact, otherwise $e = t_1 \cdots t_m$ implies $m = 1$, while

$e \notin S^{(n)}$. Therefore $S = \bigcup_{n \geq 1} S^{(n)}$ is not a group.

Now, consider the following faithful representation of F into 2×2 matrices over rationals, [57], Theorem 14.2.1,

$$x \to \begin{pmatrix} 1 & 2 \\ 0 & 1 \end{pmatrix}, \quad y \to \begin{pmatrix} 1 & 0 \\ 2 & 1 \end{pmatrix}.$$

We claim that the image of S in $M_2(\mathbf{Q})$ is an absolutely irreducible semigroup. If the subalgebra $A = K\{S\}$ spanned by S in $M_2(\mathbf{Q})$ has dimension ≤ 3, then the group of units $U(A)$ of A is solvable. But $G^{(n)} \subseteq U(A)$, a contradiction because $G^{(n)}$ is free nonabelian.

We continue with two results on finiteness of irreducible semigroups.

Proposition 4.8 *Assume that $S \subseteq M_n(K)$ is irreducible. Let j be the minimal nonzero rank of matrices in S. If $S \cap D$ is finite for every maximal subgroup D of $M_n(K)$ consisting of matrices of rank j, then S is finite.*

Proof. Let $I = \{a \in S \mid \text{rank}(a) \leq j\}$. By Proposition 4.3, I is an ideal uniform component of S and it is irreducible. The hypothesis implies that $I \cap H = S \cap H$ is finite for every $\mathcal{H}$-class of $M_n(K)$ consisting of matrices of rank j. More precisely, I is a completely 0-simple semigroup over a finite group G. By Lemma 4.1 $\sum_{i=1}^k \lambda_i a_i = 1$ for some $k \geq 1, \lambda_i \in K$ and $a_i \in I$. Clearly, these can be chosen so that $k \leq n^2$. If $x \in S$, then

$$x = \left(\sum_i \lambda_i a_i\right) x \left(\sum_i \lambda_i a_i\right) = \sum_{i,j} \lambda_i \lambda_j a_i x a_j.$$

But $a_i S a_j \subseteq H_{ij} \cup \{0\}$, where H_{ij} is the intersection of I with an $\mathcal{H}$-class of $M_n(K)$. It follows that $|S| \leq (|G| + 1)^{n^2}$. $\square$

Proposition 4.9 *Let $S \subseteq M_n(K)$ be an irreducible semigroup over an algebraically closed field K. If the set $\{\text{tr}(s) \mid s \in S\}$ is finite, then S is finite.*

Proof. Since $K = \overline{K}$, irreducibility of S implies that $K\{S\} = M_n(K)$. Hence, there exist $a_1, \ldots, a_{n^2} \in S$ which form a basis of $M_n(K)$. Let $\Lambda = \{\lambda_1, \ldots, \lambda_k\} \subseteq K$ be the set of all traces of the matrices in S. Then $\text{tr}(a_1 x), \ldots, \text{tr}(a_{n^2} x)$ is an n^2-tuple of elements of the set Λ. We will show that the system of linear equations $\text{tr}(a_1 x) = \alpha_1, \ldots, \text{tr}(a_{n^2} x) = \alpha_{n^2}$, $\alpha_i \in K$, has at most one solution in $M_n(K)$. This will prove that $|S| \leq k^{n^2}$. It is enough to consider the corresponding homogeneous system $\text{tr}(a_i x) = 0$, $i = 1, \ldots, n^2$. By the choice of a_i and linearity of the trace, $\text{tr}(ax) = 0$ for every $a \in M_n(K)$ whenever x is a solution of the latter system. Since the trace is a non-degenerate bilinear form, we must have $x = 0$. $\square$

We note that the above is no longer true if K is not algebraically closed. However, it remains true under certain assumptions on the algebra $K\{S\}$, which are satisfied in particular if $\text{ch}(K) = 0$. We give a simplified version of the approach presented in [85]. Recall that a finite dimensional simple algebra R is separable over K if the centre of R is a separable field extension of K. If A is a simple subalgebra of $M_n(K)$ and $A \simeq M_k(D)$ for a division algebra D, then $n = kr[D : K]$ for some $r \geq 1$. (It can be shown that r is the dimension of the centralizer of A in $M_n(K)$ over its centre, cf. [44], Theorem VI.4.2).

Proposition 4.10 *Let $S \subseteq M_n(K)$ be an irreducible semigroup. Assume that the simple algebra $K\{S\} \simeq M_k(D)$ is separable over K and $\text{ch}(K)$ does not divide $r = n/(k[D : K])$. If $\{\text{tr}(s) \,|\, s \in S\}$ is a finite set, then S is finite.*

Proof. Separability of $K\{S\}$ implies that $R = K\{S\} \otimes_K \overline{K}$ is a semi-simple $\overline{K}$-algebra, cf.[21], Chapter 6. But $1 \in \overline{K}\{S\} \subseteq M_n(\overline{K})$ and R can be identified with $\overline{K}\{S\}$. Every $a \in K\{S\}$ can be viewed as a map $D^k \longrightarrow D^k$. But D, as a K-space, can be identified with $K^{[D:K]}$. Hence, $D^k \simeq K^{k[D:K]} \subseteq \overline{K}^{k[D:K]}$. Let $\text{tr}_R(a)$ be the trace of $a \in R$ as a linear map $\overline{K}^{k[D:K]} \longrightarrow \overline{K}^{k[D:K]}$. Since $rk[D : K] = n$, we have a natural embedding

$$M_k(D) \simeq K\{S\} \xrightarrow{\phi} M_{k[D:K]}(K) \xrightarrow{\psi} M_n(K),$$

which is constant on K. Here ψ maps any matrix c to a block diagonal matrix with r diagonal blocks, each equal to c. Moreover $\text{tr}(\psi\phi(a)) = r \cdot \text{tr}_R(a)$ for $a \in K\{S\}$. By the Noether - Skolem theorem [36], Theorem 4.3.1, there exists $g \in GL_n(K)$ such that $\psi\phi(a) = g^{-1}ag$ for $a \in K\{S\}$. Then $\text{tr}(a) = r \cdot \text{tr}_R(a)$ for every $a \in K\{S\}$. Since $\text{ch}(K)$

does not divide r by the hypothesis, it follows that $\{\mathrm{tr}_R(a) \,|\, a \in S\}$ is a finite set.

Suppose $\mathrm{tr}_R(ax) = 0$ for some $a \in R$ and every $x \in R$. Choose x so that ax is an idempotent of rank one in a simple block of R. Then $\mathrm{tr}_R(ax) = 1$. Therefore, tr_R is a non-degenerate bilinear form on R. An argument as in Proposition 4.9 allows to prove that S is finite. $\square$

4.2 Munn algebras and representations

Our aim in this section is to describe irreducible representations of completely 0-simple semigroups. We introduce an important class of semigroup algebras arising from completely 0-simple semigroups. They are crucial for investigating local properties of arbitrary semigroup algebras. They first appeared in the independent work of Munn and Ponizovskii, cf. [15], providing one of the main tools in the representation theory of semigroups. Our presentation is in the spirit of [79], [87]. For another approach to irreducible representations and many further results in this direction we refer to [15],[80],[106].

Let R be an associative K-algebra. Let X, Y be nonempty sets and $P = (p_{yx})$ a generalised $Y \times X$ - matrix with $p_{yx} \in R$. Consider the set $\mathcal{M}(R, X, Y, P)$ of all generalised $X \times Y$ - matrices over R with finitely many nonzero entries. For any $A = (a_{xy}), B = (b_{xy}) \in \mathcal{M}(R, X, Y, P)$ addition and multiplication are defined as follows $A + B = (c_{xy})$ where $c_{xy} = a_{xy} + b_{xy}$ for $x \in X, y \in Y$, $AB = A \circ P \circ B$, where $\circ$ stands for the usual product of matrices. Also let $\alpha A = (\alpha a_{xy})$ for $\alpha \in K$. Then $\tilde{R} = \mathcal{M}(R, X, Y, P)$ subject to these operations becomes an associative K-algebra, called an algebra of matrix type over R. If not stated otherwise, we consider only the case where R has an identity and every row and every column of P contains a unit of R. Then $\tilde{R}$ is referred to as a Munn algebra over R.

The crucial motivating example comes from the following observation.

Lemma 4.11 *Let G be a group, X, Y - nonempty sets and P - an $Y \times X$-matrix over a group with zero $G \cup \{\theta\}$. Then the contracted semigroup algebra $K_0[\mathcal{M}(G, X, Y, P)]$ of the semigroup of matrix type $\mathcal{M}(G, X, Y, P)$ is naturally isomorphic to the algebra of matrix type $\mathcal{M}(K[G], X, Y, P)$ over the group algebra $K[G]$.*

Proof. Define a mapping $\phi : \mathcal{M}(G, X, Y, P) \longrightarrow \mathcal{M}(K[G], X, Y, P)$ by $((g, x', y')) \mapsto (a_{xy})$, where $a_{xy} = g$ if $x = x', y = y'$ and $a_{xy} = 0$ otherwise. From the definitions it follows that ϕ is a homomorphism into the multiplicative semigroup of the algebra $\mathcal{M}(K[G], X, Y, P)$. Since the images of the nonzero elements of $\mathcal{M}(G, X, Y, P)$ are linearly independent, it is clear that the extension of ϕ to a homomorphism of K-algebras $K_0[\mathcal{M}(G, X, Y, P)] \longrightarrow \mathcal{M}(K[G], X, Y, P)$ is an isomorphism. $\square$

We will use the above lemma by identifying the contracted semigroup algebra of a completely 0-simple semigroup with an appropriate Munn algebra.

The following notation is motivated by our convention on completely 0-simple semigroups, see Section 1.1. A matrix $(a_{xy}) \in \tilde{R}$ such that $a_{x'y'} = r, a_{xy} = 0$ for $x \neq x', y \neq y'$ is denoted also by (r, x', y'). For any $x \in X, y \in Y$ we put $\tilde{R}_{(x)}^{(y)} = \{(r, x, y) \mid r \in R\}$. If $X' \subseteq X, Y' \subseteq Y$ are nonempty subsets, then $\tilde{R}_{(X')}^{(Y')} = \sum_{x \in X', y \in Y'} \tilde{R}_{(x)}^{(y)}$. Thus $\tilde{R}_{(X')}^{(Y')} = \mathcal{M}(R, X', Y', P_{Y'X'})$, where $P_{Y'X'}$ is the $Y' \times X'$-submatrix of P.

Let $\phi : R \longrightarrow R'$ be a homomorphism of algebras. Then $\phi(P)$ stands for the $Y \times X$-matrix $(\phi(p_{yx}))$. Further,

$$\tilde{\phi} : \tilde{R} \longrightarrow \tilde{R}' = \mathcal{M}(R', X, Y, \phi(P))$$

denotes the induced homomorphism, that is, $\tilde{\phi}(A) = (\phi(a_{xy}))$, where $A = (a_{xy}) \in \tilde{R}$. If T is a subset of R, then, for any sets X, Y, an $X \times Y$-matrix A over R is said to lie over T if all entries of A are in T. We put $\mathcal{M}(T, X, Y, P) = \{A \in \tilde{R} \mid A \text{ lies over } T\}$. Thus, it clear that $\ker(\tilde{\phi}) = \mathcal{M}(\ker(\phi), X, Y, P)$ and $\tilde{\phi}(\tilde{R}) = \mathcal{M}(\phi(R), X, Y, \phi(P))$.

We start with an important auxiliary result providing necessary and sufficient conditions for a Munn algebra to have an identity. The following well known linear algebra argument will be used, cf.[15], Theorem 5.11.

Lemma 4.12 *Let R be a finite dimensional K-algebra and let P be an $m \times n$-matrix over R for some integers $m, n \geq 1$. If $m > n$, then $Q \circ P = 0$ for a nonzero $n \times m$-matrix Q over R. If $n > m$, then $P \circ Q' = 0$ for a nonzero $n \times m$-matrix Q' over R.*

Proposition 4.13 *Let R be an algebra with a nonzero finite dimensional homomorphic image. Then the following conditions are equivalent for any Munn algebra $\tilde{R} = \mathcal{M}(R, X, Y, P)$*

1. *the algebra $\tilde{R}$ has an identity,*

2. X, Y *are finite sets of the same cardinality and P is invertible as a matrix in* $M_{|X|}(R)$.

Moreover, if 1),2) hold, then $\tilde{R} \simeq M_{|X|}(R)$.

Proof. Assume that $\tilde{R}$ has an identity E. Then there exist finite subsets $U \subseteq X, Z \subseteq Y$ such that $E \in \tilde{R}_{(U)}^{(Z)}$. Let $x \in X$. We know that for any $y \in Y$

$$(*) \qquad (1, x, y) = E(1, x, y) = E \circ P \circ (1, x, y) \in \tilde{R}_{(U)}^{(Y)}.$$

This implies that $x \in U$ and so $X = U$ is a finite set. Moreover, from $(*)$ it follows that the x-th column of the matrix $E \circ P$ consists of zeros except the (x, x)-th entry which is equal to 1. Therefore, $E \circ P$ is the $X \times X$ identity matrix. Similarly, the fact that E is a right identity of $\tilde{R}$ implies that $Y = Z$ is a finite set and the $Y \times Y$ matrix $P \circ E$ is the identity matrix. To establish 2) it is enough to show that $|X| = |Y|$ since then E is the inverse of P in the algebra $M_{|X|}(R)$. Let $\phi : R \longrightarrow R'$ be a homomorphism onto a nonzero finite dimensional algebra and let $\tilde{\phi} : \tilde{R} \longrightarrow \tilde{R}' = \mathcal{M}(R', X, Y, \phi(P))$ be the induced homomorphism. Then $\tilde{R}'$ has an identity $\phi(E)$ and is finite dimensional since X, Y are finite sets. Now $A = \phi(E) \circ \phi(P) \circ A = A \circ \phi(P) \circ \phi(E)$ for all $A \in \tilde{R}'$ and from Lemma 4.12 it follows that $|X| = |Y|$. This proves 2).

Let $\psi : \tilde{R} \longrightarrow M_{|X|}(R)$ be given by $\psi(A) = A \circ P$. For any $A, B \in \tilde{R}, \lambda \in K$, we have

$$\psi(AB) = \psi(A \circ P \circ B) = A \circ P \circ B \circ P = \psi(A)\psi(B),$$

$$\psi(A + B) = (A + B) \circ P = (A \circ P) + (B \circ P) = \psi(A) + \psi(B),$$

$$\psi(\lambda A) = (\lambda A) \circ P = \lambda(A \circ P) = \lambda\psi(A).$$

If 2) holds, then ψ is an isomorphism because P is invertible in $M_{|X|}(R)$. This proves 1), as well as the remaining assertion. $\square$

Corollary 4.14 *Let* $S = \mathcal{M}(G, X, Y, P)$ *be a semigroup of matrix type over a group G. Then the following conditions are equivalent*

1. *the algebra $K_0[S]$ has an identity,*

2. X, Y are finite sets of the same cardinality and P is an invertible matrix in $M_{|X|}(K[G])$.

Moreover, if 1),2) hold, then $K_0[S] \simeq M_{|X|}(K[G])$, and S is a completely 0-simple semigroup. If S is an inverse semigroup, then 1),2) are equivalent to the fact that S has finitely many idempotents.

Proof. Since the augmentation homomorphism maps $K[G]$ onto K, Proposition 4.13, in view of Lemma 4.11 establishes the equivalence of 1) and 2), and the fact that $K_0[S] \simeq M_{|X|}(K[G])$. Moreover, the invertibility of P implies that P has no zero columns or rows, so that S is completely 0-simple.

Assume now that S is an inverse semigroup. It is well known that $S \simeq \mathcal{M}(G, X, X, Q)$, where Q is the identity $X \times X$-matrix, cf.[15], Theorem 3.9. Hence, by Lemma 1.1, S has finitely many idempotents if and only if X is a finite set. This completes the proof. $\square$

Thus, in the case described in Corollary 4.14, irreducible representations of S are in one-to-one correspondence with irreducible representations of the maximal subgroup G of S. An important example of such a situation was presented in Theorem 2.34. Our main aim in this section is to prove an extension of this result to the case of arbitrary completely 0-simple semigroups.

Recall that, if $T \subseteq R$ is a nonempty subset of an algebra R, then by $l_R(T), r_R(T)$ we mean the left, respectively right, annihilator of T in R. We write $l(T), r(T)$, if unambiguous.

Lemma 4.15 For any $x \in X, y \in Y$ we have $l(\tilde{R}) = \{A \in \tilde{R} \mid A \circ P = 0\} = l_{\tilde{R}}(\tilde{R}^{(y)})$. If the rows of P are left R-independent as elements of the left R-module R^Y, then $l(\tilde{R}) = 0$.

Proof. Assume that $A\tilde{R}^{(y)} = 0$ for some $A \in \tilde{R}, y \in Y$. Then $A \circ P \circ \tilde{R}^{(y)} = 0$, so $A \circ P \circ (r, x, y) = 0$ for any $r \in R, x \in X$. Hence r annihilates, on the right, all columns of the matrix $A \circ P$. Since $r = 1 \in R$, we get $A \circ P = 0$. Hence, the first assertion follows.

If $A \circ P = 0$ for some $0 \neq A = (a_{xy}) \in \tilde{R}$, then for any $z \in X$, $A_z \circ P = 0$ where $A_z = (b_{xy})$ with $b_{xy} = a_{xy}$ if $x = z$ and $b_{xy} = 0$ if $x \neq z$. It follows that a nontrivial left R-combination of rows of P is zero. $\square$

Let $\text{Row}(P) \subseteq R^Y$ be the left R-submodule generated by the rows of the matrix P. In other words $\text{Row}(P) = \sum_{y \in Y} R P_y$, where P_y is the y-th row of P. Moreover, for any set Z, denote by $M_Z^{row}(R)$ the algebra of all $Z \times Z$-matrices over R with finitely many nonzero rows subject to the natural addition and multiplication. The following is a direct consequence of the definitions and of Lemma 4.15.

Lemma 4.16 *The rule* $\phi(A) = A \circ P$ *defines a homomorphism of K-algebras* $\phi : \tilde{R} \longrightarrow M_X^{row}(R)$ *such that*

1. $\phi(\tilde{R})$ *is the subalgebra of $M_X^{row}(R)$ consisting of all matrices the rows of which lie in* $\text{Row}(P)$,

2. $\ker(\phi) = l(\tilde{R})$.

The submodule $\text{Col}(P)$ of the right R-module R^X is defined dually to $\text{Row}(P)$, that is, $\text{Col}(P)$ is the submodule of R^X generated by the columns $P^x, x \in X$, of P. The following result shows that $l(\tilde{R})$ is determined by some specific subsets of $\tilde{R}$, which are also essential when representing the algebra $\tilde{R}$ modulo $r(\tilde{R})$.

Lemma 4.17 *Let* $\tilde{R} = \mathcal{M}(R, X, Y, P)$ *be a Munn algebra over an algebra R, and let $Z \subseteq X$ be a subset such that the right R-submodules* $\text{Col}(P) = \sum_{x \in X} P^x R$ *and* $\sum_{x \in Z} P^x R$ *of R^X coincide. Then* $\tilde{R} = r(\tilde{R}) + \tilde{R}_{(Z)}^{(Y)}$ *and* $l(\tilde{R}) = l_{\tilde{R}}(\tilde{R}_{(Z)}^{(Y)})$.

Proof. Let $(r, x, y) \in \tilde{R}$. From the hypothesis it follows that the column P^x of P may be written as $\sum_{k=1}^n P^{x_k} r_k$ for some $n \geq 1, r_1, \dots, r_n \in R, x_1, \dots, x_n \in Z$. Define $A \in \tilde{R}$ by $A = \sum_{k=1}^n (r_k r, x_k, y) - (r, x, y)$. It is easy to see that $P \circ A = 0$, so $A \in r(\tilde{R})$. Since $\sum_{k=1}^n (r_k r, x_k, y) \in \tilde{R}_{(Z)}^{(Y)}$, we get $(r, x, y) \in r(\tilde{R}) + \tilde{R}_{(Z)}^{(Y)}$, which proves the first equality. Now $l(\tilde{R}) = l_{\tilde{R}}(r(\tilde{R}) + \tilde{R}_{(Z)}^{(Y)}) = l_{\tilde{R}}(r(\tilde{R})) \cap l_{\tilde{R}}(\tilde{R}_{(Z)}^{(Y)}) = \tilde{R} \cap l_{\tilde{R}}(\tilde{R}_{(Z)}^{(Y)}) = l_{\tilde{R}}(\tilde{R}_{(Z)}^{(Y)})$. $\square$

It is clear that the right-left symmetric analogues of Lemma 4.15, Lemma 4.16, and Lemma 4.17 may be proved. For this, one has to interchange the roles of the left and right annihilators, the roles of the mappings $A \longrightarrow A \circ P, A \longrightarrow P \circ A$, and those of the R-modules $\text{Row}(P), \text{Col}(P)$.

For any ideal J of R we denote by $\mathcal{B}(J)$, or $\mathcal{B}_{\tilde{R}}(J)$ if ambiguous, the set $\{A \in \tilde{R} \,|\, P \circ A \circ P \text{ lies over } J\}$. It is called the basic ideal of $\tilde{R}$ determined by J. The ideal $\mathcal{B}(0)$ may be characterized through some annihilators arising from $\tilde{R}$.

Lemma 4.18 *Let $\tilde{R}$ be an algebra of matrix type over an algebra R with identity. Then $\mathcal{B}(0)$ is the middle annihilator of $\tilde{R}$ and*

$$\tilde{R}/\mathcal{B}(0) \simeq (\tilde{R}/l(\tilde{R}))/r(\tilde{R}/l(\tilde{R})) \simeq (\tilde{R}/r(\tilde{R}))/l(\tilde{R}/r(\tilde{R})).$$

Proof. Let $A \in \tilde{R}$. Then A lies in the kernel of the natural homomorphism $\tilde{R} \longrightarrow (\tilde{R}/l(\tilde{R}))/r(\tilde{R}/l(\tilde{R}))$ if and only if $\tilde{R}A \in l(\tilde{R})$. The latter is equivalent to the fact that $\tilde{R}A\tilde{R} = 0$. Since R has an identity, this happens if and only if $P \circ A \circ P = 0$, that is $A \in \mathcal{B}(0)$.

The second isomorphism is established similarly. $\square$

We will describe a connection between the classes of modules over R and $\tilde{R}$, which exploits the basic ideals of $\tilde{R}$ introduced above.

Let V be a left R-module. Then there is a natural induced $M_X^{row}(R)$-module structure on the direct sum V^X, written as column X-tuples. Thus, V^X may be regarded as an $\tilde{R}$-module through the homomorphism of algebras $\phi : \tilde{R} \longrightarrow M_X^{row}(R)$ defined by $\phi(A) = A \circ P$. This module will be denoted by $V(P)$. In other words, the action of $\tilde{R}$ on $V(P)$ is given by $Av = A \circ P \circ v$ for $v \in V(P), A \in \tilde{R}$. Let $V_0(P) = \{v \in V(P) \,|\, \tilde{R}v = 0\}$. Then $V_0(P)$ is an $\tilde{R}$-submodule of $V(P)$ and $V_0(P) = \{v \in V(P) \,|\, A \circ P \circ v = 0 \text{ for all } A \in \tilde{R}\} = \{v \in V(P) \,|\, P \circ v = 0\}$ because $r_R(R) = 0$.

Lemma 4.19 *Let V be a left R-module. Then*

1. $\text{ann}_{\tilde{R}}(V(P)/V_0(P)) = \mathcal{B}(\text{ann}_R(V))$,

2. *if V is an irreducible R-module, then $V(P)/V_0(P)$ is an irreducible $\tilde{R}$-module,*

3. *the rule $W \mapsto W(P)$ defines an embedding of the lattice of R-submodules of V into the lattice of $\tilde{R}$-submodules of $V(P)$.*

Proof. 1) We have $\text{ann}_{\tilde{R}}(V(P)/V_0(P)) = \{A \in \tilde{R} \,|\, AV(P) \subseteq V_0(P)\} = \{A \in \tilde{R} \,|\, A \circ P \circ v \in V_0(P) \text{ for all } v \in V(P)\} = \{A \in \tilde{R} \,|\, P \circ A \circ P \circ$

$v = 0$ for all $v \in V(P)\} = \{A \in \tilde{R} \mid P \circ A \circ P$ lies over $\mathrm{ann}_R(V)\} = \mathcal{B}(\mathrm{ann}_R(V))$.

2) Let $v = (v_x)_{x \in X} \in V(P) \setminus V_0(P)$. Then $P \circ v$ is a nonzero element of V^X, so there exists $y \in Y$ such that the y-th entry z of $P \circ v$ is nonzero. Let $u = (u_x) \in V(P)$. Since V is an irreducible R-module, for any $x \in X$ there exists $r_x \in R$ such that $r_x z = u_x$. Let $A \in \tilde{R}$ be defined as the matrix with zeros everywhere except the y-th column, where it has r_x at the (x, y)-th entry, for all $x \in X$. It is easy to see that $u = A \circ P \circ v = Av \in \tilde{R}v$. This proves 2). 3) is straightforward. $\square$

We note that $\mathcal{B}_{\tilde{R}}(0)^3 = 0$, so $\mathcal{B}_{\tilde{R}}$ is contained in the annihilator of any irreducible $\tilde{R}$-module V. Hence V is an irreducible $\tilde{R}/\mathcal{B}(0)$-module. If T is a ring with a nonzero idempotent e and W is a left T-module, then eW is a left eTe-module which carries some important properties of W. In particular eW is irreducible or faithful whenever so is W. In the special case of Munn algebras, we derive the following result.

Lemma 4.20 *Let V be a left $\tilde{R}/\mathcal{B}(0)$-module. If p_{yx} is a unit of R, for some $y \in Y, x \in X$, then $E = (p_{yx}^{-1}, x, y)$ is an idempotent of $\tilde{R}$ and EV has a natural structure of a left R-module. Moreover*

1. *EV is an irreducible R-module if V is an irreducible $\tilde{R}/\mathcal{B}(0)$-module,*

2. *EV is a faithful R-module if V is a faithful $\tilde{R}/\mathcal{B}(0)$-module.*

Proof. We may treat V as a left $\tilde{R}$-module with $\mathcal{B}(0)V = 0$. It is clear that $E^2 = E$ and $R \simeq E\tilde{R}E$ via the map $r \mapsto (rp_{yx}^{-1}, x, y)$. Hence $EV \subseteq V$ may be regarded as an R-module. Thus 1) is a direct consequence of the foregoing remark. If $\mathrm{ann}_{\tilde{R}}(V) = \mathcal{B}(0)$, and $(EAE)EV = 0$ for some $A \in \tilde{R}$, then $EAE \in \mathcal{B}(0)$. Thus $EAE = E \circ (P \circ EAE \circ P) \circ E = 0$, which proves 2). $\square$

We now consider a special case which is of interest when looking at finite dimensional representations of Munn algebras. Assume that $\tilde{R} = \mathcal{M}(R, X, Y, P)$ is a Munn algebra over a simple Artinian algebra R. Let $R = M_r(D)$ for some $r \geq 1$ and a division algebra D. If T, Z are nonempty sets and $A = (a_{tz})$ is a $T \times Z$-matrix over R, then we write $\overline{A}$ for the $(T.r) \times (Z.r)$-matrix obtained from A by erasing the matrix brackets of all entries a_{tz} of A. Here $T.r, Z.r$ denote the disjoint unions of r copies of sets of cardinality $|T|, |Z|$ respectively. Similarly, treating elements of $(D^r)^T$ as $\{1\} \times T$-matrices over D^r, an element $\overline{v} \in D^{T.r}$ is associated with any element $v \in (D^r)^T$. With this notation we have the following result.

Lemma 4.21 *The algebras* $\mathcal{M}(M_r(D), X, Y, P)$ *and* $\mathcal{M}(D, X.r, Y.r, \overline{P})$ *are isomorphic.*

Proof. Let $A, B \in \mathcal{M}(M_r(D), X, Y, P)$. It is easy to see that $\overline{A} \circ \overline{B} = \overline{A} \circ \overline{P} \circ \overline{B} = \overline{A \circ P \circ B} = \overline{AB}$. Since $A \mapsto \overline{A}$ is a linear map, it is an isomorphism of algebras. $\square$

We define the rank of an $Y \times X$-matrix P over $M_r(D)$, and write rank(P), as $\dim_D \phi_P((D^r)^Y)$, where ϕ_P is the homomorphism of left D-spaces $(D^r)^Y \longrightarrow (D^r)^X$ determined by P. So $\phi_P(v) = (v^t \circ P)^t$, with w^t denoting the transpose of w. If $r = 1$ and $X = Y$ is finite, then our notation for the rank agrees with that used before for matrices over a field. Note that, under the natural isomorphisms of D-spaces $(D^r)^Y \simeq D^{Y.r}, (D^r)^X \simeq D^{X.r}$, ϕ_P corresponds to the isomorphism $\phi_{\overline{P}} :$ $D^{Y.r} \longrightarrow D^{X.r}$ given by $\phi_{\overline{P}}(w) = (w^t \circ \overline{P})^t$. Thus, $\dim_D \phi_{\overline{P}}(D^{Y.r}) = \dim_D(\phi_P(D^r)^Y)$. Since the former is equal to $\dim_D \text{Row}(\overline{P})$, as a direct consequence we get the following corollary.

Corollary 4.22 rank$(P) = $ rank$(\overline{P}) = \dim_D \text{Row}(\overline{P})$.

Our next auxiliary result is well known and easy to prove.

Lemma 4.23 *Let* P *be an* $Y \times X$-*matrix over a division algebra* D. *Then*

$$\dim_D \text{Col}(P) = \sup\{k \mid P \text{ has an invertible } k \times k \text{ submatrix}\}$$
$$= \dim_D \text{Row}(P).$$

This shows that rank(P) may be equivalently defined as the dimension of the image of $(D^r)^X$ under the homomorphism of right D-spaces defined dually to ϕ_P by the action of P on the left.

Proposition 4.24 *Let* $\tilde{D} = \mathcal{M}(D, X, Y, P)$ *be a Munn algebra over a division algebra* D. *Assume that* rank$(P) = t < \infty$. *Then* $\tilde{D}/\mathcal{B}(0) \simeq M_t(D)$.

Proof. Since $\text{rank}(P) = \dim_D \text{Col}(P)$, there exist $x_1, \ldots, x_t \in X$ such that the corresponding columns $P^{x_1}, \ldots, P^{x_t}$ of P are right D-independent and $\text{Col}(P) = \sum_{i=1}^{t} P^{x_i} D$. Put $Z = \{x_1, \ldots, x_t\}$, and let $\phi : \tilde{D}_{(Z)}^{(Y)} \longrightarrow M_Z^{row}(D)$ be the homomorphism defined by $\phi(A) = A \circ P_{YZ}$. Clearly, the latter algebra can be identified with $M_t(D)$. In view of Lemma 4.23 the matrix P_{YZ} has t rows which are left D-independent. Thus, from Lemma 4.16 it follows that $\phi(\tilde{D}_{(Z)}^{(Y)})$ has dimension t^2 as a left vector space over D. Therefore ϕ maps $\tilde{D}_{(Z)}^{(Y)}$ onto $M_Z^{row}(D)$, and hence again Lemma 4.16 yields

$$(*) \qquad \tilde{D}_{(Z)}^{(Y)}/l(\tilde{D}_{(Z)}^{(Y)}) \simeq \phi(\tilde{D}_{(Z)}^{(Y)}) \simeq M_Z^{row}(D) \simeq M_t(D).$$

On the other hand, from Lemma 4.15 (cf. the remark following it) it follows that $r(\tilde{D}) = r_{\tilde{D}}(\tilde{D}_{(Z)}^{(Y)})$, and consequently $r(\tilde{D}) \cap \tilde{D}_{(Z)}^{(Y)} = r(\tilde{D}_{(Z)}^{(Y)})$. Since by Lemma 4.17 $\tilde{D} = \tilde{D}_{(Z)}^{(Y)} + r(\tilde{D})$, we come to

$$\tilde{D}/r(\tilde{D}) = (\tilde{D}_{(Z)}^{(Y)} + r(\tilde{D}))/r(\tilde{D}) \simeq \tilde{D}_{(Z)}^{(Y)}/(r(\tilde{D}) \cap \tilde{D}_{(Z)}^{(Y)}) = \tilde{D}_{(Z)}^{(Y)}/r(\tilde{D}_{(Z)}^{(Y)}).$$

Thus, from Lemma 4.18 it follows that

$$\tilde{D}/\mathcal{B}_{\tilde{D}}(0) \simeq (\tilde{D}/r(\tilde{D}))/l(\tilde{D}/r(\tilde{D})) = (\tilde{D}_{(Z)}^{(Y)}/r(\tilde{D}_{(Z)}^{(Y)}))/l(\tilde{D}_{(Z)}^{(Y)}/r(\tilde{D}_{(Z)}^{(Y)})).$$

Applying Lemma 4.18 once again with respect to the algebra $\tilde{D}_{(Z)}^{(Y)}$ we get

$$\tilde{D}/\mathcal{B}_{\tilde{D}}(0) \simeq (\tilde{D}_{(Z)}^{(Y)}/l(\tilde{D}_{(Z)}^{(Y)}))/r(\tilde{D}_{(Z)}^{(Y)}/l(\tilde{D}_{(Z)}^{(Y)})).$$

The latter algebra is in view of $(*)$ isomorphic to $M_t(D)$. This completes the proof of the proposition. $\square$

Corollary 4.25 *Let* $\tilde{R} = \mathcal{M}(R, X, Y, P)$ *be a Munn algebra over a simple Artinian algebra* $R = M_r(D), r \geq 1$, *for a division algebra* D. *Then* $\tilde{R}/\mathcal{B}(0) \simeq M_t(D)$ *provided that* $t = \text{rank}(P) < \infty$.

Proof. By Lemma 4.21 $\tilde{R} \simeq \mathcal{M}(D, I.r, M.r, \overline{P})$. Thus Proposition 4.24 and Corollary 4.22 imply that $\tilde{R}/\mathcal{B}(0) \simeq M_{\text{rank}(\overline{P})}(D) = M_t(D)$. $\square$

We are now ready for the main result of this section.

Theorem 4.26 *Let $S = \mathcal{M}(G, X, Y, P)$ be a completely 0-simple semi-group. Assume that $\phi : G \longrightarrow M_r(K)$ is an irreducible representation of G such that* $\mathrm{rank}(\phi(P)) = t < \infty$. *Then the induced map $\overline{\phi}$*

$$\mathcal{M}(G, X, Y, P) \longrightarrow \tilde{R} = \mathcal{M}(M_r(K), X, Y, \phi(P)) \longrightarrow \tilde{R}/\mathcal{B}(0) \simeq M_t(K)$$

is an irreducible representation of S. Moreover, every irreducible representation of S arises in this way and two such representations are equivalent if and only if they are equivalent on G.

Proof. $V = K^r$ is an irreducible $K[G]$-module. So $V(\phi(P))/V_0(\phi(P))$ is an irreducible $\tilde{A} = \mathcal{M}(\phi(K[G]), X, Y, \phi(P))$-module by Lemma 4.19. Hence, an irreducible $K_0[S]$-module. Since

$$V_0(\phi(P)) = \{v \in V(\phi(P)) \mid \phi(P) \circ v = 0\},$$

the action of $\tilde{A}$ extends to the action of $\tilde{R} = \mathcal{M}(M_r(K), X, Y, \phi(P))$ on $V(\phi(P))/V_0(\phi(P))$. This module has dimension $\mathrm{rank}(\phi(P))$ by Corollary 4.22. Its annihilator is $\mathcal{B}_{\tilde{R}}(0)$ by Lemma 4.19. This implies that the resulting representation of S is accomplished by the natural homomorphism $\overline{\phi}$.

Let $\psi : \mathcal{M}(K[G], X, Y, P) \longrightarrow M_t(K)$ be the extension of an irreducible representation of S to its Munn algebra. Then ψ factors through the natural map $\mathcal{M}(K[G], X, Y, P) \longrightarrow \tilde{C} = \mathcal{M}(\psi(K[G]), X, Y, \psi(P))$. From Lemma 4.20 we know that ψ determines an irreducible representation $\psi_{|G}$ of G. So $\psi(K[G]) \simeq M_k(D)$ for a division algebra D by Lemma 4.1. Suppose the rank of $\psi(P)$ over D is not finite. Then Proposition 4.13 implies, in view of Lemma 4.23, that for every $n \geq 1$ $\tilde{C}$ contains a subalgebra of matrix type isomorphic to $M_{kn}(D)$. Hence, the inverse image of such an algebra in $\mathcal{M}(K[G], X, Y, P)$ intersects S nontrivially and lies in the kernel of ψ. This is a contradiction since S is completely 0-simple. Therefore $\psi(P)$ has a finite rank over D. Since $\mathcal{B}_{\tilde{C}}(0)$ is nilpotent, Corollary 4.25 implies that $\tilde{C}$ has only one irreducible module. So, this module must be induced from $\psi_{|G}$ in the way described in the first paragraph of the proof. It follows that ψ is equivalent to the representation $\overline{\phi}$ for $\phi = \psi_{|G}$. This completes the proof of the theorem. $\square$

4.3 Irreducible representations

It is well known that irreducible representations of a finite semigroup come from irreducible representations of the completely 0-simple principal factors of S. We have seen in Section 4.2 that every irreducible representation of a principal factor is determined by an irreducible representation of its maximal subgroup. A beautiful instance of this phenomenon is given in Theorem 2.35. More generally, it turns out that irreducible representations of a linear semigroup S come from the representations of the uniform components U of S. However, there is no natural one-to-one correspondence with irreducible representations of $S \cap D$ for a maximal subgroup D of the completely 0-simple closure $\widehat{U}$. The situation, in general, is much more complicated than that in the classical case. One of the reasons being that representations of $S \cap D$ do not always extend to representations of the group $\mathrm{gp}(S \cap D) \subseteq D$.

We start with a technical result of independent interest.

Lemma 4.27 *The relation $\prec$ defined on the set of uniform components of a semigroup $S \subseteq M_n(K)$ by $U \prec V$ if $SVS \cap U \neq \emptyset$ is a partial order.*

Proof. It is clear that $\prec$ is reflexive. If $U \prec V$ and $V \prec U$, then U, V consist of matrices of the same rank. Therefore, the structure theorem (Theorem 3.5) implies that we must have $U = V$. Assume that $U \prec V$ and $V \prec W$. Consider the ideal SWS of S. Then the Zariski closure $\overline{SWS}$ is an ideal of $\overline{S}$ and it intersects V. Since $\overline{S}$ is π-regular, from Remark iv) after Theorem 3.5 it follows that $\overline{SWS}$ contains every uniform component of $\overline{S}$ which it intersects nontrivially. Since V is contained in a uniform component of $\overline{S}$ by Proposition 3.13, we must have $V \subseteq \overline{SWS}$. Suppose that $SWS \cap U = \emptyset$. Then $U(SWS)U \subseteq M_{j-1}$, where j is the common rank of matrices in U. Since M_{j-1} is a closed set, it follows that $U\overline{SWS}U \subseteq M_{j-1}$. But $U \prec V$ implies that $SVS \cap U \neq \emptyset$, so that $\emptyset \neq U(SVS)U \cap U \subseteq U\overline{SWS}U$. This contradiction shows that $U \prec W$, whence $\prec$ is transitive. $\square$

Recall that, for convenience sake, a uniform component of a semigroup $S \subseteq M_n(K)$ is sometimes viewed as a subsemigroup of the corresponding principal factor M_j/M_{j-1} of $M_n(K)$, and sometimes as a subset of $M_j \setminus M_{j-1}$.

Theorem 4.28 *Let $S \subseteq M_n(K)$ be a semigroup. Then*

1. *If $\phi : S \longrightarrow M_r(L)$ is an irreducible representation over a field L, then there exists a unique uniform component U of S such that $\phi(I) = 0$ for $I = \{x \in S \mid SxS \cap U = \emptyset\}$. Moreover $\phi(U)$ is an ideal of $\phi(S)$. Hence ϕ factors through S/I and it determines an irreducible representation of U.*

2. *Conversely, every irreducible representation $\eta : U \longrightarrow M_r(L)$ of a uniform component U of S determines a unique irreducible representation $\phi : S \longrightarrow M_r(L)$ such that $\phi_{|U} = \eta$.*

Proof. 1) We may assume that $0 \in S$. Choose a minimal, with respect to the order $\prec$, uniform component U of S which is not contained in the ideal $I' = \{x \in S \mid \phi(x) = 0\}$ of S. Such a component exists because otherwise $\phi(S)$ is a nil semigroup, hence it is nilpotent by Proposition 2.14, which contradicts its irreducibility. We claim that U is uniquely determined. Suppose that U' is another component with this property. Then from Theorem 3.5 it follows that $SUS \setminus U$ and $SU'S \setminus U'$ are ideals of S and they do not intersect uniform components of S that are not contained in I'. Therefore, $\phi(SUS \setminus U)$ is a nil ideal of $\phi(S)$, so it must be zero because $\phi(S)$ is irreducible. Hence $\phi(SUS) = \phi(U)$, and similarly $\phi(SU'S) = \phi(U')$. Now $\phi(SUS)\phi(SU'S) = \phi(UU') = 0$ because UU' does not intersect U, U'. Irreducibility of $\phi(S)$ implies that $\phi(SUS) = 0$ or $\phi(SU'S) = 0$. But this contradicts the choice of U and U', proving our claim.

It is clear that I is an ideal of S. If $I \not\subseteq I'$, then as above we see that I intersects a uniform component of S. This leads to a uniform component $U' \neq U$ minimal over I', which is not possible. Hence $I \subseteq I'$. Since $\phi(U)$ is an ideal of $\phi(S)$, it is irreducible. Hence 1) follows.

2) From Chapter 3 we know that U can be viewed as an ideal of a Rees factor S/J. If $\eta : U \longrightarrow M_n(L)$ is an irreducible representation, then by Lemma 4.1 there exists $e \in L[U]$ such that $\eta(e) = 1$. Let $\phi(x) = \eta(e\overline{x})$ for $x \in L[S]$, where $\overline{x} \in L[S/J]$ denotes the image of x under the map $L[S] \longrightarrow L[S/J]$. It is clear $\eta(e\overline{x}) = \eta(e\overline{x})\eta(e) = \eta(e\overline{x}e) = \eta(\overline{x}e)$. Hence $\phi(xy) = \eta(e\overline{xy}e) = \eta(e\overline{x})\eta(e\overline{y}) = \phi(x)\phi(y)$, so that ϕ determines a representation of S which coincides with η on U. If $\phi' : S \longrightarrow M_n(L)$ is another representation with this property, then $\phi'(x) = \phi'(x)\eta(e) = \phi'(x)\phi'(e) = \phi'(xe) = \eta(\overline{x}e) = \phi(x)$. The result follows. $\square$

Remark Let $S \subseteq M_n(K)$ be a π-regular semigroup. The above theorem leads in view of the material in Section 4.2 to the following classification of irreducible representations. Let $\mathcal{G}$ be the set of maximal subgroups of S, one from each regular $\mathcal{J}$-class of S. For $G \in \mathcal{G}$, let $\mathcal{V}_G$ be the set of all nonequivalent irreducible $F[G]$-modules. If F is a field, then the irreducible $F[S]$-modules W are in one-to-one correspondence (up to equivalence) with the ordered pairs (G, V), where $G \in \mathcal{G}$ and $V \in \mathcal{V}_G$. If W is given, then there is a unique $\mathcal{J}$-class J of S such that $JW \neq 0$ and $IV = 0$, where $I = \{x \in S \mid J \not\subseteq SxS\}$. Moreover, the dimension of W is equal to $\text{rank}(\phi(P))$ for a sandwich matrix corresponding to J.

It is clear that, in general, a representation ϕ of a uniform semigroup U with a completely 0-simple closure $\hat{U}$ does not extend to a representation of $\hat{U}$. Namely, if $\phi(a) = 0$ for some $a \in U \cap D$ and a maximal subgroup D of $\hat{U}$, then we would have $\phi = 0$. However, even if $\phi(a) \neq 0$ for all $a \neq 0, a \in U$, extending ϕ may not be possible. Indeed, let $S \subseteq M_n(K), n > 1$, be an irreducible semigroup consisting of matrices of the same rank $j < n$. For example, consider $S = \{a = (a_{ij}) \in M_n(\mathbf{R}) \mid a_{ij} > 0, \text{rank}(a) = 1\}$. Let X be a free semigroup such that there exists an onto homomorphism $\phi : X \longrightarrow S$. Then ϕ does not extend to a representation ϕ' of the group G freely generated by the generating set of X. Indeed, otherwise $S \subseteq \phi'(G)$ and the latter is contained in a maximal subgroup of $M_n(K)$. Since it is irreducible, it follows that $\phi'(G) \subseteq GL_n(K)$, which contradicts the choice of j. Clearly, S can be replaced by a finitely generated irreducible subsemigroup T. Then $U = X$ is a finitely generated free semigroup, so it can be viewed as a linear semigroup $U \subseteq GL_2(\mathbf{Q})$ with $\hat{U} = \text{gp}(U) \subseteq GL_2(\mathbf{Q})$ (see Example in Section 4.1).

For the above reasons we do not get a natural correspondence between irreducible representations of U and of $U \cap D$ for a maximal subgroup D of $\hat{U}$.

Our next aim is to give an abstract characterization of the bottom layer of an irreducible semigroup, proved in [106]. We say that rows y, y' of the sandwich matrix $P = (p_{yx})$ of a completely 0-simple semigroup $S = \mathcal{M}(G, X, Y, P)$ are similar if there exists $g \in G$ such that $p_{yx} = gp_{y'x}$ for every $x \in X$. Similarity of columns x, x' is defined by the condition $p_{yx} = p_{yx'}g$ for some $g \in G$ and every $y \in Y$.

Proposition 4.29 *Let U be a uniform subsemigroup of a completely 0-simple semigroup S with completely 0-simple closure $\widehat{U} \subseteq S$ with a maximal subgroup G. Then U has an irreducible representation ϕ : $U \longrightarrow M_n(K)$ extending to a faithful representation of $\widehat{U}$ if and only if the following conditions are satisfied*

1. *the sandwich matrix of $\widehat{U} = \mathcal{M}(G, X, Y, P)$ (under some Rees matrix presentation) has no similar rows and no similar columns,*

2. *there exist $r \geq 1$ and a faithful irreducible representation $\eta : G \longrightarrow GL_r(K)$ such that the matrix $\phi(P)$ (viewed as a matrix over K with $|Y|r$ rows and $|X|r$ columns) has rank n.*

Moreover, if $U \cap G$ is a right Ore semigroup, then it is enough to assume in 2) that a faithful representation $\eta : U \cap G \longrightarrow GL_r(K)$ is given.

Proof. If 1),2) hold, then the extension of η to a homomorphism of Munn algebras $\tilde{R} = \mathcal{M}(K[G], X, Y, P) \longrightarrow \mathcal{M}(M_r(K), X, Y, \phi(P)) \longrightarrow M_n(K)$ yields an irreducible representation ϕ of $\widehat{U}$, see Theorem 4.26. Suppose that $\phi(a) = \phi(b)$ for some $a, b \in \widehat{U}$. From the construction of ϕ we know that $(a - b) \in \mathcal{B}_{\tilde{R}}(0)$. Hence $\widehat{U}(a - b)\widehat{U} = 0$. Choose an idempotent $e \in \widehat{U}$ such that $ea = a$. Then $(a-eb) \circ P \circ \widehat{U} = (a-eb)\widehat{U} = 0$. Therefore $(a - eb) \circ P = 0$. In view of 2) this implies that we must have $a \mathcal{H} eb$. Clearly $a = eb$ in this case. Hence $a \mathcal{L} b$. Similarly one shows that $a \mathcal{R} b$. Then $a \mathcal{H} b$, and $P \circ (a - b) \circ P = 0$ imply that $a = b$. Therefore ϕ is faithful.

Assume now that $\phi : \widehat{U} \longrightarrow M_n(K)$ is a faithful irreducible representation. Then $\phi(\widehat{U})$ is an irreducible, completely 0-simple subsemigroup of $M_n(K)$. From Theorem 4.26 we know that $\text{rank}(\phi(P)) = n$ and the restriction of ϕ to G is an irreducible representation. Suppose P has two similar rows. Say, $p_{yx} = gp_{y'x}$ for some $y, y' \in Y$, some $g \in G$ and for every $x \in X$. Fix some $z \in X, h \in G$. Let $a = (h, z, y), b = (hg, z, y')$. Then $ac = bc$ for every $c \in \widehat{U}$. This is impossible by Lemma 4.2, unless $y = y'$. A symmetric argument works for the columns of P.

Finally, it is easy to see that $G = (U \cap G)(U \cap G)^{-1}$ implies that the unique extension of a faithful representation of $U \cap G$ to G is also faithful. The result follows. $\square$

We conclude with a result on a representation of an irreducible semigroup S it terms of the translational hull $\Omega(S)$. Let Λ, Θ be the sets of left, respectively right translations of a semigroup S. That is, all $\lambda \in \Lambda, \rho \in \Theta$ are maps $S \longrightarrow S$ satisfying $\lambda(xy) = \lambda(x)y$ and $\rho(xy) = x\rho(y)$ for every $x, y \in S$. Recall that the translational hull $\Omega(S)$ of S is the set of all pairs

$$\{(\lambda, \rho) \,|\, \lambda \in \Lambda, \rho \in \Theta, x\lambda(y) = \rho(x)y \text{ for } x, y \in S\}$$

with operation $(\lambda, \rho) \cdot (\lambda', \rho') = (\lambda'\lambda, \rho\rho')$, [15]. Let $I(S)$ be the idealizer in $M_n(K)$ of a semigroup $S \subseteq M_n(K)$. That is, $I(S) = \{a \in M_n(K) \,|\, aS, Sa \subseteq S\}$. Clearly, every element $a \in I(S)$ determines an element $\omega(a) = (\lambda_a, \rho_a) \in \Omega(S)$, which represents the left and right multiplications λ_a, ρ_a by a. This semigroup is of interest because of the following result obtained in [106].

Proposition 4.30 *Assume that $S \subseteq M_n(K)$ is an irreducible semigroup. Then*

1. *if S is the bottom layer of a semigroup $T \subseteq M_n(K)$, then $T \subseteq I(S)$; moreover the bottom layer of $I(S)$ is equal to S if S is completely 0-simple,*

2. *the natural map $\omega : I(S) \longrightarrow \Omega(S)$ is an isomorphism.*

Proof. 1) Since S is an ideal of T, the inclusion $T \subseteq I(S)$ is clear. So, assume that S is completely 0-simple. Let $a \in I(S)$ be of rank equal to the common rank of nonzero matrices in S. Since S is irreducible, from Lemma 4.1 we know that $1 \in K\{S\}$. Hence $aS \neq 0$. Therefore $aS \subseteq S$ implies that $a\mathcal{R}b$ in $M_n(K)$ for some $b \in S$. Similarly, $a\mathcal{L}c$ for some $c \in S$. Since S is completely 0-simple, it follows that $a\mathcal{H}d$ for some $d \in S$. Then $aS \subseteq S$ implies that $a \in S$.

2) Let $\omega(s) = (\lambda_s, \rho_s)$ for $s \in I(S)$. Suppose that $(\lambda_s, \rho_s) = (\lambda_t, \rho_t)$ for some $s, t \in S$. Then $(s - t)S = 0$. Since $1 \in K\{S\}$, we must have $s - t = 0$. This means that $\omega_{|S}$ is an embedding. Thus, $\delta : \omega(S) \longrightarrow I(S) \subseteq M_n(K)$ given by $\delta(\omega(s)) = s$ is an irreducible representation of $\omega(S)$. We have $\lambda\lambda_s = \lambda_{\lambda(s)}$ and $\rho_s\rho(x) = \rho(x)s = x\lambda(s)$ whenever $(\lambda, \rho) \in \Omega(S)$ and $x \in S$. Therefore

$$(*) \qquad (\lambda_s, \rho_s)(\lambda, \rho) = (\lambda_{\lambda(s)}, \rho_{\lambda(s)}) \in \omega(S).$$

Similarly $(\lambda, \rho)(\lambda_s, \rho_s) \in \omega(S)$, so that $\omega(S)$ is an ideal of $\Omega(S)$. Hence, δ extends to a representation $\overline{\delta} : \Omega(S) \longrightarrow M_n(K)$. Then S is an ideal of $\overline{\delta}(\Omega(S))$ because $\delta(\omega(S)) = S$. Consequently $\overline{\delta}(\Omega(S)) \subseteq I(S)$. Let $y \in I(S)$. Then for every $x \in S$ we have $yx, xy \in S$ and $yx = \delta(\omega(yx)) = \overline{\delta}(\omega(y))\overline{\delta}(\omega(x)) = \overline{\delta}(\omega(y))x$. Hence $(y - \overline{\delta}(\omega(y)))S = 0$, so that $y = \overline{\delta}(\omega(y))$. Therefore $I(S) = \overline{\delta}(\Omega(S))$. It remains to show that $\overline{\delta}$ is injective. Suppose that $\overline{\delta}(a) = \overline{\delta}(b)$ for some $a, b \in \Omega(S)$. Then $\delta(ua) = \overline{\delta}(ua) = \overline{\delta}(ub) = \delta(ub)$ and $\delta(au) = \overline{\delta}(au) = \overline{\delta}(bu) = \delta(bu)$ for every $u \in \omega(S)$. This implies that $ua = ub$ and $au = bu$ for every $u \in \omega(S)$. If $a = (\lambda, \rho), b = (\lambda', \rho')$, then in view of $(*)$ this means that $\lambda_{\lambda(x)} = \lambda_{\lambda'(x)}$ for every $x \in S$. A symmetric argument shows that $\rho_{\rho(x)} = \rho_{\rho'(x)}$. So $(\lambda(x) - \lambda'(x))S = 0 = S(\rho(x) - \rho'(x))$, which implies that $\lambda = \lambda'$ and $\rho = \rho'$. So $a = b$, as desired. The result follows. $\square$

4.4   Triangularizable semigroups

Our next aim is to describe triangularizable semigroups in terms of the structure theorem of Chapter 3. The obtained conditions reduce the problem to the case of groups of matrices. It follows also that the global requirements concerning all elements of S can be replaced by 'local' conditions concerning the sets of elements of S of the same rank.

The triangularizability problem for a subsemigroup S of the monoid $M_n(K)$ has been considered by several authors, with a special emphasis on the case of the field of complex numbers, [56],[71],[112],[113],[133]. Here S is called triangularizable if there exists $g \in GL_n(K)$ such that the semigroup $g^{-1}Sg$ is upper triangular. Several sufficient conditions are known. In particular, in the case of characteristic zero or $> n/2$, a semigroup with constant trace must be triangularizable, [56],[133]. Permutability of the trace is also sufficient, [113]. The effect of certain spectral conditions on triangularizability was studied in [71],[112].

Note that, in the case of groups of matrices, several necessary and sufficient conditions are known over fields K of 'good' characteristic, cf. [71], [113]. Since we only set up the problem in terms of the structural approach, our results do not require any restriction on the field K.

Recall that every maximal subgroup D of $M_n(K)$ consists of units of the monoid $eM_n(K)e$ for some $e = e^2 \in M_n(K)$, so it can be identified with $GL_t(K)$, where $t = \mathrm{rank}(e)$. Since every idempotent of $M_n(K)$ is conjugate to a diagonal idempotent, triangularizability of $\mathrm{gp}(S \cap D)$ in $M_n(K)$ is equivalent to its triangularizability in $GL_t(K)$.

If T is a uniform component of S, then let E_T denote the set of all idempotents that are the identities of maximal subgroups of $M_n(K)$ intersecting T. In other words, $e = e^2 \in M_n(K)$ is an element of E_T if and only if there exists $t \in T$ such that $t = ete$ and $\mathrm{rank}(t) = \mathrm{rank}(e)$. Let $\langle E_T \rangle$ denote the semigroup generated by E_T.

$S \subseteq M_n(K)$ is called unipotent if the eigenvalues of every $s \in S$ lie in the set $\{0, 1\}$. Some sufficient conditions for the triangularizability of unipotent semigroups were obtained in [56],[112]. These results will be improved in Corollary 4.33. We are now ready for our main result.

Theorem 4.31 *The following conditions are equivalent for a semigroup* $S \subseteq M_n(K)$

1. *S is triangularizable,*

2. *every uniform component T of S is a subsemigroup of S, every $\langle E_T \rangle$ is a unipotent semigroup and every nonempty intersection $S \cap D$ with a maximal subgroup D of $M_n(K)$ is triangularizable,*

3. *for every uniform component T of S $\langle E_T \rangle$ is unipotent, for every $a, b \in S$, if $\mathrm{rank}(a) = \mathrm{rank}(b) = \mathrm{rank}(ab) = \mathrm{rank}((ab)^2)$, then $\mathrm{rank}(a^2) = \mathrm{rank}(a)$, and the groups associated to S are triangularizable.*

Proof. Conjugating in $M_n(K)$ we can assume that S is in the block upper triangular form of Proposition 4.4. We use the notation of the proof of this proposition. Let $\pi : R \longrightarrow Z = \sum_{i=1}^r e_i M_n(K) e_i$ be the corresponding block diagonal projection. Assume first that 1) holds. Then Z consists of diagonal matrices. Let $a, b \in S$ be such that $\mathrm{rank}(a) = \mathrm{rank}(b) = \mathrm{rank}(ab) = \mathrm{rank}((ab)^2)$. From Theorem 3.5 it follows that a, b, ab lie in a uniform component of S. Therefore, from Lemma 4.6 we know that $\pi(a), \pi(b), \pi(ab)$ have rank equal to the rank of a. Since these are diagonal matrices, we also have $\mathrm{rank}(\pi(a^2)) = \mathrm{rank}(\pi(a))$. Therefore $\mathrm{rank}(a^2) \geq \mathrm{rank}(\pi(a^2)) = \mathrm{rank}(\pi(a)) = \mathrm{rank}(a)$, which implies that $\mathrm{rank}(a^2) = \mathrm{rank}(a)$. Since every nonempty intersection $S \cap D$ with a maximal subgroup D of $M_n(K)$ is upper triangular, so is the group $\mathrm{gp}(S \cap D)$. $\langle E_T \rangle$ is unipotent because E_T is a triangular set of unipotent matrices. This proves 3).

Assume that 3) holds. Every uniform component T of S intersects all nonzero $\mathcal{H}$-classes of a completely 0-simple semigroup $\hat{T} \subseteq$

$M_j/M_{j-1}, j \geq 1$. We know that $\hat{T}$ is of the form $\hat{T} \simeq \mathcal{M}(G, X, Y, P)$. Let $a \in T$. Since $a\hat{T}$ contains a nonzero idempotent $e \in \hat{T}$, there exists $b \in T$ such that $ab\mathcal{H}e$. Hence $\mathrm{rank}(ab) = \mathrm{rank}((ab)^2)$, and clearly $\mathrm{rank}(a) = \mathrm{rank}(ab) = \mathrm{rank}(b)$. The assumption implies that $a^2 \in T$. This shows that all entries of the sandwich matrix P are nonzero. Consequently T, viewed as a subset of S, is a subsemigroup of S. Therefore 2) is a consequence of 3).

Finally, assume that 2) holds. Let π_i be the projection onto a diagonal block $Z_i \simeq M_m(K)$ such that $\pi_i(S)$ is irreducible as a subsemigroup of Z_i. We aim to show that $m = 1$. By Lemma 4.7 there exists a uniform component T of S such that $\pi_i(T) \cup \{0\}$ is an ideal of $\pi_i(S)$ that intersects the same $\mathcal{H}$-classes of $M_n(K)$ as a uniform component V of $\pi_i(S)$. Moreover, by the hypothesis $\pi_i(T)$ is a subsemigroup of $\pi_i(S)$. Since $\pi_i(T) \cup \{0\}$ is an ideal of $\pi_i(S)$, $\pi_i(T)$ is irreducible in Z_i. Let D, D', e be chosen as in Lemma 4.7. Then $\mathrm{gp}(\pi_i(T) \cap D) = \mathrm{gp}(\pi_i(T \cap D'))$ is an irreducible subgroup of the corresponding matrix ring $eM_n(K)e$. Because $T \cap D'$ is triangularizable, $\pi_i(T \cap D')$ is triangularizable and consequently $\mathrm{gp}(\pi_i(T \cap D'))$ is triangularizable. Therefore we must have $\mathrm{rank}(e) = 1$. This implies that $\pi_i(T) \subseteq M_1$, the ideal of matrices of rank at most 1 in $M_n(K)$. Hence $\pi_i(T) \cup \{0\}$ meets the same $\mathcal{H}$-classes of $M_n(K)$ as a completely 0-simple semigroup $T' \simeq \mathcal{M}(G, X, Y, P)$, where G is a subgroup of the multiplicative group K^* of the field K. But T is a subsemigroup of S, so it is contained in a union of maximal subgroups of $M_n(K)$ and E_T, T must meet the same $\mathcal{H}$-classes of $M_n(K)$. Hence $\pi_i(E_T)$ meets all $\mathcal{H}$-classes of $M_n(K)$ that are intersected by $\pi_i(T)$ (in particular $\pi_i(T)$ is contained in a union of maximal subgroups of $M_n(K)$ consisting of matrices of rank 1). Also, since $\pi_i(\langle E_T \rangle)$ is unipotent by 2), it must consist of idempotents. This means that $\pi_i(\langle E_T \rangle) = \pi_i(E_T)$ is a completely simple subsemigroup of T' over the trivial subgroup of G. Hence, if $a, b \in \pi_i(E_T)$ are in the same $\mathcal{R}$-class of $M_n(K)$, then $as = s = bs$ for every $s \in \pi_i(E_T)$. This implies that $at = bt$ for every $t \in \pi_i(T)$. Therefore $(a - b)\pi_i(T) = 0$. But $\pi_i(T)$ is irreducible. It follows that $a = b$, and hence $|X| = 1$. Similarly, one shows that $|Y| = 1$. Therefore $\pi_i(E_T)$ is a singleton, and consequently $m = 1$. This shows that 1) holds, completing the proof of the theorem. $\square$

We note that, in conditions 2),3) of Theorem 4.31, the assumption that all semigroups $\langle E_T \rangle$ are unipotent can be replaced by the requirement that they are triangularizable (clearly, triangularizability of the sets E_T is sufficient). This is an easy consequence of the proof.

Corollary 4.32 *A semigroup $S \subseteq M_n(K)$ is triangularizable if and only if its all uniform components are triangularizable, or equivalently, each of the nonempty sets $S_j \setminus S_{j-1}, j = 1, \ldots, n$, where $S_j = S \cap M_j$, is triangularizable.*

Proof. If $T \subseteq S_j \setminus S_{j-1}$ is a uniform component of S, then the semigroup $\langle T \rangle$ generated by T is contained in $T \cup S_{j-1}$. Therefore, T is its uniform component. If every T is triangularizable, then so is $\langle T \rangle$ and Theorem 4.31 applied to $\langle T \rangle$ implies that condition 2) of this theorem is satisfied. The assertion follows. $\square$

Corollary 4.32 shows that certain known sufficient conditions, for example permutability of the trace in the case of fields of 'good' characteristic, [113], Theorem 1, can be weakened by requiring only that they hold for each of the sets $S_j \setminus S_{j-1}$.

The theorem and the corollary remain valid in the more general context of subsemigroups of $M_n(F)$, where F is a division algebra. (Here, $a \in M_n(F)$ is called unipotent if $a - e$ is nilpotent for the idempotent $e \in M_n(F)$ such that $a^n \mathcal{H} e$ in $M_n(F)$). This is a consequence of our proof and of the fact that a similar structure theorem holds in this case. (The only difference, not essential for the proof, is that the number of uniform components may be infinite and the nilradicals N of $S_j \setminus S_{j-1}$ are nil modulo S_{j-1} but do not satisfy $N^{\binom{n}{j}} \subseteq S_{j-1}$ in general. However, if S is triangular, then there are at most 2^n uniform components and each N is nilpotent modulo the corresponding S_{j-1}, see Section 9.1.)

If S is triangular, then a simple description of the uniform components of S can be given. Namely, every component T is the collection of matrices in S with a fixed location of nonzero diagonal entries and such that $\text{rank}(a) = \text{rank}(\pi(a))$, where π is the diagonal projection. This is an easy consequence of the structure theorem from Section 3.2.

Corollary 4.33 *Assume that $S \subseteq M_n(K)$ is a unipotent semigroup. Then S is triangularizable if and only if the uniform components of S are subsemigroups of S.*

Proof. The necessity follows from Theorem 4.31. Since unipotent groups are triangularizable, the sufficiency follows as in the proof of the implication 2) $\implies$ 1) in Theorem 4.31 (the proof works also if one assumes that $\pi_i(T)$ is unipotent in place of the hypothesis that $\pi_i(\langle E_T \rangle)$ is unipotent). $\square$

Chapter 5

Identities

In this chapter we study the influence of identities of certain types on the structure and properties of a linear semigroup. Identities in the coordinate ring $K[x_{ij}|i,j = 1,\ldots,n]$ of $M_n(K)$, semigroup identities for $S \subseteq M_n(K)$, and also polynomial identities of the semigroup algebra $K[S]$ are considered. In what follows, if $\mathcal{N}$ is a class of groups, then we say that a group H is an almost $\mathcal{N}$-group, if H has a normal subgroup of finite index, which is an $\mathcal{N}$-group. If every finitely generated subgroup of a group G is in $\mathcal{N}$, then G is called a locally $\mathcal{N}$-group.

5.1 Semigroup identities

Let f be an element of the polynomial ring $K[t_1,\ldots,t_m]$, where $m \geq 1$. If A is a subset of $M_n(K)$, then we say that the identity $f = 0$ holds in A if for every matrices $a_k = (a_{ij}^{(k)}) \in A, k = 1,\ldots,m$, we have $f(a_1,\ldots,a_m) = 0$ in $M_n(K)$.

Lemma 5.1 *Assume that the identity $f = 0$ holds in a subset $A \subseteq M_n(K)$, where $f \in K[t_1,\ldots,t_m]$. Then $f = 0$ holds in the Zariski closure $\overline{A}$.*

Proof. For every $i = 1,\ldots,m$ define

$$A_i = \{a \in M_n(K)|\ f(a_1,\ldots,a_m) = 0$$
$$\text{for all } a_j \in \overline{A}, j < i, \text{ and all } a_k \in A, k > i\}.$$

It is clear that each A_i is a closed subset of $M_n(K)$ and $A \subseteq A_1 \subseteq \overline{A}$. It follows that $A_1 = \overline{A}$. Next, if $A_i = \overline{A}$ for some $i \geq 1$, then $A \subseteq A_{i+1} \subseteq \overline{A}$. Therefore we must have $A_{i+1} = \overline{A}$. By induction it follows that $A_m = \overline{A}$. This means that $f = 0$ holds in $\overline{A}$. $\square$

Let S be a semigroup. By an identity in S we mean an identity of the form $v(x_1, \ldots , x_m) = w(x_1, \ldots , x_m)$, where v, w are any two different words in a collection of indeterminates $x_1, \ldots , x_m$. In particular, if S is a group, then we shall consider only identities of this type, for emphasis saying that S satisfies a semigroup identity. It is well known that existence of such an identity in S implies that there exists an indentity in two variables that is satisfied in S. This is an easy consequence of the fact that the free semigroup of finite rank $m, m \geq 2$, embeds in the free semigroup $\langle x, y \rangle$.

As an example, consider the class of nilpotent semigroups introduced via a semigroup identity, independently in [77] and [86], see also [87]. Let $x, y, u_1, u_2, \ldots$ be elements of a semigroup S. Consider the sequences of elements defined inductively as follows

$$x_0 = x, \quad y_0 = y$$

$$x_{m+1} = x_m u_{m+1} y_m, \quad y_{m+1} = y_m u_{m+1} x_m$$

We will sometimes write $x_m(x, y, u_1, \ldots , u_m), y_m(x, y, u_1, \ldots , u_m)$ for x_m, y_m. S is a nilpotent semigroup if there exists $m \geq 1$ such that $x_m = y_m$ for all $x, y, u_1, \ldots , u_m \in S$. We will then say that the identity $X_m = Y_m$ holds in S. The minimal n with this property is called the nilpotency class of S and S is called m-nilpotent. If S is a group, this agrees with the classical definition. More generally, if S is cancellative, then S is m-nilpotent if and only if it has a group of quotients that is m-nilpotent, [77],[86],[87]. This class will be studied in detail in the next section.

Corollary 5.2 *Assume that a semigroup $S \subseteq M_n(K)$ satisfies an identity. Then $\overline{S}$ also satisfies this identity.*

Proof. Let $v(x_1, \ldots , x_m) = w(x_1, \ldots , x_m)$ be an identity satisfied in S. Lemma 5.1 applies to show that the latter identity is satisfied for all matrices in $\overline{S}$. $\square$

Next, we show that satisfying an identity is a local property of a semigroup $S \subseteq M_n(K)$.

Lemma 5.3 *The following conditions are equivalent for a semigroup* $S \subseteq M_n(K)$

1. *S satisfies an identity,*

2. *every nonempty intersection $S \cap D$ with a maximal subgroup D of $M_n(K)$ satisfies an identity,*

3. *every group associated to S satisfies an identity.*

Proof. Clearly, 2) is a consequence of 1). From Corollary 5.2 it follows, in view of Corollary 1.5, that 2) implies 3). Assume that 3) holds. First, we show that each factor $T = S_j/S_{j-1}$ arising from Theorem 3.5 satisfies an identity. Let $s, t \in T$. Then either $stss \in D$ for a maximal subgroup $D \subseteq M_j \setminus M_{j-1}$ or it is zero in T. In the former case, $s \in D$. Let $v(x,y) = w(x,y)$ be an identity satisfied in $\mathrm{gp}(S \cap D) \subseteq D$. Hence, in both cases we have $v(stss, s) = w(stss, s)$ in T. It is clear that the words $v(xxyx, x), w(xyxx, x)$ are different, so $v(xxyx, x) = w(xyxx, x)$ is an identity in T. Therefore, it is enough to show that, if R and R/I satisfy some identities for an ideal I of a semigroup R, then R also satisfies an identity. Thus, assume $f_1(x,y) = g_1(x,y)$ is an identity in R/I and $f_2(x,y) = g_2(x,y)$ is an identity in I. We can assume that f_2, g_2 are words in x, y of equal length, replacing them by $f_2 g_2, g_2 f_2$, if necessary. If $s, t \in R$ and $f_1(s,t) = g_1(s,t) \in R \setminus I$, then the choice of f_2, g_2 implies that $f_2(f_1(s,t), g_1(s,t)) = g_2(f_1(s,t), g_1(s,t))$. It is easy to see that the two respective words in s, t are different, so $f_2(f_1(x,y), g_1(x,y)) = g_2(f_1(x,y), g_1(x,y))$ is an identity in R. $\square$

We will need the following fundamental result on linear groups, referred to as Tits alternative. The proof can be found in the original paper [132] or in [83], see also [78], Appendix B.

Theorem 5.4 *Let $G \subseteq GL_n(K)$ be a subgroup. Then either G has a solvable normal subgroup N such that G/N is periodic or G has a free nonabelian subgroup. If $\mathrm{ch}(K) = 0$ or K is finitely generated, then either G is almost solvable or it has a free nonabelian subgroup.*

The following theorem is due to Rosenblatt [118]. It is an extension of results on solvable groups of polynomial growth due to Milnor and Wolf, see [67].

Theorem 5.5 *Let G be a finitely generated solvable group. Then either G is almost nilpotent or it has a free noncommutative subsemigroup.*

We shall only prove this theorem in the special case where G is a linear group. To this end we need the following result.

Lemma 5.6 *Let A be an abelian normal subgroup of a group G. Assume that G has no free noncommutative subsemigroups. For every $g \in G$ and $a \in A$, let $a_k = g^k a g^{-k}$ for $k \in \mathbf{Z}$. Then there exists a natural number z and integers $q_1, \ldots, q_z \in \{0, 1, -1\}$, not all equal 0 such that $a_1^{q_1} \cdots a_z^{q_z} = 1$. Moreover, the group generated by the set $\{a_k \mid k \in \mathbf{Z}\}$ is finitely generated.*

Proof. Clearly, we may assume that $a \neq 1$. The hypothesis implies that for every $x, y \in G$ there exist two different words $w(x, y), v(x, y)$ in x, y such that $w(x, y) = v(x, y)$ in G. Let $w(x, y) = x^{n_1} y^{m_1} \cdots x^{n_s} y^{m_s}$ and $v(x, y) = x^{k_1} y^{l_1} \cdots x^{k_t} y^{l_t}$ where all exponents are nonnegative and $n_i, m_i, k_j, l_j > 0$ except possibly n_1, k_1, m_s, l_t. Cancellation on the left allows us to assume that $n_1 > 0$ and $k_1 = 0$. Next, replacing $w(x, y)$ by $w(x, y)v(x, y)$ and $v(x, y)$ by $v(x, y)w(x, y)$, we can assume that the words w, v have equal length. Applying this to the elements g, ga we get a relation of the form

$$g^{n_1}(ga)^{m_1}g^{n_2}(ga)^{m_2}\cdots g^{n_s}(ga)^{m_s} = (ga)^{l_1}g^{k_2}(ga)^{l_2}g^{k_3}\cdots g^{k_t}(ga)^{l_t}$$

where $\sum_{i=1}^{s}(n_i + m_i) = \sum_{j=1}^{t}(k_j + l_j)$ with $k_1 = 0$. Since $g^k(ga)^l = a_{k+1}a_{k+2}\cdots a_{k+l}g^{k+l}$, this relation can be rewritten as

$$(a_{n_1+1}a_{n_1+2}\cdots a_{r_1})(a_{r_1+n_2+1}a_{r_1+n_2+2}\cdots a_{r_2})\cdots(a_{r_{s-1}+n_s+1}\cdots a_{r_s})g^{r_s}$$
$$= (a_{k_1+1}a_{k_1+2}\cdots a_{p_1})(a_{p_1+k_2+1}a_{p_1+k_2+2}\cdots a_{p_2})\cdots(a_{p_{t-1}+k_t+1}\cdots a_{p_t})g^{p_t},$$

where $p_j = k_1 + l_1 + \cdots + k_j + l_j$ for $j = 1, 2, \ldots, t$ ad $r_i = n_1 + m_1 + \cdots + n_i + m_i$ for $i = 1, 2, \ldots, s$. Then $r_s = p_t$, so the factors g^{r_s}, g^{p_t} can be cancelled on both sides. Since A is abelian, $0 = k_1 \neq n_1$ and the sequences of indices appearing on both sides of this equality are increasing, we get a relation of the desired form: $a_1^{q_1} \cdots a_z^{q_z} = 1$, with $z = r_s$ and $q_1 \neq 0$.

Choose z as small as possible. Then $q_z \neq 0$. Clearly, $z \geq 2$ because otherwise $a_1 = 1$, and hence $a = 1$.

We have shown that a_z is a word (allowing inverses) in $a_1, \ldots, a_{z-1}$. Conjugating by g we see that a_{z+1} is a word in $a_2, \ldots, a_z$, and therefore a word in $a_1 \ldots, a_{z-1}$. An easy induction implies that every $a_k, k \geq 1$, is a word in $a_1, \ldots a_{z-1}$. Similarly, it follows that a_1 is a word in $a_2, \ldots, a_z$, so that a_0 is a word in $a_1, \ldots, a_{z-1}$. Then a_{-1} is a word in $a_0, \ldots, a_{z-2}$, and hence in $a_1, \ldots, a_{z-1}$. Again, by induction it follows that every $a_{-k}, k \geq 1$, is a word in $a_1, \ldots, a_{z-1}$. This means that the latter elements generate the group generated by the set $\{a_k | k \in \mathbf{Z}\}$. $\square$

Proof of Theorem 5.5 for $G \subseteq GL_n(K)$.

Suppose that G has no free noncommutative subsemigroups. From Malcev's theorem we know that G has a normal subgroup T of finite index which is triangularizable in $M_n(\overline{K})$, see [134], Theorem 3.6. Then T is finitely generated, so replacing G by T and K by $\overline{K}$, we can assume that $G \subseteq GL_n(K)$ is upper triangular. We proceed by induction on n, the case $n = 1$ being clear. Let $e_1, \ldots, e_n$ be the diagonal idempotents of rank one and let I be the identity matrix. Then the induction hypothesis implies that the projections $G_1 = (I - e_1)G(I - e_1)$ and $G_2 = (I - e_n)G(I - e_n)$ are almost nilpotent (as they can be considered as subgroups of $GL_{n-1}(K)$). Then $N = (I + e_1 M_n(K)e_n) \cap G$ is the intersection of the kernels of these projections. Since G/N embeds into $G_1 \times G_2$, there exists a normal subgroup H of finite in G such that $N \subseteq H$ and H/N is nilpotent. Thus, we may assume that N is nontrivial. First consider the case where there exists a matrix $c \in G$ such that the element $b = c_1 c_n^{-1}$ is not a root of unity in K, where $c_1, \ldots, c_n$ is the diagonal of c. Let $a \in N, a \neq I$. Then there exists a nonzero $x \in K$ such that $c^i a c^{-i} = I + (c_1 c_n^{-1})^i x e_{1n}$ for every $i \in \mathbf{Z}$, where e_{1n} is the $(1, n)$-th matrix unit. From Lemma 5.6 we know that the subgroup of N generated by all $a_i = c^i a c^{-i}$ is finitely generated. If $\mathrm{ch}(K) = p > 0$, then it must be finite. Then $c_1 c_n^{-1}$ is a root of unity, a contradiction. Therefore $\mathrm{ch}(K) = 0$. Let L be a finitely generated subfield of K such that $G \subseteq GL_n(L)$. We can assume that L is a subfield of the field $\mathbf{C}$ of complex numbers. It has been shown that there exist $r \geq 1$ such that $b^i x \in \sum_{j=1}^r b^j x \mathbf{Z}$ for every $i \geq 1$. This implies that b is integral over $\mathbf{Z}$.

Then $|b| \neq 1$, since every element of $\mathbf{C}$ which is integral over $\mathbf{Z}$ and has absolute value 1 is a root of unity, see [8], p.104-105. Now, replacing c by c^i or by c^{-i} for i big enough, we may assume that $|b| > 2$. From Lemma 5.6 we know that for a natural number z and some integers $q_1, \ldots, q_z \in \{0, 1, -1\}$ with $q_z \neq 0$, we have $a_1^{q_1} \cdots a_z^{q_z} = I$. Therefore $q_1 b + q_2 b^2 + \cdots + q_z b^z = 0$. This contradicts the fact that $|b| > 2$.

Hence, suppose that $c_1 c_n^{-1}$ is a root of unity for every $c \in G$. The set A of roots of unity in a finitely generated field $L \subseteq K$ such that $G \subseteq GL_n(L)$ is finite by Lemma 1.7. It follows easily that $F = \{c \in G | c_1 c_n^{-1} = 1\}$ is a subgroup of finite index in G. It is clear that N lies in the centre of F. But $F/N \subseteq G/N$, so it is almost nilpotent. This implies that F is almost nilpotent. Therefore G is almost nilpotent. This completes the proof of the theorem. $\square$

Recall that, if every finitely generated subgroup of a group G is almost solvable, then we say that G is locally almost solvable. It is well known that a linear group of this type has a solvable normal subgroup H such that G/H is a periodic linear group, see [134], Theorem 10.8. In particular, G is a periodic extension of a solvable group. Similarly, every locally almost nilpotent group $G \subseteq GL_n(K)$ is a periodic extension of a nilpotent group. This will be used in the following result, coming from [5]. The proof requires standard facts on the structure of algebraic groups. All of these results are stated in Section 6.1.

Proposition 5.7 *Let K be an algebraically closed field which is not the closure of a finite field. Assume that G is a closed connected subgroup of $GL_n(K)$. Then the following conditions are equivalent*

1. *G satisfies a semigroup identity,*

2. *G is nilpotent,*

3. *G has no free noncommutative subsemigroups.*

Proof. As mentioned after Lemma 5.1, nilpotent groups satisfy semigroup identities of some special type (see also Section 5.3). Hence 1) is a consequence of 2). Since 1) implies 3), it is enough to prove the implication 3) $\Rightarrow$ 2). Thus, assume that G has no free noncommutative subsemigroups. Then G has no free nonabelian subgroups, so from Theorem 5.4 it follows that G has a solvable normal subgroup H such that

G/H is periodic. Let R be the solvable radical of G (the largest closed connected solvable normal subgroup). Suppose that $H \neq G$. Then G/R is a semisimple algebraic group and $H/(H \cap R)$ is its solvable normal subgroup. The identity component of the Zariski closure of $H/(H \cap R)$ is a closed connected solvable normal subgroup of G/R, so it must be trivial. Therefore $H/(H \cap R)$ is finite. It follows that G/R is periodic. In particular, a maximal torus of G/R is periodic. This is impossible by the hypothesis on K. Hence we must have $H = G$ and G is solvable. Now, Theorem 5.5 implies in view of the hypothesis that G is locally almost nilpotent. Therefore G has a nilpotent normal subgroup N such that G/N is periodic. Replacing N by the identity component of its closure we may assume that N is closed and connected. On the other hand, since G is connected, it is well known that the set G_u of unipotent elements of G forms a closed nilpotent subgroup of G and G is a direct product of G_u and a maximal torus T of G, see Theorem 6.3. Then $G_u N$ is a normal subgroup of G which is also closed, see Corollary 6.5. The algebraic group $G/G_u N$ is a periodic torus, hence it is trivial by the hypothesis on K. The sets N_s, N_u of semisimple, respectively unipotent elements of N are subgroups of N and $N = N_s \times N_u$ by Theorem 6.3. Since N is a normal subgroup of G, N_s is a normal subgroup of G. But $G = G_u N = G_u N_s$ implies then that $G = N_s \times G_u$ is nilpotent. This completes the proof. $\square$

It is clear that the assertion of Proposition 5.7 does not hold for $K = \overline{\mathbf{F}}_p$. Namely, consider $G = GL_n(\overline{\mathbf{F}}_p)$.

Corollary 5.8 *Let $S \subseteq M_n(K)$ be a semigroup. The following conditions are equivalent*

1. *S satisfies an identity,*

2. *every group associated to S is almost nilpotent.*

Proof. Extending K, if necessary, we may assume that K is algebraically closed and it is not the closure of a finite field. If 1) holds, then by Corollary 5.2 $\overline{S}$ satisfies an identity. From Corollary 1.5 it follows that every group H associated to S is contained in a maximal subgroup G of $\overline{S}$. But G is the group of matrices of maximal rank in $eM_n(K)e \cap \overline{S}$, where $e = e^2 \in G$ (see Remark iv) before

Proposition 3.6). Therefore G is a closed subgroup of a maximal subgroup of $M_n(K)$. Hence Proposition 5.7 implies that G is almost nilpotent. Therefore H is almost nilpotent and 2) follows. On the other hand, every almost nilpotent group satisfies an identity of the form $x_m(x^k, y^k, u_1^k, \ldots, u_m^k) = y_m(x^k, y^k, u_1^k, \ldots, u_m^k)$, where x_m, y_m are the words defined after Lemma 5.1. Therefore, the implication 2) $\Rightarrow$ 1) follows from Lemma 5.3. $\square$

We note that the results of this section will be strengthened in Chapter 6.

5.2   Almost unipotent semigroups

Our aim in this section is to describe two classes of semigroups $S \subseteq M_n(K)$ defined by identities over the coordinate ring $K[x_{ij}]_{i,j=1,\ldots,n}$, as considered in Lemma 5.1, of some special types. We follow [93] and [100].

Let $\overline{K}$ be the algebraic closure of the field K. An element $s \in M_n(K)$ is called unipotent if the eigenvalues of the matrix $s \in M_n(\overline{K})$ lie in the set $\{0, 1\}$. If there exists $k \geq 1$ such that s^k is unipotent, then s is called almost unipotent.

A semigroup $S \subseteq M_n(K)$ is almost unipotent if every group associated to S is an almost unipotent group. Clearly, this condition is stronger than that considered in Corollary 5.8. Hence, every semigroup of this type satisfies an identity. This property can be described by a number of equivalent conditions.

Proposition 5.9 *The following conditions are equivalent for a linear semigroup $S \subseteq M_n(K)$*

1. *S is almost unipotent,*

2. *for some $m \geq 1$ the identity $x^m(I - x^m)^n = 0$ holds for $x \in S$, where I denotes the identity matrix.*

3. *the set of eigenvalues of elements of S is finite,*

4. *S is conjugate to a subsemigroup of $M_n(\overline{K})$ that is block upper triangular with finite projections onto the diagonal blocks.*

Moreover, if K is finitely generated, these conditions are satisfied if and only if.

5. *every element $s \in S$ is almost unipotent.*

Proof. Assume that 1) holds. Since S has at most 2^n uniform components, Corollary 3.4 implies that there exists $m \geq 1$ such that for each $s \in S$ the matrix s^m lies in a subgroup D_s of $M_n(K)$ and is unipotent. Let $e = e^2 \in D_s$. Conjugating e to a diagonal idempotent we see that $(I - s^m)^n = I - e$. Since $s^m = s^m e$, the identity of 2) follows. Therefore 1) implies 2).

It is clear that 3) is a consequence of 2).

Assume that 3) holds. By Lemma 4.4 S is conjugate to a block upper triangular subsemigroup T of $M_n(\overline{K})$ whose projections $T_{1,1}, \ldots, T_{r,r}$, $r \leq n$, onto the diagonal blocks are absolutely irreducible or zero. From Proposition 4.9 it follows that each $T_{i,i}$ must be finite if 3) holds. Hence 3) implies 4).

Assume that 4) holds. Let G be the group generated by $T \cap D$ for a maximal subgroup D of $M_n(K)$. The block diagonal projections $(T \cap D)_{i,i}$, $i = 1, \ldots, r$, of $T \cap D$ are finite subsemigroups of the respective projections $G_{i,i}$ of G. Since $(T \cap D)_{i,i}$ generates $G_{i,i}$, we must have $(T \cap D)_{i,i} = G_{i,i}$. Hence each $G_{i,i}$ is finite and G must be a finite extension of a unipotent group. This implies that 1) is satisfied.

If $s \in M_n(K)$ is almost unipotent, then the nonzero eigenvalues of s are roots of unity. It is known that there are finitely many roots of unity that are roots of polynomials of degree n over a given finitely generated field, see Lemma 1.7. Therefore, for such a field K, 5) is equivalent to 3). This completes the proof. $\square$

Note that condition 2) of the above proposition implies that the Zariski closure $\overline{S}$ of S also is an almost unipotent semigroup, because it satisfies the same identities as S by Lemma 5.1.

Remark Let $Z_1, \ldots, Z_r$ be the $\overline{K}$-subalgebras of $M_n(\overline{K})$ corresponding to the diagonal blocks determined by $T_{1,1}, \ldots, T_{r,r}$ of the above proof. That is, Z_i is spanned by $T_{i,i}$ if $T_{i,i}$ is irreducible and Z_i is the appropriate 1-dimensional diagonal block if $T_{i,i}$ is a zero semigroup. The block diagonal projection $\pi : T \longrightarrow Z = Z_1 \times \ldots \times Z_r$ is a homomorphism that gives a Z-gradation on T. The components of idempotents

$T_{(e)} = \pi^{-1}(e)$, $e = e^2 \in Z$, are subsemigroups of T. Since each e can be diagonalized by conjugating within Z, $T_{(e)}$ is conjugate to a semigroup V such that each projection $V_{i,i}$ is a diagonal idempotent in Z_i. Therefore, V is an upper triangular semigroup which is unipotent. Let A be the semigroup of all upper triangular matrices with diagonal equal to the diagonal of e. Then $J = \{a \in A|\ \mathrm{rank}(a) = \mathrm{rank}(e)\}$ is an ideal of A. From Theorem 3.5 it follows that J is a uniform component of A. If D is the maximal subgroup of $M_n(K)$ containing e, then $J \cap D$ is a group. Hence J is a completely simple semigroup, which is the principal factor of e in A (see Proposition 3.1). Thus, from Theorem 3.5 it follows that $I = J \cap V$ is a uniform subsemigroup of J, whose associated group is unipotent. Therefore $T_{(e)}$ has an ideal $I_{(e)}$, which is its uniform component, whose associated group is unipotent and such that $(T_{(e)})^k \subseteq I_{(e)}$ for some $k \geq 1$. All this should be compared to the semilattice decomposition of upper triangular subsemigroups of $M_n(K)$, [108], Corollary 1.16, Theorem 5.12, Corollary 6.32.

Assume now that $S = \langle a_1, \ldots, a_t \rangle \subseteq M_n(K)$ is an almost unipotent semigroup that is in the block upper triangular form of 4) in Proposition 5.9. Let e_j be the identity of the diagonal block $Z_j, j = 1, \ldots, r$. For $k \leq j$ write $x_{kj}^{(i)} = e_k a_i e_j$, so that $a_i = \sum_{k,j=1,\ k\leq j}^r x_{kj}^{(i)}$. We define the sets

$$C_{i,i} = e_i S e_i, \quad C_{i,i+1} = C_{i,i}^1 X_{i,i+1} C_{i+1,i+1}^1$$

where $X_{i,j} = \{x_{ij}^{(k)} \mid k = 1, \ldots, t\}$, and inductively, for $i < j - 1$,

$$C_{i,j} = C_{i,i+1}C_{i+1,j} \cup \ldots \cup C_{i,j-1}C_{j-1,j} \cup C_{i,i}^1 X_{i,j} C_{j,j}^1.$$

It is clear that each $C_{i,j}$ is a finite set (induction). Let $D_{i,j}, i \leq j$, be the R-submodule of the ring $M_n(K)$ generated by $C_{i,j}$, where $R = \mathbf{Z}$ if $\mathrm{ch}(K) = 0$ and R is the prime subfield of K otherwise.

Lemma 5.10 *With the above notation we have*

1. *If $s \in S$, then $e_i s e_j \in D_{i,j}$ for $i \leq j$.*

2. *If $s = a_{i_1} \ldots a_{i_m}, i_q \in \{1, \ldots, t\}$, and $i < j$, then $e_i s e_j$ is a sum of at most $h_{j-i}(m)$ elements of the set $C_{i,j}$, where h_k is a polynomial in m of degree k, for $k = 1, \ldots, n-1$.*

Proof. 1) follows from the block multiplication of the matrices in S. Namely, for $s \in S$, by induction on the length m of s in the generators $a_1, \ldots, a_t$ of S, we show that, if $i \leq j$, then

$$e_i s e_j \in D_{i,i+1} D_{i+1,j} + \ldots + D_{i,j-1} D_{j-1,j} + C^1_{i,i} X_{i,j} C^1_{j,j} \subseteq D_{i,j}.$$

For $m = 1$ the assertion is clear. Let $q \in \{1, \ldots, t\}$. Then

$$e_i (s a_q) e_j = \sum_{l=i}^{j} (e_i s e_l)(e_l a_q e_j).$$

From the definitions it follows easily that $C_{p,p} C_{p,k} \subseteq C_{p,k}$ for $p \leq k$, and also $C_{p,k} X_{k,r} \subseteq C_{p,k} C_{k,r} \subseteq C_{p,r}$, for $p \leq k \leq r$. Since by the induction hypothesis

$$(e_i s e_l)(e_l a_q e_j) \in (D_{i,i+1} D_{i+1,l} + \ldots + D_{i,l-1} D_{l-1,l} + C^1_{i,i} X_{i,l} C^1_{l,l}) X_{l,j},$$

the inductive claim follows. This implies that 1) holds.

The above block multiplication pattern easily implies that to get the desired bound one can take $h_0(m) = 1, h_k(1) = 1$ for $k = 1, \ldots, n-1$, and inductively $h_k(m) = h_k(m-1) + h_{k-1}(m-1) + \ldots + h_0(m-1)$ for $k > 0, m > 1$. By induction on k we show that $h_k(m)$ is a polynomial in m of degree k. For $k = 0$ this is clear. Let $k > 0$. The induction hypothesis implies, in view of the definition, that $h_k(m) - h_k(m-1)$ is a polynomial of degree $k - 1$. It is well known, and easy to check, that in such a case $h_k(m)$ is a polynomial of degree k, see [67], Lemma 1.5. $\square$

Corollary 5.11 *If* $\mathrm{ch}(K) > 0$, *then every almost unipotent semigroup* $S \subseteq M_n(K)$ *is periodic. Moreover, if it is finitely generated, it must be finite.*

Proof. If $a \in S$, then there exists $k \geq 1$ such that $a^k \in D$ for a maximal subgroup D of $M_n(K)$ and a^k is unipotent. Therefore a^{kr} is the identity of D, where r is a power of the characteristic of K, so S is periodic. Assume that S is finitely generated. Since that sets $C_{i,j}$ are finite, Lemma 5.10 implies that S must be finite. Note that the latter follows also from the more general assertion of Theorem 7.3. $\square$

Proposition 5.12 *Let $S \subseteq M_n(K)$ be a finitely generated almost unipotent semigroup in the block upper triangular form of 4) in Proposition 5.9. If $C_{i,j}, D_{i,j}$ have the earlier meaning, then*

1. *the set $E = \{a \in M_n(K)| e_i a e_i \in C_{i,i}, e_i a e_j \in D_{i,j} \text{ for } i < j\}$ is a π-regular almost unipotent semigroup containing S.*

2. *the groups associated to S are finitely generated.*

Proof. We have seen that E is a semigroup containing S. From Proposition 5.9 it follows that E is almost unipotent. Let D be a maximal subgroup of $M_n(K)$ such that $E \cap D \neq \emptyset$. We know that the group G generated by $E \cap D$ has a subgroup V of finite index that is unipotent. The finiteness of the index easily implies that $E \cap V$ generates V as a group, see [87], Lemma 7.4. Let $u \in E \cap V$. Then $(u-e)^n = 0$, where e is the identity of V. Let $R = \mathbf{Z}$ if $\mathrm{ch}(K) = 0$ and $R = \mathbf{F}_p$ if $\mathrm{ch}(K) = p > 0$. Hence e is an R-combination of $u, \ldots, u^n$. If u^{-1} denotes the inverse of u in D, then we get $e_i u^{-1} e_j \in \sum_{l=1}^{n} e_i u^l e_j R \subseteq D_{i,j}$ for $i < j$. Since every $C_{i,i} = e_i S e_i$ is a finite semigroup, the projection $e_i G e_i$ is a finite group, so we also have $e_i u^{-1} e_i \in C_{i,i}$. Therefore $u^{-1} \in E$, which shows that $V = \mathrm{gp}(E \cap V) \subseteq E$. Since $E \cap D$ generates G and $[G : V] < \infty$, it follows that $G \subseteq E$. This proves 1).

From Lemma 4.6 it follows that, conjugating in the K-subalgebra of $M_n(K)$ that is block upper triangular with the block pattern of S, we can assume that e is a diagonal idempotent. We can also assume that $S = E$ because each group of the form G as above is a subgroup of E and a subgroup of a finitely generated almost nilpotent group is finitely generated. If $e_1 e e_1 = 0$ then $G \subseteq (I - e_1)M_n(K)(I - e_1)$, so that one can proceed by induction on the number r of diagonal blocks. (G is finite if $r = 1$.) This allows to assume that $e_1 e e_1 \neq 0$. Similarly, we can assume that $e_r e e_r \neq 0$. It is enough to show that the block upper triangular unipotent subgroup $W = \{g \in G| e_i g e_i = e_i e e_i \text{ for } i = 1, \ldots, r\}$ of G is finitely generated (note that $[G : W] < \infty$.) Let $F = e + e D_{1,r} e$. Since $D_{1,r}$ is a finitely generated R-module, the same is true of $e D_{1,r} e$, whence F is a finitely generated group. Therefore $F \cap W$ also is finitely generated. On the other hand, induction shows that $W/(F \cap W)$ is finitely generated, because it is isomorphic to a subgroup of $G_1 \times G_2$, where $G_1 = (I - e_1)G(I - e_1), G_2 = (I - e_r)G(I - e_r)$ are the corresponding projections of G. Hence G is finitely generated, as desired. $\square$

Our next aim is to discuss a more general class of semigroups, those that are 'almost solvable'. It turns out that this class also is determined by identities over $K[x_{ij}]_{i,j=1,...}$ of some special type.

Proposition 5.13 *Assume that the associated groups of a linear semigroup $S \subseteq M_n(K)$ are almost solvable. Then the maximal subgroups of $\overline{S}$ are almost solvable.*

Proof. Let $U_j, W_j, j = 1, 2 \dots$, be words in x, y defined inductively as follows

$$U_1 = x, \quad W_1 = y,$$
$$U_{j+1} = U_j^3, \quad W_{j+1} = U_j W_j U_j \text{ for odd } j,$$
$$U_{j+1} = W_j U_j W_j, \quad W_{j+1} = W_j^3 \text{ for even } j.$$

We claim that, if $x, y \in M_n(K)$, then $u = U_{2n+1}, w = W_{2n+1}$ lie in a maximal subgroup D of $M_n(K)$. Since $\mathrm{rank}(U_{i+1}) \le \mathrm{rank}(U_i), \mathrm{rank}(U_{i+1}) \le \mathrm{rank}(W_i)$, and $\mathrm{rank}(W_{i+1}) \le \mathrm{rank}(W_i), \mathrm{rank}(W_{i+1}) \le \mathrm{rank}(U_i)$, it is clear that either $U_{2n+1} = W_{2n+1} = 0$ or there exists $i \le 2n - 1$ such that $\mathrm{rank}(U_i) = \mathrm{rank}(W_i) = \mathrm{rank}(U_{i+1}) = \mathrm{rank}(W_{i+1})$. In view of the definition, this implies that U_{i+1}, W_{i+1} lie in a maximal subgroup of $M_n(K)$, and consequently U_j, W_j lie in this subgroup provided that $j > i$. This proves the claim. By the hypothesis and Malcev's theorem, [134], Theorem 3.6, the group H generated by $S \cap D$ is almost triangularizable. From Corollary 3.4, we know that there are finitely many conjugacy classes of linear groups arising from S in this way. Therefore, there exists $N \ge 1$ such that if $x, y \in S$, then

$$[c(ab - ba)]^n = 0 \text{ for every } a, b, c \in \langle u^N, w^N \rangle.$$

If a, b, c are fixed words in x, y and x, y run over S, then this is a polynomial identity which is satisfied in S. Therefore, by Lemma 5.1 this identity must be also satisfied for every $x, y \in \overline{S}$. Let G be a maximal subgroup of $\overline{S}$ and let $e = e^2 \in G$. Suppose that $x, y \in G$ generate a free nonabelian subgroup. Then u^N, w^N also generate a free nonabelian subgroup X (as, clearly, these two words in x, y do not commute, and every subgroup of a free group is free). The above implies that the commutator ideal J of the K-subalgebra A generated by u^N, w^N in $M_n(K)$

has a basis consisting of nilpotents. It is well known that J must be then nilpotent (for example, use the fact that the trace of every $a \in J$ is zero and apply Lemma 4.1 and Proposition 4.4). From Corollary 1.5 it follows that $e \in A$ is the identity of A and X lies in the group of units A^* of A. However, the nilpotency of J easily implies that A^* is solvable, a contradiction. In view of Theorem 5.4 this shows that G has a solvable normal subgroup H such that G/H is periodic. Since K can be chosen nonalgebraic over its prime subfield, the connected component G^c of G is solvable (otherwise, the semisimple group $G^c/\mathcal{R}(G^c)$, with $\mathcal{R}(G^c)$ denoting the radical of G^c, would have a periodic torus, which contradicts the choice of K). Hence G is almost solvable, as desired. $\square$

We shall see in Chapter 6 that, if a linear semigroup S has no free noncommutative subsemigroups and $\text{ch}(K) = 0$ or K is finitely generated, then the groups associated to S are almost solvable. Hence, the maximal subgroups of $\overline{S}$ are almost solvable by Proposition 5.13.

Let $G = GL_n(\overline{\mathbf{F}}_p), n > 1$, where $\overline{\mathbf{F}}_p$ is a proper subfield of K. Then G is periodic, but the Zariski closure G' of G in $GL_n(K)$ is not a periodic extension of a solvable group. Namely, G' contains the group $B' \subseteq GL_n(K)$ of upper triangular matrices, because B' is the closure of $B' \cap GL_n(\overline{\mathbf{F}}_p)$, so $G' = GL_n(K)$ by Theorem 2.7. Therefore, 'almost solvable' cannot be replaced by 'a periodic extension of a solvable group' in the statement of Proposition 5.13.

5.3   Nilpotent semigroups

In this section we discuss nilpotent semigroups, as defined in Section 5.1 via an identity $X_m = Y_m$ for some $m \geq 1$. Thus, the reader should be warned that the term 'nilpotent' used here is more general than that in the foregoing chapters, where it was used for a semigroup S with zero θ such that $S^m = \theta$ for some $m \geq 1$. (To make a distinction, the semigroups of the latter type will be sometimes called power nilpotent.) It is known that every nilpotent cancellative semigroup S has a group of classical quotients G which is nilpotent and of the same nilpotency class as S. Groups, and linear groups in particular, satisfying certain related semigroup identities, introduced in [124], have been recently studied in [6], [105], [123]. In particular, finitely generated residually finite groups satisfying a semigroup identity must be almost nilpotent, [123]. On the other hand, the results of Section 5.1, which will be strengthened in Chapter 6, show that a linear semigroup $S \subseteq M_n(K)$ satisfying any identity has a strong flavour of nilpotency.

A natural question that arises here is to decide which of these semigroups are nilpotent. Because of the powerful classical theory of nilpotent linear groups, cf.[134], one can also ask whether such semigroups can be approached via group theoretical methods. We note that the very special case of nilpotent connected algebraic monoids has been recently considered in [39].

The aim of this section is to develop a structural approach to nilpotent semigroups of matrices. First, we find a structural characterization of such semigroups. Next, we show that the smallest π-regular subsemigroup $\text{cl}(S) \subseteq M_n(K)$ containing S is very close to S. Namely, $\text{cl}(S)$ is also nilpotent, $S, \text{cl}(S)$ intersect the same maximal subgroups of $M_n(K)$ (even more: S intersects all $\mathcal{H}$-classes of $\text{cl}(S)$ contained in regular $\mathcal{J}$-classes of $\text{cl}(S)$) and $H = \text{gp}(S \cap H)$ for every maximal subgroup H of $\text{cl}(S)$. Moreover, there are at most 2^n maximal subgroups of $M_n(K)$ intersected by S. These are very special properties, which allow in view of the general structure theorem discussed in Section 3.2, to apply group theoretical tools in the study of an arbitrary nilpotent semigroup $S \subseteq M_n(K)$. The results of this section come from [94].

Recall that for a subset A of a semigroup S we denote by $E(A)$ the set of idempotents of A. First, we collect a few simple observations on nilpotent semigroups.

Lemma 5.14 *1. Let J be an ideal of a semigroup S with zero θ. If there exists $k \geq 1$ such that $(aS)^k = \theta$ for every $a \in J$, and S/J is nilpotent, then S is nilpotent.*

2. if S is a uniform subsemigroup of a completely 0-simple semigroup M and $\widehat{S}$ is the completely 0-simple closure of S in M, then S is nilpotent if and only if $\widehat{S}$ is a Brandt semigroup over a nilpotent group, that is $\widehat{S} \simeq \mathcal{M}(G, Z, Z, I)$ for a nonempty set Z, a nilpotent group G and the $Z \times Z$ identity matrix I.

3. if $S \subseteq M_n(K)$ is nilpotent, then $\text{cl}(S) \subseteq \overline{S}$ are nilpotent.

Proof. 1) Assume that S/J is m-nilpotent. So, if $x, y, u_1, u_2, \ldots \in S$, then either $x_m = y_m$ or $x_m, y_m \in J$. If $q \geq 1$, then the word X_{m+q}

contains 2^{q-1} copies of the word X_m separated by some u_i, Y_i. Therefore, in the latter case $x_{m+q}, y_{m+q} = \theta$ whenever $2^{q-1} \geq k$. This proves 1).

2) If $ef = f, fe = e, f \neq e$ for some $e, f \in E(\widehat{S}) \setminus \{\theta\}$, then, for every $m \geq 1$, $x_m(e, f, e, \dots, e), y_m(e, f, e, \dots, e)$ are not in the same $\mathcal{H}$-class of $\widehat{S}$. Therefore, choosing $s, t \in S$ such that $s\mathcal{H}e$ and $t\mathcal{H}f$, we see that $x_m(s, t, s, \dots, s)$ and $y_m(s, t, s, \dots, s)$ are not in the same $\mathcal{H}$-class, hence they are not equal. This implies that each $\mathcal{R}$-class of $\widehat{S}$ has only one idempotent if S is nilpotent. A similar argument works for the $\mathcal{L}$-classes. By [77], every maximal subgroup G of $\widehat{S}$ is nilpotent because it is generated by the nilpotent semigroup $G \cap S$. It is well known that S is of the desired form in this case, [15]. Conversely, assume that $\widehat{S}$ is a Brandt semigroup over a nilpotent group G. Let $r = \max(m, 2)$, where m is the nilpotency class of G. Choose $x, y, u_1, \dots, u_r \in \widehat{S}$. If $xu_1y = \theta$ or $yu_1x = \theta$, then $x_2 = \theta = y_2$. Suppose that xu_1y, yu_1x are nonzero. Write $x = (g, w, z), y = (h, w', z')$ for $g, h \in G$ and $w, w', z, z' \in Z$. The assumptions imply that $w = w'$ and $z = z'$. Hence, we must have $x_i\mathcal{H}y_i$ for every $i \geq 1$. If $x_r \neq \theta$, it follows also that all u_i are in the same $\mathcal{H}$-class H of $\widehat{S}$ – the only one that satisfies $xHy \neq \theta$. Therefore $u_i = (f_i, z, w)$ for some $f_i \in G$. Since G is m-nilpotent, this easily leads to $x_r = y_r$ (because this equality reduces to the corresponding equality in G with g, h in place of x, y and f_i in place of u_i). It follows that $\widehat{S}$ is nilpotent of class at most $\max(m, 2)$. This proves 2).

3) From Corollary 1.5 we know that $\mathrm{cl}(S) \subseteq \overline{S}$. The latter semigroup satisfies the same identities as S by Corollary 5.2. $\square$

Since $M_n(K)$ has no infinite chains of idempotents, the principal factors of every regular semigroup $S \subseteq M_n(K)$ are completely 0-simple (or completely simple). Hence, if S is also nilpotent, then Lemma 5.14 implies that S is inverse, so that $E(S)$ is a commutative semigroup. However, if S is not regular, then $E(S)$ need not be a subsemigroup of S. For example, $S = \{ \begin{pmatrix} 1 & 0 \\ 0 & 0 \end{pmatrix}, \begin{pmatrix} 0 & 0 \\ 1 & 1 \end{pmatrix}, \begin{pmatrix} 0 & 0 \\ 1 & 0 \end{pmatrix}, 0 \} = \langle E(S) \rangle \subseteq M_2(K)$ is nilpotent by assertion 1) of Lemma 5.14.

If I is a principal right ideal of a subsemigroup S of M_n/M_{j-1} such that $I \subseteq S_j/S_{j-1}$ and I is nil, then it is easy to see that $I^2 = \theta$. In this context assertion 1) of Lemma 5.14 (with $k = 2$) will be later applied.

Note that it was mistakenly stated in [69], Proposition 2.3, that a nilpotent regular semigroup must be a semilattice of nilpotent groups (though, a slightly different definition was used there).

Our first aim is to look at the closure of a nilpotent semigroup S. We will prove that $\mathrm{cl}(S)$ is very close to S. In fact, Theorem 5.15 shows that they are as close as possible in terms of the uniform components and the associated groups. Namely, every uniform component of $\mathrm{cl}(S)$ is the completely 0-simple closure of a uniform component of S.

Theorem 5.15 *Assume that the completely 0-simple closure of every uniform component of a semigroup $S \subseteq M_n(K)$ is a Brandt semigroup. Then*

1. *every uniform component of $\mathrm{cl}(S)$ intersects the same $\mathcal{H}$-classes of $M_n(K)$ as a uniform component of S,*

2. *for every maximal subgroup G of $\mathrm{cl}(S)$ we have $G = \mathrm{gp}(S \cap G)$,*

3. *S intersects at most 2^n maximal subgroups of $M_n(K)$.*

This applies in particular to every nilpotent semigroup $S \subseteq M_n(K)$.

Proof. Let E_j be the set of nonzero idempotents of all Brandt semigroups mentioned above and contained in M_j/M_{j-1}. From Proposition 3.6 we know that E_j is a triangular set of idempotents. Hence $|E_j| \leq \binom{n}{j}$ by Lemma 2.11. This implies that 3) holds. Now, assertions 1),2) follow from Proposition 3.15, Proposition 3.14 and its left-right dual. The last assertion is a direct consequence of Lemma 5.14. $\square$

Now we will focus on irreducible nilpotent semigroups. We say that $a \in M_n(K)$ is a monomial matrix if each row and each column of a contains at most one nonzero entry. $S \subseteq M_n(K)$ is a monomial semigroup if S is conjugate to a semigroup consisting of monomial matrices. More generally, if $1 = e_1 + \ldots + e_r$ for some mutually orthogonal idempotents $e_i \in M_n(K)$, and for every i there exists at most one j with $e_i s e_j \neq 0$, and for every j there exists at most one i with $e_i s e_j \neq 0$, then we say that s is a block monomial matrix with respect to $e_1, \ldots, e_r$. This allows us to define block monomial semigroups $S \subseteq M_n(K)$.

Proposition 5.16 *Assume that $S \subseteq M_n(K)$ is an irreducible nilpotent semigroup. Then*

1. $\mathrm{cl}(S)$ *is block monomial with respect to idempotents* $e_1, \ldots, e_r$ *such that* $E(\mathrm{cl}(U)) \setminus \{0\} = \{e_1, \ldots, e_r\}$ *for an ideal* U *of* S. *Moreover, if* D *is an algebraically closed field, then* S *is monomial.*

2. S *and* $\mathrm{cl}(S)$ *intersect the same* $\mathcal{H}$-*classes of* $M_n(K)$.

3. $E(\mathrm{cl}(S))$ *is a commutative subsemigroup of* $\mathrm{cl}(S)$ *and it contains at most* $\binom{r}{q}$ *idempotents of rank* qn/r *for every* $q \in \{1, \ldots, r\}$, *but no nonzero idempotents of other ranks. Consequently* $|E(\mathrm{cl}(S))| \leq 2^r$ *and* S *intersects at most* 2^r *maximal subgroups of* $M_n(K)$.

Proof. First note that the set U of matrices of the least nonzero rank in S (with zero if it is in S) is an ideal of S that is a uniform component of S, see Proposition 4.3. Clearly, U also is irreducible. $\hat{U}$ can be viewed as a subsemigroup of $M_n(K)$ and the proof of assertion 2) of Lemma 5.14 shows that it is a Brandt semigroup or a group (if $0 \notin S$). In the latter case, the irreducibility of U implies that $U \subseteq GL_n(K)$, so that $S \subseteq GL_n(K)$. Hence, adjoining 0 to S if necessary, we may assume that $U \subseteq \hat{U} = \mathrm{cl}(U) \subseteq M_n(K)$ is such that $H = \mathrm{gp}(H \cap U)$ for every maximal subgroup H of $\mathrm{cl}(U)$, U intersects all $\mathcal{H}$-classes of $\mathrm{cl}(U)$, and $\mathrm{cl}(U) \simeq \mathcal{M}(G, Z, Z, I)$ for a set Z and a nilpotent group G. Since the idempotents of $\mathrm{cl}(S)$ are orthogonal in $M_n(K)$, we have $|Z| = r \leq n$. Let $e = e_1 + \ldots + e_r$ be the sum of all nonzero idempotents of $\mathrm{cl}(U)$. Since U is irreducible, $\mathrm{cl}(U)$ also is irreducible. This implies that $e = 1$ because $\mathrm{cl}(U) = e\,\mathrm{cl}(U)e$. Hence $1 = e_1 + \ldots + e_r$ is a decomposition into a sum of orthogonal idempotents. Conjugating in $M_n(K)$ we can assume that $e_1, \ldots, e_r$ are diagonal.

Let $s \in S$. Since $S' = S \cup \mathrm{cl}(U)$ is a subsemigroup of $M_n(K)$ by Lemma 3.11, $e_i s, se_i \in \mathrm{cl}(U)$ for every i. If $e_i s \neq 0$, then $e_i s = e_i se_j$ for a unique j. Therefore, in the block decomposition $s = \sum_{i,j=1}^r e_i se_j$, for every i there exists at most one j such that $e_i se_j \neq 0$. Similarly, for every j there exists at most one i with $e_i se_j \neq 0$. That is, s is block monomial with respect to $e_1, \ldots, e_r$. If additionally D is an algebraically closed field, then the irreducibility of $\mathrm{cl}(U)$ implies by Proposition 4.3 that $e_i \mathrm{cl}(U)e_i$ is irreducible as a subsemigroup of $e_i M_n(K)e_i \simeq M_t(K)$, $t = \mathrm{rank}(e_i)$. But it is a group with zero, so it is monomial by [134], Theorem 1.14. Since $e_1, \ldots, e_r$ are diagonal, after an appropriate conjugation in $M_n(K)$, all diagonal blocks $e_i \mathrm{cl}(U)e_i$ consist of monomial matrices. Assume that $x = e_i xe_j \in \mathrm{cl}(U)$ is nonzero. Let $e_{ji} = e_j e_{ji} e_i$ be the matrix whose only nonzero block is the identity $t \times t$ matrix. Then $e_{ji} \in \mathrm{cl}(U)$ and $0 \neq xe_{ji} \in \mathrm{cl}(U)$. This implies that x is a monomial matrix. Hence, $\mathrm{cl}(U)$ is monomial.

Let A be the set of all matrices $a \in M_n(K)$ that are block monomial with respect to $e_1, \ldots, e_r$ and with $e_i a e_j \in e_i \mathrm{cl}(U) e_j$ for every i, j. Denote by $B \subseteq GL_t(K)$ the group corresponding to the maximal subgroup $e_1 \mathrm{cl}(U) e_1 \setminus \{0\}$ of $\mathrm{cl}(U)$. Clearly $S \subseteq A$ and $k = qt$, for $q = 1, \ldots, r$, are the only ranks of nonzero matrices of A. It is easy to see that $(A \cap M_k)/(A \cap M_{k-1}) \simeq \mathcal{M}(G^q, \binom{r}{q}, \binom{r}{q}, I)$ where G^q is isomorphic to the group of block monomial matrices of rank qt contained in $(e_1 + \ldots + e_q) M_n(K)(e_1 + \ldots + e_q)$ and with nonzero blocks contained in B. Therefore A is π-regular (even regular) and so $\mathrm{cl}(S) \subseteq A$. This completes the proof of 1).

Moreover, $A = RM$, where M is the group of invertible matrices in the monoid $\sum_{i=1}^{r} e_i A e_i$ and R is the corresponding inverse monoid (isomorphic to the full symmetric inverse submonoid in $M_r(K)$). Let $\varphi : A \longrightarrow R$ be the natural homomorphism. In other words, $\varphi(a)$ is the element of R having the same location of nonzero blocks as a and all nonzero blocks equal to the identity $t \times t$ matrix. It is easy to see that the inverse image in A of any maximal subgroup H' of R is a subgroup H of A (in fact, H is the set of all matrices $a \in A$ with the location of zero blocks identical to the location of zero blocks a matrix $a' \in H'$). In particular, every maximal subgroup of A is the inverse image of a maximal subgroup of R. For the same reason, every $\mathcal{H}$-class of A is the collection of all matrices of A with the patterns of nonzero blocks of some fixed types. Since $\phi(S)$ is finite, it follows that $\varphi^{-1}\varphi(S) \subseteq A$ is π-regular, and it contains S. Hence $\varphi^{-1}\varphi(S) \supseteq \mathrm{cl}(S)$, and consequently $\varphi(S) = \varphi\varphi^{-1}\varphi(S) \supseteq \varphi(\mathrm{cl}(S))$. Therefore $\varphi(S) = \varphi(\mathrm{cl}(S))$. This implies that S and $\mathrm{cl}(S)$ intersect the same $\mathcal{H}$-classes of A, and hence of $M_n(K)$ (use again the above description of the $\mathcal{H}$-classes of A). This proves 2).

Let $e \in E(\mathrm{cl}(S))$. From 1) it is clear that e must be block diagonal. Thus $e = e_1 e e_1 + \ldots + e_r e e_r$. Every $e_i e e_i$ is an idempotent and it lies in $\mathrm{cl}(S)$. Since by 2) $\mathrm{cl}(S)$ does not intersect nonzero maximal subgroups of $M_n(K)$ consisting of matrices of ranks $< \mathrm{rank}(e_i)$, it follows that $e_i e e_i \in \{0, e_i\}$. Hence, $E(\mathrm{cl}(S))$ is diagonal and 3) follows. $\square$

Note that the monoid A used in the above proof is an obvious generalisation of the monoid of all monomial matrices in $M_n(K)$, while M

corresponds to the group of diagonal invertible matrices.

Remark Recall that $S \subseteq M_n(K)$ is completely reducible if K^n is a direct sum of irreducible $K\{S\}$ - modules. In this case S is conjugate to a block diagonal semigroup $S' \subseteq M_n(K)$ whose diagonal blocks are irreducible. Proposition 5.16 implies that S' is block monomial with respect to the set of idempotents coming from all irreducible blocks. It can be checked that the proof of the second part of Proposition 5.16 extends to this case, showing that $S, \mathrm{cl}(S)$ intersect the same $\mathcal{H}$-classes of $M_n(K)$.

If $a \in M_n(K)$ is a monomial matrix, then $a^{n!}$ is diagonal. This is clear for $a \in GL_n(K)$. If $\mathrm{rank}(a) < n$, then $b = a^n \in G$ for a maximal subgroup G of $M_n(K)$. Then $b^{j!}$ is diagonal, where $j = \mathrm{rank}(b)$ (compare Lemma 2.6). It follows that indeed $a^{n!}$ is diagonal.

Corollary 5.17 *If $S \subseteq M_n(K)$ is a nilpotent semigroup, then S is a triangularizable - by - (periodic of bounded index) subsemigroup of $M_n(\overline{K})$.*

Proof. S is conjugate to a semigroup $S' \subseteq M_n(\overline{K})$ which is block upper triangular and whose diagonal blocks are irreducible or zero. By Proposition 5.16 1) we may assume that S consists of monomial matrices. Therefore, for any $s \in S$, every projection of $s^{n!}$ onto a diagonal block is a diagonal matrix. Hence, the assertion follows. $\square$

Let T be a uniform component of a semigroup $S \subseteq M_n(K)$ such that the completely 0-simple closure $\widehat{T}$ of T is a Brandt semigroup with finitely many idempotents. Let $e_1, \dots, e_r$ be the idempotents of the maximal subgroups $H_1, \dots, H_r$ of $M_n(K)$ intersected by T and k the rank of matrices in T. We know that the semigroup $S_T = \langle S, \mathrm{gp}(S \cap H) \rangle$ intersects the same $\mathcal{H}$-classes of $M_k \setminus M_{k-1}$ as S, and $\widehat{T}$ is a uniform component of S_T, see Lemma 3.11 and the remark following it. Then $\widetilde{S}_T = (S \setminus S_{k-1}) \cup \widehat{T}$ has a natural semigroup structure and it is a Rees factor of S_T. Moreover $\widehat{T}$ is an ideal of $\widetilde{S}_T$. It is well known that the sum of idempotents of $\widehat{T}$ is the identity of the contracted semigroup ring $K_0[\widehat{T}]$ and $K_0[\widehat{T}] \simeq M_r(K[G])$ for the maximal subgroup G of $\widehat{T}$, see Section 4.2. Consider the mapping $\alpha_T : K[S_T] \longrightarrow K_0[\widehat{T}]$ defined for $s \in S_T$ by $\alpha_T(s) = \overline{e_T s}$, where $e_T = e_1 + \dots + e_r \in K[S_T]$

and $\bar{a}$ is the image of $a \in K[S_T]$ in $K_0[\tilde{S}_T]$. It is a homomorphism because $\overline{e_T}$ is central in $K_0[\tilde{S}_T]$. It is easy to see that $\alpha_T(s)$ has a monomial pattern with respect to $\overline{e_1}, \ldots, \overline{e_r}$, so that the nonzero entries of $\alpha_T(s)$ as an $r \times r$ matrix determine a 'quasi permutation' of the set $A = \{1, \ldots, r\}$ (a map $\sigma : A \longrightarrow A \cup \{0\}$ that is one-to-one on $\sigma^{-1}(A)$). We say that $\alpha_T(S) \subseteq K_0[\widehat{T}]$ has no inversions if for every $s, t \in S$ and $e_i, e_j, e_p, e_q \in E(\widehat{T}) \setminus \{\theta\}$ with $e_i \neq e_j, e_p \neq e_q$, at least one of the elements $\alpha_T(e_q s e_j), \alpha_T(e_q t e_i), \alpha_T(e_p t e_j), \alpha_T(e_p s e_i)$ is zero (or equivalently, one of the matrices $e_q s e_j, e_q t e_i, e_p t e_j, e_p s e_i$ has rank $< k$).

Recall that we often identify a uniform component T of S with the subset of the corresponding $S_k \setminus S_{k-1}$. We are now ready for a structural description of nilpotent semigroups.

Theorem 5.18 *The following conditions are equivalent for a semigroup $S \subseteq M_n(K)$*

1. S is nilpotent,

2. the groups associated to S are nilpotent and for every $x, y, u, z \in S$, if x, y, xuy, yux, xzx, yzy lie in the same uniform component of S, then $x \mathcal{H} y$ in $M_n(K)$,

3. the groups associated to S are nilpotent, the completely 0-simple closure of every uniform component T of S is a Brandt semigroup with finitely many idempotents and $\alpha_T(S)$ has no inversions.

Moreover, if k denotes the maximum of the nilpotency classes of the groups associated to S, then there exists a function $f(n, k)$ such that the nilpotency class of S is bounded by $f(n, k)$.

Proof. Assume that S is nilpotent and there exists a uniform component T of S containing x, y, xuy, yux, xzx, yzy for some $x, y, u, z \in S$. Let $u_{2m} = z$ and $u_{2m-1} = u$ for $m = 1, 2, \ldots$. Since in a completely 0-simple semigroup $ab \neq \theta, bc \neq \theta$ imply $abc \neq \theta$, it follows that $x_m = x_m(x, y, u_1, \ldots, u_m) \in T$ and $y_m = y_m(x, y, u_1, \ldots, u_m) \in T$. Therefore $x_m \mathcal{H} x$ and $y_m \mathcal{H} y$ if m is even. But the identity $X_m = Y_m$ holds in S for some m. Hence $x \mathcal{H} y$ in $M_n(K)$. Since $H = \mathrm{gp}(S \cap H)$ for every group H associated to S, H must be nilpotent, [77]. Therefore 1) implies 2).

Assume that 2) holds. Suppose first that a uniform component T of S intersects two different maximal subgroups H, G of $M_n(K)$ that are in the same $\mathcal{R}$-class of $M_n(K)$. Let $x, u, z \in S \cap H, y \in S \cap G$. Then $xuy, yzy \in G, yux, xzx \in H$. This contradicts 2). A similar argument shows that T does not intersect different $\mathcal{L}$-related maximal subgroups of $M_n(K)$. Thus, in view of Theorem 5.15, the second part of 3) is satisfied.

Suppose that $\alpha_T(S)$ has an inversion. Let $\{e_1, \ldots, e_r\}$ be the idempotents of the maximal subgroups of $M_n(K)$ intersecting T and let $l = \operatorname{rank}(e_1)$. Then there exist $i, j, p, q, i \neq j, p \neq q$, such that $\operatorname{rank}(e_q s e_j) = \operatorname{rank}(e_q t e_i) = \operatorname{rank}(e_p t e_j) = \operatorname{rank}(e_p s e_i) = l$ for some $s, t \in S$. Choose $x, y \in T$ such that $x = e_i x e_p, y = e_j y e_q$. Then $ysy = y e_q s e_j y \in T, ytx = y e_q t e_i x \in T, xty = x e_p t e_j y \in T, xsx = x e_p s e_i x \in T$. This contradicts 2). Therefore 3) follows.

Assume that 3) holds. To prove 1) we may assume that $0 \in S$, adjoining 0 to S if necessary. From the general structure theorem described in Section 3.2 we know that S has an ideal chain $S = T_1 \supset \ldots \supset T_m$, $m < 2^n + n$, such that T_m and each factor T_i/T_{i+1} is either a nonzero nil subsemigroup of a completely 0-simple semigroup or it is a uniform component of S with a completely 0-simple closure that is a Brandt semigroup over a nilpotent group and with finitely many idempotents. We show by induction on j that every factor S/T_j is nilpotent (under the remaining assumption of 3)). Since the linearity of S/T_j will not be needed, we simply assume that S/T_m is nilpotent (this takes care of the case $m = 1$ as well). If T_m is a nil semigroup, then assertion 1) of Lemma 5.14 applies with $k = 2$, showing that S is nilpotent. Thus, assume that $T = T_m$ is a uniform component of S. As mentioned above, it is known that $S' = S \cup \widehat{T}$ has a semigroup structure extending that of S. Then $K_0[\widehat{T}] \simeq M_r(K[G]), r \geq 1$, is a ring with an identity element and it is an ideal of $K_0[S']$. Therefore it is a ring direct summand of $K_0[S']$ and the centre of $K_0[\widehat{T}]$ is contained in the centre of $K_0[S']$. Identifying $K_0[\widehat{T}]$ with $M_r(K[G])$, we see that the set Z of invertible matrices in the centre of $M_r(K[G])$ lies in the centre of the algebra $K_0[S']$. Let ρ be the relation on S' defined by $s\rho u$ if and only if $s \in uZ, s, u \in \widehat{T}$, or $s = u$. It is easy to see that ρ is a congruence on S'. Moreover, an easy calculation shows that S' is nilpotent of class at most $t + 1$ if S'/ρ is nilpotent of class t. Let $\overline{Z} = \{gI \mid g \in G\} \subseteq M_r(K[G])$, where I is the identity matrix. Let $\overline{\rho}$ be the congruence defined by $s\overline{\rho}u$ if

$s \in \overline{Z}u, s, u \in \widehat{T}$, or $s = u$. Since G is q-nilpotent for some q and Z contains the scalar matrices over the centre of G, an induction on q shows that S' is nilpotent of class at most $t + q$ if $S'/\overline{p}$ is nilpotent of class $t - q$. The location of nonzero blocks in every $\alpha_T(s), s \in S$, is preserved under the map $S' \longrightarrow S'/\overline{p}$, so the additional assumption of 3) carries over to $S/\overline{p}$. Therefore, we will further assume that G is trivial, so that the $\mathcal{H}$-classes of $\widehat{T}$ are singletons (hence $\widehat{T} = T$) and $K_0[\widehat{T}] = M_r(K)$.

It is easy to see that S embeds into the direct product $K_0[S/T] \oplus K_0[T]$ via the map $s \mapsto (s', es)$, where s' is the image of s in $K_0[S/T]$ and e is the identity of $K_0[T]$. By the induction hypothesis, S/T is nilpotent, say it satisfies the identity $X_p = Y_p$. Therefore, we only need to show that $eS \subseteq K_0[T]$ is nilpotent. Choose $x, y, u_1, u_2, \ldots \in eS$. If $x_p, y_p \notin eT = T \subseteq K_0[T]$, then we have $x_p = y_p$ because eS/eT is a homomorphic image of S/T. Hence, we only need to consider the case where $x_p, y_p \in T$.

Now, if $x_p \mathcal{H} y_p$ in $T = \mathcal{M}(1, r, r, I)$, then they are equal. Thus, suppose that x_p, y_p are not in the same $\mathcal{H}$-class. If $x_{p+2} = \theta$ or $y_{p+2} = \theta$, then $x_{p+3} = \theta = y_{p+3}$. Hence, assume that $x_{p+2} \neq \theta \neq y_{p+2}$. Thus, $aubzbua, buazaub$ are nonzero for $a = x_p, b = y_p \in T$ and some $u, z \in eS$. If $a\mathcal{R}b$ in T then $u(bzbua)\mathcal{R}u(azaub)$ because the left multiplication by u permutes the $\mathcal{R}$-classes of T. Since $aubzbua, buazaub$ are nonzero and T is a Brandt semigroup, it follows that $a\mathcal{L}b$. Hence $a\mathcal{H}b$, a contradiction. Thus a, b are not in the same $\mathcal{R}$-class. Similarly one shows that they are not in the same $\mathcal{L}$-class. Hence $a = e_i a e_q, b = e_j b e_p$ for some $e_i, e_j, e_p, e_q \in E(T)$ with $i \neq j, p \neq q$. Since aza, bzb, aub, bua are nonzero, this contradicts 3) (u, z would give an inversion). This shows that the identity $X_{p+3} = Y_{p+3}$ holds in S, so that S is nilpotent.

The existence of a function $f(n, k)$ is an easy consequence of the proof. $\square$

Example Let $S = E_3 \cup G_3 \subseteq M_3(K)$ and $S' = E_3 \cup \{a, b\} \subseteq M_3(K)$, where

$$a = \begin{pmatrix} 0 & 0 & 1 \\ 0 & 1 & 0 \\ 0 & 0 & 0 \end{pmatrix}, \quad b = \begin{pmatrix} 0 & 1 & 0 \\ 0 & 0 & 1 \\ 0 & 0 & 0 \end{pmatrix},$$

E_3 is the semigroup of 3×3 matrix units with zero and G_3 is the subgroup of order 3 in $GL_3(K)$ consisting of permutation matrices. Then S is nilpotent, but S' is not nilpotent, though both 'layers' $S_2'/S_1', S_1'$ of

S' are nilpotent.

Corollary 5.19 *Let $S \subseteq M_n(K)$ be a semigroup such that every non-empty intersection $S \cap G$ with a maximal subgroup G of $M_n(K)$ is nilpotent. Then S is nilpotent if and only if every finitely generated subsemigroup of S is nilpotent.*

Proof. Let P be a finitely generated subsemigroup of S. Every group associated to P is a subgroup of a group H associated to S and $H = \mathrm{gp}(S \cap H)$ is nilpotent by the hypothesis and [77]. From Theorem 5.18 it follows that P is nilpotent of class at most $f(n,k)$, where k is the maximal nilpotency class of a group associated to S. Therefore S also satisfies the identity $X_{f(n,k)} = Y_{f(n,k)}$. $\square$

Corollary 5.20 *Assume that every uniform component of S is contained in a subgroup of $M_n(K)$. Then S is nilpotent if and only if all uniform components of S are nilpotent. The assumption is satisfied in particular if S is π-regular and $E(S)$ is contained in the centre of S.*

Proof. The first assertion is a direct consequence of condition 3) in the above theorem because, with the notation used there, $\widehat{T} \setminus \{\theta\}$ is a group, so $E(\widehat{T}) \setminus \{\theta\}$ is a singleton, in this case. If S is π-regular with central idempotents, then every regular $\mathcal{J}$-class of S is a group, so the assertion follows. $\square$

This result covers the case of nilpotent connected algebraic monoids M, recently considered in [39]. Note that for such monoids the Zariski closure in $M_n(K)$ of the group of units G is equal to M. Therefore, in view of Lemma 5.14, the definition of nilpotency used in [39] (the nilpotency of the group G) is equivalent to our definition.

Theorem 5.18 can be used to show that the class of semigroups considered in [45] (where $u_1, u_2, \ldots$ used in our definition are allowed to be chosen from S^1) coincides with the class of nilpotent semigroups, when restricted to subsemigroups of $M_n(K)$.

Corollary 5.21 *If $S \subseteq M_n(K)$ is nilpotent, then S^1 is nilpotent.*

Proof. We show that condition 2) of Theorem 5.18 holds for S^1 if it holds for S. Assume that $x, y, xuy, yux, xzx, yzy \in T$ for some $x, y, u, z \in$

S^1 and a uniform component T of S^1. Suppose that x, y are not in the same $\mathcal{H}$-class of $M_n(K)$. Then $T \subseteq M_{n-1}$, so that $x, y \in S$ and T is a uniform component of S. We use the fact that the completely 0-simple closure of T is a Brandt semigroup. This easily implies that either $u \in S$ or $z \in S$. First assume that $z = 1$. Then $x = e_i x e_i, y = e_j y e_j$ for some $e_i, e_j \in E(\widehat{T})$. By the supposition $i \neq j$. The quasi permutation corresponding to $\alpha_T(u)$ switches i, j because $\alpha_T(e_i u e_j) \neq 0, \alpha_T(e_j u e_i) \neq 0$ (since $xuy, yux \in T$). Therefore, the quasi permutation corresponding to $\alpha_T(u^2)$ fixes i and j. This means that $e_i u^2 e_i \mathcal{H} e_i$, hence $xu^2 x \mathcal{H} x$, and similarly $yu^2 y \mathcal{H} y$. Thus, we can replace $z = 1$ by $u^2 \in S$, coming to a contradiction with the assumption on S. Thus, assume that $u = 1$. Then $xy, yx \in T$ implies that $xy, yx, yxuxy, xyuxy, yxzxy, xyzyx \in T$ and $xy = e_i x e_i, yx = e_j y e_j$ for some $e_i, e_j \in E(\widehat{T})$. Moreover xy, yx are not in the same $\mathcal{H}$-class because of the supposition on x, y. As above, we can replace u by $z^2 \in S$, coming to $xy, yx, yxz^2 yx, xyz^2 xy, yxzxy, xyzyx \in T$. This again contradicts the assumption on S. Thus, the assertion follows. $\square$

We note that the nilpotency class of S^1 can exceed that of S. For example, this is so for the Brandt semigroup $S = \mathcal{M}(\{1\}, n, n, I), n > 1$, over the trivial group, which is 2-nilpotent.

Using the functions α_T, we can also extend the comment following Lemma 5.14 as follows.

Proposition 5.22 *Assume that $S \subseteq M_n(K)$ is a nilpotent semigroup. Then $\langle E(\mathrm{cl}(S)) \rangle$ is a finite semigroup.*

Proof. $E(\mathrm{cl}(S))$ is finite by Theorem 5.15. Let $F = \langle E(\mathrm{cl}(S)) \rangle$. Choose a uniform component T of $\mathrm{cl}(S)$. Then the semigroup $\alpha_T(F)$ is generated by idempotents $\alpha_T(e), e \in E(\mathrm{cl}(S))$. But $\alpha_T(\mathrm{cl}(S))$ has a block pattern that is monomial with respect to the decomposition $e = e_1 + \ldots + e_r$ of the identity e of $K_0[T]$, where $E(T) = \{e_1, \ldots, e_r\}$. Therefore, the elements of $\alpha_T(E(\mathrm{cl}(S)))$ are block diagonal idempotents. Since each nonzero block is an invertible matrix, this implies that these idempotents commute, so that $\alpha_T(F)$ is a finite semigroup. Moreover, $T \cap F$ embeds into $\alpha_T(F)$. The ideal chain used for S in the proof of the implication 3) $\Rightarrow$ 1) in Theorem 5.18 yields now an ideal chain of F, whose factors are either finite or nil subsemigroups of completely 0-simple semigroups. Hence, each factor is locally finite, so F must be locally finite, see [87], Proposition 2.2 or Theorem 7.3. Since it is finitely generated, it must be finite. □

Our last aim is to show that the class of semigroups $S \subseteq M_n(K)$ whose all 'layers' S_j/S_{j-1} are nilpotent is not very far from the class of nilpotent semigroups. Namely, such a semigroup is periodic modulo a nilpotent subsemigroup of a very special type, described in Corollary 5.20.

Theorem 5.23 *Assume that the groups associated to $S \subseteq M_n(K)$ are nilpotent and the completely 0-simple closure of every uniform component of S is a Brandt semigroup. Let M be the set of all $x \in S$ such that for every uniform component T of S and every $f \in E(\widehat{T}) \setminus \{\theta\}$ either $xf = fxf$ or $xf \notin T$. Then*

  1. *M is a nilpotent semigroup,*

  2. *S is periodic of bounded index modulo M,*

  3. *$\mathrm{cl}(S), S, M$ intersect the same maximal subgroups of $M_n(K)$.*

Moreover, if S is π-regular (regular respectively), then M is π-regular (regular).

Proof. If T is a uniform component of S, then let $E(\widehat{T}) \setminus \{\theta\} = \{e_1, \ldots, e_r\}$, see Theorem 5.15. Let $x \in M$. The definition of M is equivalent to saying that $\alpha_T(x) \in e_1\widehat{T}e_1 + \ldots + e_r\widehat{T}e_r \subseteq K_0[\widehat{T}]$ (for all T). In other words, $\alpha_T(x)$ has a block diagonal form with respect to the blocks arising from the decomposition $1 = e_1 + \ldots + e_r \in K_0[\widehat{T}]$. Since α_T is a homomorphism, it follows that M is a subsemigroup of S.

The argument above shows also that $M \cap T$ is contained in the union of maximal subgroups of $M_n(K)$ intersecting T. Therefore, every uniform component of M must be contained in a maximal subgroup of $M_n(K)$. From Corollary 5.20 it follows that M is nilpotent.

It is clear that, if $s \in S$ and $r!$ divides q, then $\alpha_T(s)^q$ is block diagonal. Since $r \leq |E(S)|$, this implies that S is periodic of bounded index modulo M. Consequently, S, M intersect the same maximal subgroups of $M_n(K)$. Hence 3) follows from Theorem 5.15.

Finally, assume that $S = \mathrm{cl}(S)$. Let $a \in H \cap M$ for a maximal subgroup H of $\mathrm{cl}(S)$. Let $b \in H$ be such that $ab = e = e^2 \in H$. It is easy to see that, for every uniform component T of S, $\alpha_T(b)$ has a block diagonal form because $\alpha_T(a)$ has such a form. Therefore, $b \in M$. This implies that $M = \mathrm{cl}(M)$. A similar argument also shows that M is regular whenever S is regular. Namely, if $aba = a, bab = b$ for some $a \in M, b \in S$, then it follows easily that every $\alpha_T(b)$ also is block diagonal. This completes the proof. $\square$

If $S \subseteq M_n(K)$ is periodic of bounded index modulo a nilpotent subsemigroup, then S satisfies an identity of the form

$$X_m(x^k, y^k, u_1^k, \dots, u_m^k) = Y_m(x^k, y^k, u_1^k, \dots, u_m^k).$$

As in the proof of Lemma 5.14 this implies that the completely 0-simple closures of uniform components of S are Brandt semigroups. The groups associated to S must be almost nilpotent by Corollary 5.8.

In [45] the structure of semigroup algebras of nilpotent semigroups was studied, in particular via prime Goldie homomorphic images, leading naturally to nilpotent subsemigroups of the matrix monoids $M_n(F)$ over division rings F. The results of this section go over to this more general setting, though the proofs require more work because we are unable to use $\overline{S}$ and to immediately derive the nilpotency of $\mathrm{cl}(S)$, [94].

5.4 PI algebras

In this section we study conditions on a linear semigroup $S \subseteq M_n(K)$ which ensure that the semigroup algebra $F[S]$ over a field F satisfies a polynomial identity. In fact, we find necessary and sufficient conditions in terms of the sandwich matrices of the uniform components of S and also we prove that they are equivalent to a combinatorial property for S. The results come from [97]. Throughout this section, our basic reference to PI - algebras is [117]. We will also need certain results on semigroup algebras [87].

A semigroup S has the property $\mathcal{P}_m, m \geq 2$, if for every $a_1, \dots, a_m \in S$ we have $a_1 \cdots a_m = a_{\sigma(1)} \cdots a_{\sigma(m)}$ for a nontrivial permutation σ. S has the permutation property $\mathcal{P}$ if S satisfies $\mathcal{P}_m$ for some m.

Our starting point is the following observation.

Lemma 5.24 *Assume that S is a semigroup such that the semigroup algebra $F[S]$ over a field F satisfies a polynomial identity of degree m. Then S satisfies $\mathcal{P}_m$.*

Proof. It is well known that $F[S]$ satisfies a multilinear identity of degree m, [117]. Therefore, there exist $\alpha_1, \ldots, \alpha_m \in F$ such that $x_1 \cdots x_m = \sum_\sigma \alpha_\sigma x_{\sigma(1)} \cdots x_{\sigma(m)}$ for every $x_1, \ldots, x_m \in F[S]$, where the summation runs over nontrivial permutations of $\{1, \ldots, m\}$. Choosing x_i from S only we see that $\alpha_\sigma \neq 0$ for some σ. The F-independence of the elements of S implies that S satisfies $\mathcal{P}_m$. $\square$

The class of groups with permutation property was described in [19]. This was extended by the author to arbitrary cancellative semigroups as follows, see [87], Theorem 19.8.

Lemma 5.25 *Let S be a cancellative semigroup. Then S satisfies the permutation property if and only if S has a group of quotients G which has a subgroup H of finite index such that the commutator subgroup $[H, H]$ is finite. If G is finitely generated, this is equivalent to G being almost abelian.*

We note that the converse to Lemma 5.24 is not true in general, even for group algebras, due to the following fundamental result, see [102].

Proposition 5.26 *The group algebra $F[G]$ over a field F satisfies a polynomial identity if and only if one of the following holds*

1. *$\mathrm{ch}(F) = 0$ and G is almost abelian,*

2. *$\mathrm{ch}(F) = p > 0$ and G has a normal subgroup H of finite index such that $[H, H]$ is a finite p-group.*

We will focus on linear semigroups with permutation property.

Lemma 5.27 *If $S \subseteq M_n(K)$ is a semigroup having $\mathcal{P}_m$, then so does its Zariski closure $\overline{S}$.*

Proof. We proceed as in the proof of Lemma 5.1. For every nontrivial permutation σ of the set $\{1, \dots, m\}$ and every $i = 1, \dots, m$, put

$$
\begin{aligned}
X_i(\sigma) \;=\; &\{a_i \in \overline{S} \,|\, a_1 \cdots a_m = a_{\sigma(1)} \cdots a_{\sigma(m)} \\
&\text{for all } a_j \in \overline{S}, j < i, \text{ and all } a_k \in S, k > i\}
\end{aligned}
$$

Then every $Y_i = \sum_\sigma X_i(\sigma)$ is a closed subset of $\overline{S}$. Since $S \subseteq Y_1 \subseteq \overline{S}$, it follows that $Y_1 = \overline{S}$. If $Y_i = \overline{S}$ for some i, then $S \subseteq Y_{i+1} \subseteq \overline{S}$. Hence $Y_{i+1} = \overline{S}$. Therefore, by induction $Y_m = \overline{S}$. This means that $\overline{S}$ has $\mathcal{P}_m$. $\square$

We shall need some results on closed subsemigroups S of $M_n(K)$. Recall that by Theorem 1.4 such a semigroup is π-regular and the uniform components of S coincide with regular $\mathcal{J}$-classes of S, see Remark iv) preceding Proposition 3.6. In other words, they determine completely 0-simple principal factors of S. For more background in this area we refer to [108].

Lemma 5.28 *Let $S \subseteq M_n(K)$ be a Zariski closed semigroup with m irreducible components. Then*

1. *if G is a maximal subgroup of S with the identity component G^c, then $|G/G^c| \leq m$,*

2. *if J is a regular $\mathcal{J}$-class of S, then $E(J)$ is a closed subset of S with at most m^2 irreducible components.*

Proof. 1) Let f be the identity of G. Then fSf is the image of S under the morphism $x \mapsto fxf$. Hence fSf has at most m irreducible components. Since G is an open subset of fSf, the same is true of G.

2) Let k be the rank of matrices in J. If $a \in S$, then define $\delta(a)$ as the sum of all products of k eigenvalues of a. Since $\delta(a)$ is a coefficient of the characteristic polynomial of a, it follows that $\delta : S \longrightarrow K$ is a morphism. Clearly, the set $Z = \{x \in S \,|\, x^2 = x, \operatorname{rank}(x) \leq k\}$ is closed. But $X = \{x \in X \,|\, \delta(x) = 1\}$ is then also closed. Note that X coincides with the set of idempotents of rank k in S. Let $U_1, \dots, U_t$ be the uniform components of S consisting of matrices of rank k. We know that each of them is a regular $\mathcal{J}$-class of S and one of them is equal to J. Then $X = X_1 \cup \cdots \cup X_t$, where $X_i = U_i \cap X$, so one of the X_i coincides with $E(J)$. In order to show that X_i are closed it is enough to

prove that every $Y_i = \bigcup_j X_j \setminus X_i$ is closed. Let $V_i = (S \cap M_k) \setminus U_i$. From Theorem 3.5 we know that $V_i = \{x \in S \cap M_k | (xU_1)^2 \in M_{k-1}\}$. Since M_{k-1} is closed, it follows that V_i is closed. But $V_i \cap X = Y_i$. Since X is closed, so is Y_i. It follows that $E(J)$ is closed, as desired.

Let $e \in E(J)$ and let G be the maximal subgroup of S containing e. Conjugating in $M_n(K)$ we may assume that $e = \begin{pmatrix} I & 0 \\ 0 & 0 \end{pmatrix}$. Define $U = \{(x,y) | x, y \in S, eyxe \in G\}$. U is a nonempty subset of the linear closed semigroup $S \times S \subseteq M_{2n}(K)$. If $\theta(exe)$ denotes the determinant of exe treated as a $k \times k$ matrix, then $(x,y) \mapsto (\theta(eyxe))^{-1}$ is a morphism on U. Thus $(x,y) \mapsto (eyxe)^{-1}$ is also a morphism on U (where the inverse is taken in G). Therefore the map $\psi : (x,y) \mapsto x(eyxe)^{-1}y$ is a morphism from U into $E(S)$. Since $ey\psi(x,y)xe = eyxe \in G$, it follows that $\psi(x,y) \in E(J)$. On the other hand, if $f \in E(J)$, then $xey = f$ for some $x, y \in S$. Let $e_1 = eyxe$. Then $e_1 \in J \cap eSe = G$ is an idempotent. Hence $e_1 = e$. So $(x,y) \in U$ and $\psi(x,y) = f$. We have shown that ψ is a morphism of U onto $E(J)$. Since $S \times S$ has m^2 irreducible components by the hypothesis, and U is an open subset, U has at most m^2 irreducible components. Hence $E(J)$, being an image of U, has at most m^2 irreducible components. $\square$

The following 'zigzag' lemma is very useful, see [108].

Lemma 5.29 *Let $S \subseteq M_n(K)$ be a Zariski closed connected semigroup. Assume that $SeS = SfS$ for some idempotents $e, f \in S$. Then there exist $e_1, e_2, f_1, f_2 \in E(S)$ such that $e\mathcal{R}e_1\mathcal{L}f_1\mathcal{R}f$ and $e\mathcal{L}e_2\mathcal{R}f_2\mathcal{L}f$, where $\mathcal{R}, \mathcal{L}$ stand for Green's relations in S.*

Proof. Let $k = \mathrm{rank}(e)$. We claim that there exists $e_1 \in E(S)$ such that $e\mathcal{R}e_1$ and $e_1 f \mathcal{J} f$ in S. Suppose otherwise. Let H, H' be the $\mathcal{H}$-classes in S of e, f respectively. Then $eSe \setminus H = eSe \cap M_{k-1}$ is a closed set. Similarly, $fSf \setminus H'$ is closed. By the hypothesis there exist $x, y \in S$ such that $xey = f$. Since eS is closed, the set $X = \{a \in eS | fxaf \in fSf \setminus H'\}$ is closed. Also $Y = \{a \in S | ae \in eSe \setminus H\}$ is closed. Since $fxeyf = f$, we have $ey \notin X$. Clearly $e \notin Y$. Since S is connected, so is eS. It follows that $eS \not\subseteq X \cup Y$. So there exists $a \in eS$ such that $a \notin X \cup Y$. Then $ea = a, fxaf \in H'$ and $ae \in H$. Hence there exists $z \in S$ such that $zae = e$. Then $za^2 = zaea = ea = a$. Therefore $Sa^2S = SaS$, so a lies in a maximal subgroup of S. Let $e_1 \in E(S)$

be the identity of this group. Then $e \mathcal{R} e_1$ in S because $ea = a$. Hence $Se_1 f S = S a f S \supseteq S f x a f S = S f S$, and consequently $Se_1 f S = S f S$, which contradicts our supposition and proves our claim. Thus, there exists $e_1 \in E(S)$ such that $e \mathcal{R} e_1$ and $e_1 f \mathcal{J} f$ in S. Since the principal factor of S containing $e_1, f, e_1 f$ is completely 0-simple, it follows that there exists an idempotent $f_1 \in E(S)$ such that $e_1 \mathcal{L} f_1$ and $f_1 \mathcal{R} f$ in S. This proves the first assertion. The proof of the second assertion is similar. $\square$

Recall that a rectangular band B is a direct product of a right zero semigroup and a left zero semigroup. In particular, B satisfies the identity $xyzw = xzyw$.

Lemma 5.30 *Let $S \subseteq M_n(K)$ be a Zariski closed semigroup with the property $\mathcal{P}$. Then*

1. *if G is a maximal subgroup of S, then the identity component G^c of G is abelian,*

2. *if J is a regular $\mathcal{J}$-class of S, then every irreducible component A of $E(J)$ is a rectangular band.*

Proof. G^c satisfies $\mathcal{P}_m$, and hence by Lemma 5.25 it has a normal subgroup H of finite index such that the commutator subgroup $[H, H]$ is finite. It is well known that G^c has no closed subgroups of finite index, see [40], Proposition 7.3. Hence H is dense in G^c. On the other hand, we have a morphism $\phi : G^c \times G^c \longrightarrow G^c$ given by $\phi(x, y) = xyx^1 y^{-1}$. Now $G^c \times G^c$ is irreducible, $H \times H$ is dense in $G^c \times G^c$ and $\phi(H \times H)$ is finite, so $\phi(G^c \times G^c) = \phi(H \times H)$ being irreducible must be trivial. This means that G^c is abelian.

For every positive integer i $\overline{A^i}$ is a closed irreducible set (being the closure of an image of $A \times \cdots \times A$.) Hence we have an ascending chain of closed irreducible sets $A \subseteq \overline{A^2} \subseteq \overline{A^3} \subseteq \cdots$. Comparing the dimensions we see that this chain must stabilize. Therefore, for some i, $T = \overline{A^i} = \overline{A^j}$ for all $j \geq i$. Now $\bigcup_{k \geq 1} A^k$ is a semigroup with closure T. Hence T is a connected semigroup satisfying $\mathcal{P}_m$. Now the irreducible set $T \times \cdots \times T$ (m times) is a finite union of the closed sets

$$B_\sigma = \{(a_1, \ldots, a_m) | \, a_i \in T, a_1 \cdots a_m = a_{\sigma(1)} \cdots a_{\sigma(m)}\}, \quad \sigma \neq 1.$$

Hence $T \times \cdots \times T = B_\sigma$ for some $\sigma \neq 1$. Thus, T satisfies a fixed permutation identity. It is known that this implies that, for some $k \geq 1$, T^k satisfies the identity $xyzw = xzyw$, see [87], Lemma 21.5. It follows that $E(T)$ is a subsemigroup of T. Let $J_1, \ldots, J_t$ be the regular $\mathcal{J}$-classes of T. If $e, f \in J_k$ for some k, then by the connectedness of T there exist $e_1, f_1 \in E(J_k)$ such that $e \mathcal{R} e_1 \mathcal{L} f_1 \mathcal{R} f$ in T, see Lemma 5.29. Therefore $ee_1 = e_1, e_1 f_1 = e_1$ and $f_1 f = f$. Consequently $ee_1 f_1 f \in J_k$. Thus, the above identity yields $ef(e_1 f_1)f = e(e_1 f_1)f \in J_k$. It follows that $ef \in J_k$. This implies that each $E(J_k)$ is a rectangular band. Also, by Lemma 5.28, each $E(J_k)$ is closed and irreducible. Since $A \subseteq E(T) = \bigcup_k E(J_k)$, we see that $A \subseteq E(J_r) \subseteq E(J)$ for some r. So $A = E(J_r)$ is a rectangular band. $\square$

Corollary 5.31 *Assume that $S \subseteq M_n(K)$ has the permutation property. Then S satisfies an identity.*

Proof. This is a direct consequence of Lemma 5.27, Lemma 5.30 and Lemma 5.3. $\square$

Lemma 5.32 *Let $S = \mathcal{M}(G, X, Y, P)$ be a completely 0-simple semigroup and $E \subseteq S, E \neq \theta$, a rectangular band. Then*

1. *for some $X' \subseteq X, Y' \subseteq Y$ and the $Y' \times X'$ submatrix Q of P, E is the idempotent set of $\mathcal{M}(G, X', Y', Q)$,*

2. *any two rows of Q are similar, that is, for every $y, y' \in Y'$ there exists $g \in G$ such that $p_{yx} = gp_{y'x}$ for all $x \in X'$.*

Proof. Let $X' = \{x \in X \mid (g, x, y) \in E \text{ for some } g \in G, y \in Y\}$ and $Y' = \{y \in Y \mid (g, x, y) \in E \text{ for some } g \in G, x \in X\}$. Then the first assertion is clear.

Assume that $y, y' \in Y', x' \in X'$ and $g = p_{yx'} p_{y'x'}^{-1} \in G$. If $x \in X'$, then $e = (p_{yx}^{-1}, x, y), f = (p_{y'x'}^{-1}, x', y') \in E$. Since $efe = e$, we obtain $p_{yx}^{-1} p_{yx'} p_{y'x'}^{-1} p_{y'x} p_{yx}^{-1} = p_{yx}^{-1}$. Hence $p_{yx} = gp_{y'x}$, as desired. $\square$

Remark Let $S \subseteq M_n(K)$ be a π-regular semigroup with the permutation property. Assume that T is a completely 0-simple principal factor of S. Since T is contained in a principal factor of $\overline{S}$, by Lemma 5.30 $E(T) \setminus \{\theta\}$ is a union of finitely many rectangular bands $E_1, \ldots, E_r$.

We can then refine this union to write E as a disjoint union of finitely many rectangular bands. This is a consequence of the fact that $E_i \cap E_j$ is a rectangular band and $E_i \setminus (E_i \cap E_j)$ is a disjoint union of at most three rectangular bands. If $T = \mathcal{M}(G, X, Y, P)$, then it follows that X, Y can be partitioned as $X = X_1 \cup \cdots \cup X_p, Y = Y_1 \cup \cdots \cup Y_s$ so that any two rows of the submatrix P_{ij} of P corresponding to the semigroup $T_{ij} = \mathcal{M}(G, X_j, Y_i, P_{ij}) \subseteq T$ are similar. Thus T is a 0-disjoint union of the semigroups T_{ij} and each T_{ij} is either a null semigroup or $T_{ij} \setminus \{\theta\}$ is a direct product of the almost abelian group G and a rectangular band.

Let F be a field. If T is a completely simple semigroup over an almost abelian group G such that $E(T)$ is a rectangular band, then it is well known that T is a direct product $G \times E(T)$. Therefore, $F[T] \simeq F[G] \otimes_F F[E(T)]$. Now, $F[G]$ is a PI - algebra by Proposition 5.26 and $F[E(T)]$ also is a PI - algebra since $E(T)$ satisfies the multilinear identity $xyzw = xzyw$. Hence, by [117], Theorem 6.1.1, $F[T]$ is a PI - algebra.

If M is a completely 0-simple principal factor of a Zariski closed semigroup $S \subseteq M_n(K)$ satisfying $\mathcal{P}$, then by Lemma 5.28 and Lemma 5.30 the set of non-nilpotents of S can be covered by at most m^2 completely simple semigroups of the above type. This will allow us to prove that $F[M]$ is a PI - algebra. The proof of the main result of this section is based on this idea. In view of Lemma 5.27 the following theorem characterizes arbitrary linear semigroups whose semigroup algebras are PI.

Theorem 5.33 *Let $S \subseteq M_n(K)$ be a π-regular semigroup. Then the following conditions are equivalent*

1. *S has the permutation property,*

2. *$E(S)$ is a finite union of rectangular bands and every maximal subgroup of S is almost abelian,*

3. *$F[S]$ satisfies a polynomial identity for a field F,*

4. *$F[S]$ satisfies a polynomial identity for every field F.*

Moreover, if these conditions hold and $\mathcal{J}(F[S])$ denotes the Jacobson radical of $F[S]$, then $\mathcal{J}(F[S])$ is nilpotent of index at most $3^{n-1} \prod_{j=1}^{n} \binom{n}{j}$

and $F[S]/\mathcal{J}(F[S])$ satisfies an identity of degree $2m(2^{m^2}-1)+2$, where m is the number of irreducible components of the Zariski closure $\overline{S}$.

Proof. That 1) implies 2) follows from Lemma 5.27, Lemma 5.28 and Lemma 5.30. By Lemma 5.24 1) is a consequence of 3). Since 3) follows from 4), we only need to prove that 2) implies 4). It is enough to consider the case where $\mathrm{ch}(F) = 0$. In fact, an identity of the algebra $\mathbf{Q}[S]$ yields an identity in $\mathbf{F}_p[S]$, and therefore in every $F[S]$. If J is an ideal of S, then $F_0[S/J] \simeq F[S]/F[J]$. Moreover, the class of PI - algebras is closed under ideal extensions. Hence, by Theorem 3.5, it suffices to show that $F[V]$ is a PI - algebra for every completely 0-simple principal factor V of S.

By the hypothesis $V \simeq \mathcal{M}(G, X, Y, P)$ for an almost abelian group G and $E(V) \setminus \{\theta\} = E_1 \cup \cdots \cup E_r$, where each E_i is a rectangular band. We can construct subsemigroups $V_j = \mathcal{M}(G, X_j, Y_j, P_j)$ of V, where $X_j \subseteq X, Y_j \subseteq Y$, with $E(V_j) = E_j, j = 1, \ldots, r$. Since V is completely 0-simple, we have $X = X_1 \cup \cdots \cup X_r$.

Let $Z_1, \ldots, Z_t$ be all nonempty subsets of X, which are minimal in the Boolean algebra generated by $X_1, \ldots, X_r$. Then $X = Z_1 \cup \cdots \cup Z_t$. Clearly $t \leq 2^r - 1$ and this is a disjoint union. Consider the homomorphism $\phi : F_0[V] \longrightarrow M_X(F[G])$ given by $a \mapsto a \circ P$. Here $F_0[V]$ is identified with the Munn algebra $\mathcal{M}(F[G], X, Y, P)$ and $M_X(F[G])$ is the algebra of $X \times X$ matrices over $F[G]$ with finitely many nonzero rows. From Lemma 4.16 we know that the kernel of ϕ is equal to the left annihilator of $F_0[V]$. Moreover $\phi(F_0[V])$ coincides with the set of all elements of $M_X(F[G])$ whose rows lie in the left $F[G]$-submodule of $F[G]^X$ generated by the rows of P. Let $M \subseteq F[G]^X$ be the submodule generated by the set of all projections of the rows of P onto $F[G]^{Z_j} \subseteq F[G]^X, j = 1, \ldots, t$. Consider the subalgebra A of $M_X(F[G])$ consisting of all matrices whose rows lie in M. For each j choose a row z_j with a nonzero projection onto $F[G]^{Z_j}$. Let $Z = \{z_j \mid j = 1, \ldots, t\}$. From Lemma 5.32 we know that any two nonzero projections onto a given $F[G]^{Z_j}$ are similar. Since every row of P is the sum of its projections, it follows that M is a left $F[G]$-module generated by Z and $\phi(F_0[V]) \subseteq A$. Consider the Munn algebra $R = \mathcal{M}(F[G], X, Z, Q)$, where for every $z \in Z$ the z-th row of Q is equal to z. Let $\psi : R \longrightarrow M_X(F[G])$ be given by $a \mapsto a \circ Q$. Clearly, $\phi(F_0[V]) \subseteq A = \psi(R)$. In particular, we can identify $\phi(F_0[V])$ with a subalgebra of $R/l(R)$, where $l(R)$ is the left annihilator of R.

On the other hand, we have a homomorphism $R \longrightarrow M_Z(F[G]) \simeq M_t(F[G])$, determined by $a \mapsto Q \circ a$. Its image is isomorphic to $R/r(R)$, where $r(R)$ is the right annihilator of R. Moreover, G has an abelian normal subgroup H of index $q < \infty$. Hence $F[G]$ embeds into $M_q(F[H])$. Therefore $R/r(R)$ embeds into $M_{qt}(F[H])$. Consider the maps $F_0[V] \longrightarrow R/l(R) \longrightarrow R/(r(R) + l(R))$. By the Amitsur - Levitzki theorem, see [117], it follows that $R/(r(R) + l(R))$ (being an image of $R/r(R)$) satisfies the standard identity $\mathbf{S}_{2qt}(x_1, \ldots, x_{2qt}) = 0$ of degree $2qt$. Hence $\phi(F_0[V]) \subseteq R/l(R)$ satisfies the identity $x_0 \mathbf{S}_{2qt}(x_1, \ldots, x_{2qt}) = 0$. Since $\ker(\phi) F_0[V] = 0$, this implies that $F_0[V]$ satisfies the identity

$$x_0 \mathbf{S}_{2qt}(x_1, \ldots, x_{2qt}) x_{2qt+1} = 0.$$

Therefore $F[S]$ is a PI - algebra, as desired. This completes the proof of the equivalence of conditions 1) - 4).

From Lemma 5.28 and Lemma 5.30 it follows that above we can take $H = G^c$ and $q \leq m, r \leq m^2$, so that $t \leq 2^{m^2} - 1$. Let S_j, T_j be as in Theorem 3.5. Each S_j/T_j is a 0-disjoint union of completely 0-simple ideals. Hence, the algebra $F_0[S_j/T_j]$ is a direct product of the corresponding contracted semigroup algebras. If G is a maximal subgroup of S_j/T_j, then G embeds into $GL_j(K)$. Since $\mathrm{ch}(F) = 0$, it is well known that the algebra $F[G]$ is semisimple (for example, use [102], Lemma 7.1.5, Proposition 7.4.3, the fact that finitely generated linear groups are residually finite and Maschke's theorem). Therefore the Jacobson radical $\mathcal{J}(F[S_j/T_j])$ is nilpotent of index at most 3, [87], Corollary 5.18. (In fact, this is also a consequence of Theorem 4.26. Namely, since all primitive homomorphic images of $F[S_j/T_j]$ are finite dimensional simple F-algebras, because it is a PI - algebra, one only needs to look at the intersection of the kernels of all homomorphisms corresponding to finite dimensional irreducible representations of S_j/T_j over F). Since $T_j^{\binom{n}{j}} \subseteq S_{j-1}$, it follows that $\mathcal{J}(F[S])^r = 0$, where $r = 3^{n-1} \prod_{j=1}^n \binom{n}{j}$.

Let R be a simple algebra which is a homomorphic image of $F[S]$. As in Theorem 4.28 we see that R is an image of the semigroup algebra $F_0[V]$ of a completely 0-simple principal factor V of S. We have checked above that $F_0[V]$ satisfies the identity $x_0 \mathbf{S}_{2qt}(x_1, \ldots, x_{2qt}) x_{2qt+1} = 0$. Hence, it satisfies $x_0 \mathbf{S}_p(x_1, \ldots, x_p) x_{p+1} = 0$, with $p = 2m(2^{m^2} - 1)$.

Since $F[S]$ is PI, $F[S]/\mathcal{J}(F[S])$ is a subdirect product of simple algebras, [117]. Therefore it satisfies the same identity. This completes the proof of the theorem. $\square$

We conclude with the following extension of Theorem 5.33. The proof requires standard facts about Noetherian and PI - algebras, [117], results on Noetherian semigroup algebras, [87], and a representation theorem from [2].

Corollary 5.34 *Assume that S is a semigroup such that $F[S]$ is a right Noetherian algebra. If S has the permutation property, then $F[S]$ embeds into a matrix algebra over a field $K \supseteq F$. In particular, S is a linear semigroup and $F[S]$ is a PI - algebra.*

Proof. Let Q be a prime ideal of $F[S]$ and let S_Q denote the image of S in $F[S]/Q$. Then $F[S]/Q$ embeds into $M_n(E)$ for a division algebra E over F. We shall see in Section 9.2 that S_Q has an ideal U which is a uniform subsemigroup of a completely 0-simple semigroup $V \subseteq M_n(E)$. We may assume that V is the completely 0-simple closure of U. Then $H = G \cap U \neq \emptyset$ for a maximal subgroup G of V. Since $F[S]$ is right Noetherian, from [87], Lemma 7.21, it follows that the group ring $F[HH^{-1}]$ is Noetherian. Therefore $G = HH^{-1}$ is a finitely generated group and Proposition 5.26 and Lemma 5.25 imply that $F[G]$ is a PI - algebra. Since $F[S_Q]$ is right Noetherian and U intersects every $\mathcal{H}$-class of V, it follows that the number t of $\mathcal{R}$-classes of V is finite. This implies that $F_0[V]$ is a PI - algebra (as in the proof of Theorem 5.33 we see that $F_0[V]$ modulo its left annihilator embeds into the algebra $M_t(F[G])$). Therefore $F[U]$ is a PI - algebra. Since the image of $F[U]$ in $F[S]/Q$ is a nonzero ideal and $F[S]/Q$ is prime and right Noetherian, the latter also is a PI - algebra, see [117]. This implies that $F[S]$ modulo its prime radical satisfies a polynomial identity. Since the radical is nilpotent, $F[S]$ also satisfies an identity. Finally, right Noetherian semigroup algebras satisfying a polynomial identity are finitely generated, [87], Theorem 19.14. Therefore, from [2] it follows that $F[S]$ embeds into a matrix ring over a field extension of F. $\square$

Further results on PI - algebras of linear semigroups will follow from the material presented in Chapter 7. This is due to the well-known fact that finitely generated PI - algebras have polynomial growth, [67], Corollary 10.7, or [87], Theorem 19.4.

Chapter 6

Generalised Tits alternative

Tits alternative asserts that a finitely generated linear group G either is almost solvable or it contains a free nonabelian subgroup, Theorem 5.4. This, together with Theorem 5.5, implies that G is almost nilpotent or it contains a free noncommutative subsemigroup. Moreover, by Proposition 5.7, any linear group (not necessarily finitely generated) is almost nilpotent if and only if it satisfies a semigroup identity. On the other hand, in view of our structural approach, it seems natural to study the status of cancellative subsemigroups of a linear semigroup S, and in particular to ask whether a generalisation of Tits alternative holds for S.

In this chapter we show that a finitely generated linear semigroup $T \subseteq GL_n(K)$ with no free noncommutative subsemigroups generates an almost nilpotent subgroup of $GL_n(K)$. This extends the results on finitely generated linear groups and finitely generated solvable groups mentioned above. We call this result the 'generalised Tits alternative', we extend it to an arbitrary linear semigroup $S \subseteq M_n(K)$ and obtain consequences for the structure of the Zariski and π-regular closures of such an S.

Our approach comes from [100] and is based on a refinement of arguments of Tits and Rosenblatt, [132],[118]. We refer to [7] and [40] for the general material on algebraic groups, extensively used in this section. Some of the key auxiliary results are stated in the first section. A complete presentation of the ideas of Tits can be found in [83] and [78].

6.1   Auxiliary results

In this section we state several classical results, used in the proof of
the result of Tits [132] and required for the refinement of his argument
presented in the next section.

Let G be a linear algebraic group. In other words, G is a closed
subgroup of $GL_n(K)$ for a field K and some $n \geq 1$. For the proof of the
following result we refer to [134], Theorem 6.4.

Theorem 6.1 *If H is a closed normal subgroup of a linear algebraic
group G, then G/H is a linear algebraic group.*

The connected component G^c of the identity is a normal subgroup
of finite index in G. It is a connected group, and it has no nontrivial
closed subgroups of finite index. Thus, we will further assume that G
is connected. Recall that G has the largest solvable normal subgroup
$\mathcal{R}(G)$. This is a closed subgroup and it is called the radical of G. The
group G is semisimple if $\mathcal{R}(G)$ is trivial. Recall also that a matrix
$a \in M_n(K)$ is called semisimple if it is diagonalizable in $M_n(\overline{K})$. An
element $g \in G$ is semisimple if g is a semisimple matrix. Note that
this does not depend on the chosen rational linear representation of the
group G. By G_s we denote the set of all semisimple elements of G. A
torus of G is a closed connected diagonalizable subgroup of G. A torus
of an algebraic linear group G is conjugate to the product $(K^*)^r$ of r
copies of the multiplicative group of the field K, for some $r \geq 0$.

The next classical result is extracted from [40], Theorem 22.2, Corollary 26.2, Theorem 28.5.

Theorem 6.2 *Assume that G is a semisimple algebraic group over an
algebraically closed field. Then*

  *1. for every closed normal subgroup H of G, the algebraic group G/H
     is semisimple,*

  2. the commutator subgroup $[G, G]$ coincides with G,

  *3. $G_s = \bigcup_{g \in G} g^{-1} T g$ for any maximal torus T of G and G_s contains
     a nonempty open subset of G.*

The second extreme is the class of solvable connected groups. Let G_u denote the set of unipotent elements of G. Recall that the unipotent radical of a linear group G is the largest unipotent normal subgroup of G. For the following result we refer to [40], §§19.2, 19.3, 21.4.

Theorem 6.3 *Assume that G is a connected algebraic group over an algebraically closed field. Then*

1. *if G is solvable, then G is triangularizable and G is a semidirect product of G_u and a maximal torus T of G; moreover G_u is the unipotent radical of G and it is a closed subgroup,*

2. *G is nilpotent if and only if G_s is a subgroup. In this case G_s is a closed subgroup and $G = G_s \times G_u$,*

3. *if G has only one maximal torus, then G is nilpotent.*

An intersection of a closed subset of a topological space V with an open subset of V is called locally closed. A union of finitely many locally closed subsets is called a constructible subset of V, see [40], Chapter 4.4. In particular, a constructible set contains a dense open subset of its closure.

Lemma 6.4 *Let $\phi : G \longrightarrow H$ be a morphism of linear algebraic groups. If $A \subseteq G$ is a constructible set, then $\phi(A)$ is a constructible subset of H. In particular, $\phi(A)$ contains a nonempty open subset of $\phi(H)$.*

We continue with a simple application of constructible sets.

Corollary 6.5 *If H_1, H_2 are closed subgroups of a linear algebraic group G, and H_1 normalizes H_2, then $H_1 H_2$ is a closed subgroup of G.*

Proof. The product $H = H_1 H_2$ is a constructible set as the image of $H_1 \times H_2$ under the product map $G \times G \longrightarrow G$. So $U \subseteq H$ for a dense open subset U of the closure $\overline{H}$. If $h \in \overline{H}$, then hU^{-1}, U are dense open subsets of $\overline{H}$, so they intersect nontrivially. Therefore $\overline{H} = UU$. As H is a subgroup, we come to $H = \overline{H}$. $\square$

It is clear that a nonempty open subset of a connected group G is dense in G. Hence, so it the intersection of any two such subsets. We will need the following extension of this fact.

Lemma 6.6 *Assume that G is a connected group and A, B are dense subsets of G such that A contains a nonempty open subset V of G. Then $A \cap B$ is a dense subset of G.*

Proof. Since B is dense in G, we have $V \cap B \neq \emptyset$. This set is open in B, so it must be dense in B. Therefore it is dense in G and the assertion follows. $\square$

A field K is called locally compact if there is a norm $|\ |$ on K such that the normed space $K, |\ |$ is locally compact. The field $\mathbf{R}$ of real numbers, the field $\mathbf{Q}_p$ of p-adic numbers and the field $\mathbf{F}_p((t))$ of power series over the finite field $\mathbf{F}_p$ are of this type. Here these fields are considered with their standard norms. These are: the absolute value on $\mathbf{R}$, the p-adic norm on $\mathbf{Q}_p$, and for the latter field, the t-adic norm defined by $|\sum_{i=r}^{\infty} a_i t^i| = 2^{-r}$, if $a_r \in \mathbf{Z}$ is such that $a_r \neq 0$. Moreover, if K is a finite extension of one of these fields, then there is a unique extension of the norm to K and K is locally compact with respect to this norm. This, and the following converse result, may be found in [10], §9.

Theorem 6.7 *Every locally compact field is a finite extension of one of the fields $\mathbf{R}, \mathbf{Q}_p, \mathbf{F}_p((t))$.*

In particular, this implies that locally compact fields are perfect.

Our next auxiliary result turns out to be very useful. Its proof does not require Theorem 6.7, since it leads directly to the three classes of locally compact fields mentioned above, see [132], [83], 53.1.1.

Lemma 6.8 *Let α be an element of a finitely generated field K which is not a root of unity. Then there exists an embedding $\sigma : K \longrightarrow L$, for a locally compact field L with a norm $|\ |$, such that $|\sigma(\alpha)| \neq 1$.*

The space K^n and its projective space P are endowed with the topology induced from the field K, that is, the product topology on K^n and the quotient topology on P coming from the natural mapping $\pi : K^n \setminus \{0\} \longrightarrow P$. It is well known that P is compact in this topology.

Let $x = (x_1, \dots, x_n)$ be an affine coordinate system on K^n. We define $d_x : K^n \times K^n \longrightarrow \mathbf{R}$ by

$$d_x(a, b) = \sup_i |x_i(a) - x_i(b)|.$$

The equation $y_0 = 0$ defines a hyperplane H in P. Then $x_i = y_0^{-1} y_i$, for $i = 1, \ldots, n$, is an affine coordinate system on P and $D_x = P \setminus H$ is the domain of this system.

A distance function $d : P \times P \longrightarrow \mathbf{R}_+$ on the projective space P is called admissible if it defines a metric compatible with the topology on P and if, for every affine coordinate system x in P and every compact subset C of D_x there exist $m, M > 0$ such that

$$m\, d_x|_{C \times C} \leq d|_{C \times C} \leq M\, d_x|_{C \times C}.$$

It is known that an admissible distance function exists on P, see [132], 3.3, [83], 53.2.

For a linear map $g : P \longrightarrow P$ we denote by $\| g \|$ the norm of g with respect to the metric d. That is,

$$\| g \| = \sup \frac{d(g(a), g(b))}{d(a, b)}$$

where the supremum runs over all $a, b \in P$ such that $a \neq b$. It is known that we have $\| g \| < \infty$, [132], Lemma 3.5, [83], 53.2.

Any semigroup $S \subseteq GL_n(K)$ acts on the projective space P of K^n. This is accomplished via the natural map $GL_n(K) \longrightarrow PGL_n(K)$, onto the projective linear group. For a linear map $g : P \longrightarrow P$, let $\tilde{g}$ be a representative of g in $GL_n(K)$. Let $\chi(x) = \prod_{i=1}^n (x - \lambda_i) \in \overline{K}[x]$ be the characteristic polynomial of $\tilde{g}$. Put $\Omega = \{i \mid |\lambda_i| = \sup\{|\lambda_j| \mid 1 \leq j \leq n\}\}$, where $|\ |$ stands also for the extension of the norm to $\overline{K}$. Let $\chi_1(x) = \prod_{i \in \Omega}(x - \lambda_i)$ and $\chi_2(x) = \prod_{i \notin \Omega}(x - \lambda_i)$. Clearly $\chi = \chi_1 \chi_2$. Since the extension of $|\ |$ to $\overline{K}$ is unique, every K-automorphism of $\overline{K}$ preserves the norm. Therefore, the coefficients of $\chi_1(x)$ are invariant under all such automorphisms. Since K is a perfect field, this implies that $\chi_1, \chi_2 \in K[x]$. Applying the Jordan form of s we also get $\ker(\chi_1(\tilde{g})) \oplus \ker(\chi_2(\tilde{g})) = K^n$. Define $A(g), A'(g)$ as the subspaces of P that correspond to $\ker(\chi_1(\tilde{g})), \ker(\chi_2(\tilde{g}))$, respectively. In particular $A(g) \cap A'(g) = \emptyset$. Now, for any $s \in GL_n(K)$ we denote by $A(s), A'(s)$ the spaces $A(g), A'(g)$, where g is the image of s in $PGL_n(K)$. With this notation we have the following important technical result.

Lemma 6.9 ([132], Lemma 3.8) *Let $g : P \longrightarrow P$ be a linear map, $C \subseteq P$ a compact subset.*

1. *Assume that g is semisimple, $A(g)$ is a point and $C \cap A'(g) = \emptyset$. Then $\| g^n|_C \| \longrightarrow_{n \to \infty} 0$ and for every neighbourhood X of $A(g)$ there exists $N \geq 1$ such that $g^k(C) \subseteq X$ for $k \geq N$.*

2. *Assume that $g(C) \subseteq \mathrm{Int}(C)$, the interior of C in P, and $\| g|_C \| < 1$. Then $A(g)$ is a point contained in $\mathrm{Int}(C)$.*

6.2   Free subsemigroups

Recall that a cancellative semigroup T is almost nilpotent if it has a group of quotients that is a finite extension of a nilpotent group. This notion can also be given an intrinsic definition in terms of subsemigroups of finite index and semigroup identities introduced in Section 5.1. We say that a subsemigroup T of S is a subsemigroup of finite index if there exists a finite set $Z \subseteq S$ such that for every $s \in S$ there exists $z \in Z$ with $sz \in T$.

Lemma 6.10 *The following conditions are equivalent for a cancellative semigroup S*

1. *S has a nilpotent subsemigroup T of finite index,*

2. *S has an almost nilpotent group of quotients.*

Proof. Assume that 1) holds and a finite set $Z \subseteq S$ is chosen for T. If $s, t \in S$, then there exist $x, y \in Z$ such that $sx, ty \in T$. Now $sxT \cap tyT \neq \emptyset$, because T is nilpotent, so S has a group of right quotients G. Let H be the subgroup generated by T. Then H is nilpotent (see Chapter 5) and it follows that $[G : H] \leq |Z|$. Since S has no free noncommutative subsemigroups, G is a (right and left) group of quotients of G.

Conversely, if H is a nilpotent normal subgroup of finite index in the group G of quotients of S, then S intersects all cosets of H in G. Therefore, we may take as Z any set of coset representatives contained in S. $\square$

The main result of this chapter reads as follows.

Theorem 6.11 *Let $S \subseteq M_n(K)$ be a linear semigroup. Consider the following conditions*

1. S has no free noncommutative subsemigroups,

2. the associated linear groups are almost nilpotent,

3. every cancellative subsemigroup of S is almost nilpotent,

4. S satisfies a semigroup identity.

Then the following implications hold: 2) $\Leftrightarrow$ 3) $\Leftrightarrow$ 4) $\Rightarrow$ 1). *Moreover, if the field* K *is finitely generated, then* 1) $\Rightarrow$ 2).

Proof. The equivalence of conditions 2) and 4) was established in Corollary 5.8. Proposition 3.7 implies that 3) is a consequence of 2). Clearly, 2) is a consequence of 3) and 1) is a consequence of 4).

It remains to prove that, if 1) holds and K is finitely generated, then 2) is satisfied. Replacing S by its intersections with maximal subgroups of $M_n(K)$, it is enough to consider the case where $S \subseteq GL_n(K)$, $n \geq 1$. The proof will be completed by establishing the following assertion:

(∗) If $S \subseteq GL_n(K)$ and K is finitely generated, then either S has a free noncommutative subsemigroup or S is almost nilpotent.

Let $G \subseteq GL_n(K)$ be the subgroup generated by S.

First consider the case where G is almost solvable. We extend the proof of Theorem 5.5, showing that the hypothesis on the nonexistence of free subsemigroups in S is sufficient there. Using the reduction argument of that proof, and the notation used there, we may assume that G is upper triangular and there exists a normal subgroup H of finite index in G such that H/N is nilpotent, where $N \subseteq H$ is the subgroup consisting of unipotent matrices with all non-diagonal entries, except possibly the $(1,n)$-th entry, equal to zero. Replacing S by $S \cap H$ we may further assume that G/N is nilpotent, of nilpotency class r, say. Let $x, y \in S$. Then $x_r(x, y, u_1, \ldots, u_r)N = y_r(x, y, u_1, \ldots, u_r)N$ for the Malcev words x_r, y_r in x, y and any fixed $u_1, \ldots, u_r \in G$. Since G is upper triangular, the definition of x_r, y_r implies that x_r, y_r have the same diagonal, say $(c_1, \ldots, c_n)$. First, consider the case where $c_1 c_n^{-1}$ is a root of unity or $x_r = y_r$ (for every choice of $x, y, u_1, \ldots, u_r \in S$). Since K is finitely generated, there exists $k > 1$ (independent of c_1, c_n) such that $c_1^k = c_n^k$. It follows easily that

$$x_r(x, y, u_1, \dots, u_r)^k y_r(x, y, u_1, \dots, u_r)^k$$
$$= y_r(x, y, u_1, \dots, u_r)^k x_r(x, y, u_1, \dots, u_r)^k$$

because x_r^k, y_r^k may differ only in the $(1, n)$-th entry, but in this case their $(1, 1)$ and (n, n) entries are all equal. Hence S satisfies a nontrivial identity, so its Zariski closure S' in $GL_n(K)$ also satisfies this identity by Corollary 5.2. Since S' is a group by Theorem 1.4 and Corollary 1.5, $G \subseteq S'$ and so Proposition 5.7 implies that S', G are almost nilpotent.

So suppose that $c_1 c_n^{-1}$ is not a root of unity for some $x, y, u_1, \dots, u_r \in S$ and $s = x_r(x, y, u_1, \dots, u_r) \neq t = y_r(x, y, u_1, \dots, u_r)$. Let T be the subgroup generated by s, t. We know that $t = sa$ for some $a \in N$. Then we get the conclusions of Lemma 5.6 for the pair s, a (the argument used there requires only the fact that $\langle s, sa \rangle$ is not a free noncommutative semigroup). As in the proof of Theorem 5.5, this leads to a contradiction. This completes the argument in the case where G is almost solvable.

Assume now that G is not almost solvable. We will use the Zariski topology in the K-algebraic group $GL_n(F)$, where F is a sufficiently big algebraically closed extension of K, and the topology induced to certain subsets of $GL_n(F)$. It is well known that the F-closure $\overline{S}$ of S in $GL_n(F)$ is an algebraic K-group. Let $\overline{S}^c$ denote the connected component of $\overline{S}$. Since $\overline{S}^c$ has finite index in $\overline{S}$, the group $\overline{S}^c$ is generated by $S \cap \overline{S}^c$. But $S \cap \overline{S}^c$ is a closed group, so it must be equal to $\overline{S}^c$. If $S \cap \overline{S}^c$ contains a free noncommutative subsemigroup, then we are done. If $S \cap \overline{S}^c$ is almost nilpotent, then so is $\overline{S}^c = \overline{S \cap \overline{S}^c}$ by Corollary 5.2 and Corollary 5.8, and hence $\overline{S}$ and S are almost nilpotent. Therefore, it is enough to establish assertion $(*)$ for the semigroup $S \cap \overline{S}^c$. But this semigroup is dense in the connected K-group $\overline{S}^c$. This allows us to assume that $\overline{S}$ is a connected group.

Consider the natural K-homomorphism $\phi : \overline{S} \longrightarrow \overline{S}/\mathcal{R}(\overline{S})$, where $\mathcal{R}(\overline{S})$ is the solvable radical of $\overline{S}$. Since $G \subseteq \overline{S}$ is not solvable, it follows that $\overline{S} \neq \mathcal{R}(\overline{S})$. Hence $\phi(S)$ is dense in the nontrivial semisimple K-group $\overline{S}/\mathcal{R}(\overline{S})$ and it is contained in the set of K-rational points of this group. Clearly, it is enough to show that $\phi(S)$ has a free noncommutative subsemigroup. Thus, we can assume that $\overline{S}$ is nontrivial, connected and semisimple.

From Lemma 1.7 we know that there are finitely many roots of unity that satisfy a polynomial of degree n over K. Hence, the degrees of torsion semisimple elements of S divide a natural number N. Let $B = \{g \in \overline{S} \mid g^N = 1\}$. B is a closed subset and $B \neq \overline{S}$ because otherwise $\overline{S}$ is periodic of bounded index, and so almost nilpotent by Corollary 5.8, which contradicts the assumption. From Theorem 6.2 it follows that the set $A = \{g \in \overline{S} \mid g$ is semisimple$\}$ contains a nonempty open subset of $\overline{S}$. Hence, by Lemma 6.6, there exists $s \in S \cap (\overline{S} \setminus B) \cap A$. Then s is a semisimple element of infinite order. Extending K we may assume that the eigenvalues of s lie in K. One of the eigenvalues, say λ, is not a root of unity. From Lemma 6.8 we know that there exists a locally compact field K' with a norm $|\ |$ and a homomorphism $\sigma : K \longrightarrow K'$ such that $|\sigma(\lambda)| \neq 1$. We will further assume that K is locally compact and $|\lambda| \neq 1$. The unique extension of $|\ |$ to an algebraic extension of K will also be denoted by $|\ |$.

Let r be the number of the eigenvalues of s of maximal norm. Since the commutator subgroup $[\overline{S}, \overline{S}]$ coincides with $\overline{S}$ by Theorem 6.2, we must have $\det(s) = 1$. Hence $r < n$ and the maximal norm eigenvalues of s is not equal to 1. Replacing $S, \overline{S}$ by their images under the exterior power homomorphism $g \mapsto \Lambda^r(g)$, we can assume in view of Lemma 1.6 that $r = 1$. (Note that all homomorphic images of $\overline{S}$, $\Lambda^r(\overline{S})$ in particular, are semisimple by Theorem 6.2.) Passing to a finite field extension of K (it is also locally compact), we can assume that $\overline{S}$ has a block triangular form with absolutely irreducible diagonal blocks. Replacing $\overline{S}$ by the block containing an eigenvector corresponding to the eigenvalue λ, we can assume that $\overline{S}$ is absolutely irreducible. Clearly, $\overline{S}$ is nontrivial because $|\lambda| \neq 1$.

From Section 6.1 we know that there exists a metric $d : P \times P \longrightarrow \mathbf{R}_+$ inducing a topology that agrees with that coming from the field K. Moreover, on open subsets of P that can be identified with K^{n-1}, d is equivalent in the usual sense with the metrics introduced via the cartesian metrics on K^{n-1}. For a linear map $g : P \longrightarrow P$ the norm $\|g\|$ of g with respect to the metric d satisfies $\|g\| < \infty$, see Section 6.1. Also, $A(s), A'(s)$ have their earlier meaning.

Lemma 6.12 Let $S \subseteq GL_n(K)$ be a semigroup such that $\overline{S}$ is a connected group. If there exists a semisimple element $s \in S$ such that $A(s)$ is a point, then the set $Z = \{s' \in S \mid A(s')$ is a point $\}$ is dense in $\overline{S}$.

Proof. Choosing an appropriate irreducible component of the set $s^{\mathbf{N}} = \{s^k \mid k \in \mathbf{N}\}$ we can assume that there exists an infinite subset $M \subseteq \mathbf{N}$ such that s^M is K-irreducible. Let

$$U = \{u \in \overline{S} \mid uA(s) \notin A'(s)\}.$$

Then U is open in $\overline{S}$. It is easy to see that $sA(s) = A(s)$. Since $A(s) \cap A'(s) = \emptyset$, it follows that $s \in U$, so $U \cap S \neq \emptyset$. Let $u \in U \cap S$. Since $A'(s)$ is closed in the topology of P, there exists a neighbourhood X of the point $A(s)$ in P such that $u\hat{X} \cap A'(s) = \emptyset$, where $\hat{X}$ denotes the closure of X in the topology of P. From the first assertion of Lemma 6.9 applied to $C = u\hat{X}$ it follows that there exists $N_u \geq 1$ such that $s^k(u\hat{X}) \subseteq X$ and $\| s^k u|_{\hat{X}} \| < 1$ for $k \geq N_u$ (note that $\| s^k u|_{\hat{X}} \| \leq \| s^k|_{u(\hat{X})} \| \| u \|$). Take any $k \geq N_u$. The second assertion of Lemma 6.9 applied to $g = s^k u$ and $C = \hat{X}$ implies that $A(s^k u)$ is a point. Hence $s^k u \in Z$ for $u \in U \cap S$ and $k \geq N_u$. Now the set $s^{M \cap \{k \mid k \geq N_u\}}$ is dense in s^M because the latter is irreducible. Hence $s^M u \subseteq \overline{Z}$. Let $m_0 \in M$. We have shown that $s^{m_0}(U \cap S) \subseteq \overline{Z}$. Consequently $s^{m_0} \overline{U \cap S} \subseteq \overline{Z}$. Since $\overline{U \cap S} = \overline{S}$ (see Lemma 6.6), we come to $\overline{S} = s^{m_0} \overline{S} \subseteq \overline{Z}$. This shows that Z is dense in $\overline{S}$. $\square$

The element $s \in S$ found earlier has precisely one eigenvalue of maximal norm, so that $A(s)$ is a point. By Lemma 6.12 Z is dense in S. Let $T \subseteq \overline{S}$ be a maximal torus of $\overline{S}$. Then T diagonalizes in a basis $v_1, \ldots, v_n$ of F^n. By $\lambda_i(t)$ we denote the eigenvalue of $t \in T$ corresponding to v_i. Define $U_{ij} = \{t \in T \mid \lambda_i(t) \neq \lambda_j(t)\}$. The sets U_{ij} are open in T and at least one of these sets is nonempty, because otherwise T would be the only maximal torus of $\overline{S}$, which in view of Theorem 6.3 leads to a contradiction. Let $t_0 \in \bigcap_{U_{ij} \neq \emptyset} U_{ij}$. We may assume that $v_1, \ldots, v_r$ is the set of simple eigenvectors of t_0 (that is, eigenvectors corresponding to simple eigenvalues of t_0). The choice of t_0 implies that each simple eigenvector v of an element $t \in T$ lies in $\{v_1, \ldots, v_r\}$, up to a scalar multiple.

Let $w \in K^n \subseteq F^n$ be a vector corresponding to the point $A(s)$, and $W \subseteq F^n$ a hyperplane corresponding to $A'(s)$. Then w is a simple eigenvector of s. Define

$$
\begin{aligned}
L(s) \;=\; & \{g \in \overline{S} \mid \text{if } w' \text{ is a simple eigenvector of } g, \\
& W' \text{ its } g - \text{invariant complement, then } w \notin W' \text{ and } w' \notin W\}.
\end{aligned}
$$

We will show that $L(s)$ contains an open subset of $\overline{S}$. Let

$$V = \{g \in \overline{S} \mid gv_i \notin W, w \notin g \operatorname{Lin}(v_1, \dots, v_{i-1}, v_{i+1}, \dots, v_n)$$
$$\text{for } i = 1, \dots, r\}.$$

Then V is an open subset of $\overline{S}$ as an intersection of the open subsets $\{g \in \overline{S} \mid gv_i \notin W\}$ and $\{g \in \overline{S} \mid g^{-1}w \notin \operatorname{Lin}(v_1, \dots, v_{i-1}, v_{i+1}, \dots, v_n)\}$, for $i = 1, \dots, r$. Hence V is nonempty because otherwise one of these sets is empty (because $\overline{S}$ is connected), which would contradict absolute irreducibility of $\overline{S}$. Let v be a simple eigenvector of gtg^{-1}, where $g \in V, t \in T$. Then $g^{-1}v$ is a simple eigenvector of t. Hence $g^{-1}v \in \operatorname{Lin}(v_i)$ for some $i \in \{1, \dots, r\}$. Therefore $v \in \operatorname{Lin}(gv_i)$ and $v \notin W$ by the definition of V. A similar argument shows that $g \operatorname{Lin}(v_1, \dots, v_{i-1}, v_{i+1}, \dots, v_n)$ is a gtg^{-1}-invariant complement of v and it does not contain w. The definition of $L(s)$ implies now that $gtg^{-1} \in L(s)$.

Let $\phi : \overline{S} \times T \longrightarrow \overline{S}$ be given by the formula $\phi(g, t) = gtg^{-1}$ for $(g, t) \in \overline{S} \times T$. We have shown that $\phi(V \times T) \subseteq L(s)$. But $\phi(V \times T)$ is a constructible set as an image of $V \times T$. Hence it contains an open subset of $\overline{\phi(V \times T)}$, see Lemma 6.4. Since V is dense in $\overline{S}$, we have $\overline{\phi(V \times T)} = \phi(\overline{S} \times T) = \overline{S}$ by Theorem 6.2. Therefore, $L(s)$ contains an open subset of $\overline{S}$, as claimed.

The set C of elements $g \in \overline{S}$ such that w is not an eigenvector of g is open and nonempty because $\overline{S}$ is absolutely irreducible. Hence we may choose an element $t \in Z \cap L(s) \cap \{g \in \overline{S} \mid g \text{ is semisimple }\} \cap C$. In particular, $A(t)$ is a point. Since w is not an eigenvector of t by the definition of C, it follows that $A(t) \neq A(s)$. Also $A(s) \notin A'(t)$ and $A(t) \notin A'(s)$ by the definition of $L(s)$. The following lemma shows that there exists $n_0 \geq 1$ such that s^{n_0}, t^{n_0} generate a free noncommutative semigroup, completing the proof of the theorem.

Lemma 6.13 *Assume that $s, t \in GL_n(K)$ are semisimple. Assume also that $A(s), A(t)$ are points such that $A(s) \neq A(t), A(s) \notin A'(t)$ and $A(t) \notin A'(s)$. Then there exists $n_0 \geq 1$ such that s^{n_0}, t^{n_0} generate a free noncommutative semigroup.*

Proof. There exist neighbourhoods U_s, U_t of the points $A(s), A(t)$ respectively, such that $U_s \cap U_t = \emptyset$, $\hat{U}_s \cap (A'(s) \cup A'(t)) = \emptyset$ and $\hat{U}_t \cap (A'(s) \cup A'(t)) = \emptyset$, (as before, $\hat{X}$ denotes the closure of X in P). Choose

$$p \notin U_s \cup U_t \cup A'(s) \cup A'(t).$$

From Lemma 6.9, applied to $X = U_s$ and $C = \{p\} \cup \hat{U}_s \cup \hat{U}_t$, it follows that there exists $n_0 \geq 1$ such that $s^k(\{p\} \cup U_s \cup U_t) \subseteq U_s$. Similarly, adjusting n_0 if necessary, we get $t^k(\{p\} \cup U_s \cup U_t) \subseteq U_t$ for $k \geq n_0$. Suppose that $s^{n_0 a_1} t^{n_0 a_2} \cdots = t^{n_0 b_1} s^{n_0 b_2} \cdots$ for some $a_1, a_2, \ldots, b_1, b_2, \ldots \in \mathbf{N}$. An easy induction shows that $s^{n_0 a_1} t^{n_0 a_2} \cdots (p) \in U_s$ and $t^{n_0 b_1} s^{n_0 b_2} \cdots (p) \in U_t$, contradicting the fact that $U_s \cap U_t = \emptyset$. Similarly, $p \notin U_s \cup U_t$ implies that a nontrivial word in s^{n_0}, t^{n_0} cannot be the identity of $GL_n(K)$. The assertion follows. $\square$

Remark The argument of Tits can also be modified to obtain an alternative proof of the almost solvable case in the above theorem. Indeed, it is enough to consider the situation where K is finitely generated and $S \subseteq GL_n(K)$ is such that the group $\overline{S}$ is connected and solvable. Then, by Theorem 6.3, $\overline{S}$ is triangularizable in $GL_n(K')$ for a finite field extension K' of K. Hence, extending K we can assume that $\overline{S}$ is indecomposable and $\overline{S}$ is contained in the group of upper triangular matrices over K. For an element $g \in \overline{S}$ let $\lambda_1(g), \ldots, \lambda_n(g)$ be the subsequent diagonal entries of g. Define $\phi : \overline{S} \longrightarrow GL_n(F)$ by $\phi(g) = g\lambda_1(g)^{-1}$. Then ϕ is a K-homomorphism of algebraic groups. If $\phi(S)$ is almost nilpotent, then S is almost nilpotent because the kernel of ϕ is central. So, passing to $\phi(S)$, we can assume that $\lambda_1(g) = 1$ for all $g \in \overline{S}$. If for every $s \in S$ each $\lambda_i(s)$ is a root of unity, then $\{\lambda_i(s) \mid s \in S, \ i = 1, \ldots, n\}$ is a finite set. In this case the group generated by S is almost unipotent, and so we are done. Thus, assume that $\lambda_i(s)$ is not a root of unity for some $s \in S$ and some i. As in the proof of the semisimple case, Lemma 6.8 allows us to assume that K is locally compact and $|\lambda_i(s)| \neq 1$. Replacing s by s^{-1}, if necessary, we may assume that $|\lambda_i(s)| > 1$. S acts on the projective space P of K^n. Since $|\lambda_1(s)| = 1$ and $|\lambda_i(s)| \neq 1$, we have $A(s) \neq P$. Let $W, W' \subseteq K^n$ be the linear spaces corresponding to $A(s), A'(s)$ respectively. Define $W_F(s) = \mathrm{Lin}_F(W)$ and $W'_F(s) = \mathrm{Lin}_F(W')$.

Assume first that $W_F(s)$ is $\overline{S}$-invariant. Then, extending K, we can assume that $W = \mathrm{Lin}_K(e_1, \ldots, e_r), W' = \mathrm{Lin}_K(e_{r+1}, \ldots, e_n)$, where e_i is the i-th vector of the standard basis of K^n, and $\overline{S}$ is upper block triangular with respect to the decomposition $F^n = W_F(s) \oplus W'_F(s)$. Since W, W' are s-invariant, the matrix s is block diagonal subject to this decomposition. Consider the transpose $S^T \subseteq \overline{S}^T$. Suppose that $W_F(s^T)$ (defined similarly as $W_F(s)$ above) is $\overline{S}^T$-invariant. Note that $W_F(s^T) = \ker(\chi_1(s)^T)$, where $\chi_1(x)$ is the polynomial used in the definition of $A(s)$. Hence it is easy to see that $W_F(s^T) = W_F(s)$ is a direct summand of the $\overline{S}^T$-module F^n, contradicting the indecomposability of $\overline{S}$. Therefore, replacing S by S^T, if necessary, we can assume that $W_F(s)$ is not $\overline{S}$-invariant.

We apply the exterior power Λ^r to $\overline{S}$. Since $r = \dim_F W_F(s)$, we come to an element $\Lambda^r(s) \in \Lambda^r(S)$ which has only one simple eigenvalue of maximal norm (see Section 1.3). Therefore $A(\Lambda^r(s))$ is a point. This allows us to assume that $s \in S$ is such that $A(s)$ is a point, $\overline{S}$ is triangular with $\lambda_1(g) = 1$ for $g \in \overline{S}$, and $W_F(s)$ is not $\overline{S}$-invariant. Choose a subset $M \subseteq \mathbf{N}$, $|M| = \infty$, such that the set s^M is K-irreducible. Let $\lambda_i(s)$ be the (unique, because $A(s)$ is a point) eigenvalue of s of maximal norm. Then $i > 1$. It is easy to see that for $u \in S$ there exists $N_u \geq 1$ such that, for $k \geq N_u$, $\lambda_i(us^k)$ is an eigenvalue of us^k of maximal norm. Let $Z = \{s' \in S \mid \lambda_i(s')$ is an eigenvalue of s' of maximal norm$\}$. Then $us^k \in Z$ for $u \in S$ and $k \geq N_u$. An argument as that at the end of the proof of Lemma 6.12 allows us to show that Z is a dense subset of $\overline{S}$. Let $t \in Z \cap \{g \in \overline{S} \mid W_F(s)$ is not g-invariant$\}$. (The latter of the intersected sets is open in $\overline{S}$ and nonempty by the choice of s, so that the intersection is nonempty.) We have chosen elements $s, t \in S$ such that $\lambda_i(s), \lambda_i(t)$ are eigenvalues of s, t, respectively, of maximal norms and $A(s) \neq A(t)$. Replacing s, t by their restrictions to $\mathrm{Lin}(e_1, \dots, e_i)$ we can assume that $n = i$. Then $A'(s) = A'(t)$ and $\mathrm{Lin}(e_1, \dots, e_{n-1})$ is the corresponding subspace of K^n. The elements s, t satisfy the hypotheses of Lemma 6.13, except for being semisimple. However, as mentioned in [132], the assertions of Lemma 6.9 (and hence also of Lemma 6.12 and Lemma 6.13) are valid without this restriction. Therefore, $\langle s, t \rangle$ contains a free noncommutative subsemigroup.

Now we can summarize the information on the associated groups of a semigroup $S \subseteq M_n(K)$ with no free noncommutative subsemigroups.

Corollary 6.14 *Assume that $S \subseteq M_n(K)$ has no free noncommutative subsemigroups. Then every group G associated to S has a nilpotent normal subgroup H such that G/H is periodic. Moreover, G is locally almost nilpotent and it is almost solvable if $\mathrm{ch}(K) = 0$ and almost nilpotent if K is finitely generated.*

Proof. Recall from Section 5.1 that a locally almost nilpotent linear group G has a normal subgroup H with a periodic quotient G/H. G

is almost solvable if $\text{ch}(K) = 0$, by Theorem 5.4. Thus, the assertion follows from Theorem 6.11. □

If R is a finitely generated domain, then every locally almost nilpotent subgroup of $GL_n(R)$ is almost nilpotent, cf.[134], Theorem 10.14. The following extension of this result is an immediate consequence of Theorem 6.11.

Corollary 6.15 *Assume that $G \subseteq GL_n(K)$ is a linear group over a finitely generated field K. If G is locally almost nilpotent, then it is almost nilpotent.*

We note that, in contrast to the case of linear groups, finitely generated (noncancellative) linear semigroups satisfying the conditions of the theorem need not have polynomial growth, even if the associated linear groups are finitely generated. This topic will be discussed in the next chapter.

Let $S \subseteq M_n(K)$. The groups associated to the Zariski closure $\overline{S}$ of S in $M_n(K)$ are just the maximal subgroups of $\overline{S}$, see Section 3.2. Thus, $\overline{S}$ has a better structure than S. From Corollary 6.14 we know that, if S has no free noncommutative subsemigroups, then the groups associated to S are periodic extensions of nilpotent groups. Moreover, Theorem 6.11 implies that, if S is finitely generated, then $\overline{S}$ has no free noncommutative subsemigroups. The latter is no longer true if S is not finitely generated. For example, the group G of upper triangular matrices in $GL_n(K), n > 1, K \not\subseteq \overline{\mathbf{F}}_p$, with periodic diagonal is a solvable group which is a periodic extension of a nilpotent group, but its closure (= the group of all upper triangular matrices) is not of this type. Simply, the Zariski closure of a semigroup $S \subseteq M_n(K)$ is not controlled by the Zariski closures of the finitely generated subsemigroups of S. However, we have seen in Section 3.3 that this is the case for the π-regular closure $\text{cl}(S)$ of S.

Corollary 6.16 *Let $S \subseteq M_n(K)$ be a semigroup. If S has no free noncommutative subsemigroups, then $\text{cl}(S)$ has no free noncommutative subsemigroups .*

Proof. This is a direct consequence of Theorem 6.11 and Lemma 3.12. □

Note that Corollary 3.18 yields an important information on the ideal structure of a linear semigroup S with no free noncommutative subsemigroups. Namely, 0-simple principal factors of S are completely 0-simple.

If $S \subseteq M_n(K)$ is a connected algebraic semigroup (that is, Zariski closed and irreducible as an algebraic variety), then the maximal subgroups of S are connected algebraic groups, see [108]. If $K \not\subseteq \overline{\mathbf{F}}_p$, then from Theorem 6.11 and from Proposition 5.7 it follows that S has no free noncommutative subsemigroups if and only if the maximal subgroups of S are nilpotent, which is also equivalent to the fact that S satisfies an identity.

Finally, we note that some other variants of Tits alternative have been studied for groups of units of integral group rings of finite groups, [35], for skew fields,[74],[75], and domains,[64].

Chapter 7

Growth

It was shown in Chapter 6 that a finitely generated linear semigroup $S \subseteq M_n(K)$ with no free noncommutative subsemigroups 'locally' is almost nilpotent in the sense that each nonempty intersection $S \cap D$ with a maximal subgroup D of $M_n(K)$ generates an almost nilpotent subgroup of D. However, in contrast to the case of linear groups, the growth of such a semigroup S often is not polynomial.

The aim of this chapter is to look for a structural description of linear semigroups of polynomial growth. To this end we first present certain important special classes of semigroups of polynomial growth, proving in particular that unipotent semigroups are of this type. Then we discuss the second extreme case of 'almost semisimple semigroups', and we explain why semigroups of these two types should be crucial for the structure of arbitrary linear semigroups of polynomial growth. Also the class of semigroups of linear growth and its connection with a combinatorial property, called repetitivity, are studied.

Techniques resulting from the general structure theorem for linear semigroups, obtained in Chapter 3, are essential for our approach. The polynomiality of the growth of S turns out to depend not only on the properties of the uniform components of S but also on the interaction between them.

We refer to [67], [81] for the general theory of growth of associative systems.

7.1   Burnside theorem and some examples

Let S be a finitely generated semigroup. Choose a generating set $V = \{a_1, \ldots, a_t\}$. By $V^m, m \geq 1$, we denote the set of all elements of S which are of the form $v_1 \cdots v_m$ for $v_i \in V$. Then the cardinality $d_V(m)$ of the set $V \cup V^2 \cup \cdots \cup V^m$, $m \geq 1$, is called the growth function of S corresponding to the generating set V. If $d_V(m)$ is bounded by a polynomial in m, then we say that S has polynomial growth. It turns out that this is independent of the choice of the generating set V. The minimal degree of such a polynomial is called the degree of growth of S. Moreover, the Gelfand - Kirillov dimension of S is defined as $\mathrm{GK}(S) = \limsup \log_m d_V(m)$. In other words, $\mathrm{GK}(S) < \infty$ if and only if S has polynomial growth. If S is not finitely generated, then $\mathrm{GK}(S)$ is defined as the supremum of $\mathrm{GK}(T)$ where T runs over the set of finitely generated subsemigroups of S.

More generally, growth and Gelfand - Kirillov dimension can be defined for any K-algebra A, by considering the K-dimension of the linear span of $V \cup \cdots \cup V^m$ for a generating set V of A. Clearly, $\mathrm{GK}(S) = \mathrm{GK}(K[S])$ for any semigroup S.

It is known that $\mathrm{GK}(A)$ can be equal to $0, 1$ any real number ≥ 2 or ∞. Dimensions 0 and 1 are distinguished also from the structural point of view. Namely, $\mathrm{GK}(A) = 0$ if and only if A is locally finite, that is, every finitely generated subalgebra of A is finite dimensional. Also, algebras of dimension 1 are very special. If $\mathrm{GK}(A) = 1$ for a finitely generated K-algebra A, then the prime radical $\mathcal{B}(A)$ of A is nilpotent and $A/\mathcal{B}(A)$ is a finite module over its centre, which is finitely generated, [128]. In particular, A is a PI - algebra.

The following fundamental result due to Gromov,[32], and Bass,[4], is the starting point for studying the growth problem for linear semigroups.

Theorem 7.1 *Let G be a finitely generated group. Then G has polynomial growth if and only if G is almost nilpotent. Moreover, in this case $\mathrm{GK}(S) = \sum_i i r_i$, where r_i is the torsion - free rank of the i-th factor of the lower central series of a nilpotent subgroup of finite index in G.*

In view of Theorem 7.1 and of the results of Chapter 3, in case of a semigroup $S \subseteq M_n(K)$ one needs in the first place to consider the relation between $GK(S)$ and $GK(G)$ for a cancellative semigroup S generating a group G. This was done by Grigorchuk, [31], cf. [87].

Theorem 7.2 *Let S be a cancellative semigroup. Then $GK(S)$ is finite if and only if S has a group of quotients G and $GK(G)$ is finite. Moreover, in this case $GK(S) = GK(G)$.*

The next step is to see how the 'local' information (that on the level of the groups associated to a given linear semigroup $S \subseteq M_n(K)$) affects the growth of S. The first step looks rather promising. It was first proved in [82] through quite a different approach.

Theorem 7.3 *The following conditions are equivalent for a semigroup $S \subseteq M_n(K)$*

1. *S is locally finite,*

2. *S is periodic,*

3. *$GK(S) = 0$,*

4. *$GK(S \cap G) = 0$ for every maximal subgroup G of $M_n(K)$ nontrivially intersecting S.*

Proof. We know that 1),3) are equivalent. Clearly, 2) is a consequence of 1). If S is periodic, then so is every nonempty $S \cap G$. So $S \cap G$ is a periodic linear group, and therefore it is locally finite, [17], Theorem 36.2. Hence 2) implies 4). It remains to show that 1) is a consequence of 4). This is an easy consequence of the structure theorem. Namely, it is easy to see that a completely 0-simple semigroup over a locally finite group is locally finite, [87], Lemma 2.3. Also, if $T^N = \theta$ for a semigroup T, then T is locally finite. Next, if I is an ideal of a semigroup T and $I, T/I$ are locally finite, then T is locally finite, cf.[87], Proposition 2.2. But S has an ideal chain $S = I_1 \supset \cdots \supset I_t$ such that every factor T of this chain is either completely 0-simple or it satisfies $T^N = \theta$ for some $N \geq 1$, Corollary 3.8. Therefore, the assertion follows by induction on t. $\square$

It turns out that one cannot expect the equivalence of 3) and 4) in general. Polynomial growth of every group associated to a given $S \subseteq M_n(K)$ does not even imply that S has polynomial growth.

Example There exist linear semigroups of subexponential growth with all associated groups isomorphic to the infinite cyclic group. Namely, let $\mathbf{Z}$ be the ring of integers, $M_3(\mathbf{Z})$ the ring of 3×3 matrices over $\mathbf{Z}$. Denote by S the subsemigroup generated in $M_3(\mathbf{Z})$ by the matrices

$$e = \begin{pmatrix} 1 & 0 & 0 \\ 0 & 0 & 0 \\ 0 & 0 & 1 \end{pmatrix} \quad \text{and} \quad g = \begin{pmatrix} 1 & 1 & 1 \\ 0 & 2 & 1 \\ 0 & 0 & 1 \end{pmatrix}.$$

We claim that the growth of S is not polynomial. For this, first note that $g^n = \begin{pmatrix} 1 & a_n & b_n \\ 0 & 2^n & a_n \\ 0 & 0 & 1 \end{pmatrix}$, where $a_0 = 0, b_0 = 0$, and for $n \geq 1$

$$\begin{aligned} a_n &= a_{n-1} + 2^{n-1} \\ &= 1 + 2 + \cdots + 2^{n-1} \\ &= 2^n - 1 \end{aligned}$$

and

$$\begin{aligned} b_n &= b_{n-1} + a_{n-1} + 1 \\ &= a_{n-1} + a_{n-2} + \cdots + a_2 + a_1 + b_1 + n - 1 \\ &= (2^{n-1} - 1) + 2^{n-2} + \cdots + 2 + 1 = 2^n - 1 \end{aligned}$$

Consider the elements $t = eg^{n_1}eg^{n_2}\cdots eg^{n_s} \in S$. The $(1,3)$-entry of t is equal to $b_{n_1} + \cdots + b_{n_s}$. Let $A_m = \{b_{n_1} + \cdots + b_{n_s} \mid n_1 + \cdots + n_s \leq m; s \geq 1\}$ and $\overline{A}_m = \{b_{n_1} + \cdots + b_{n_s} \mid n_1 + \cdots + n_s + s \leq m; s \geq 1\}$. Then $A_m \subseteq \overline{A}_{2m}$ because if $n_1 + \cdots + n_s \leq m$, then $s \leq m$. The maximal number of the set A_m is equal to b_m. Thus, the sets $A_m, A_m + b_{m+1}, \cdots, A_m + b_{2m}$ are disjoint (because $\max(A_m + b_{m+k}) = b_m + b_{m+k} < b_{m+k+1}$) and of the same cardinality. Since all of them lie in A_{3m}, we come to

$$|A_{3m}| \geq 3^{m-1}|A_{3m-1}| \geq \cdots \geq 3^{m-1}3^{m-2}\cdots 3^0 = 3^{m(m-1)/2}.$$

Therefore

$$|\overline{A}_{2\cdot 3m}| \geq |A_{3m}| \geq 3^{m(m-1)/2}.$$

Hence

$$\log_{2\cdot 3m}|\overline{A}_{2\cdot 3m}| \geq \log_{3m+1} 3^{m(m-1)/2} = \frac{m(m-1)}{2(m+1)}.$$

It follows that the function

$$f(m) = |\{s \in S \mid s \text{ is a word of length at most } m \text{ in } e, g\}|$$

is not bounded by a polynomial, as desired.

We note that $|A_n| \leq \sum_{k \leq n} p(k)$, where $p(k)$ denotes the number of partitions of k. It is known that $p(n) \sim (1/4n\sqrt{3})e^{A\sqrt{n}}$ for $A = \pi\sqrt{2/3}$, cf. [33], Chapter 4. This easily implies that S has subexponential growth.

All matrices of rank three belong to the cyclic group generated by g. For any matrix $s \in S$ of rank two, we see that ese belongs to the maximal subgroup of $M_3(\mathbf{Q})$ containing e. This implies that matrices of rank two form a uniform component of S. Hence, all associated groups consisting of matrices of rank two are isomorphic to the cyclic group

$$\left\{ \begin{pmatrix} 1 & 0 & x \\ 0 & 0 & 0 \\ 0 & 0 & 1 \end{pmatrix} \mid x \in \mathbf{Z} \right\}.$$

Finally, S has no elements of rank one. Therefore all the associated groups of S are cyclic infinite.

Before studying the structure of arbitrary linear semigroups of polynomial growth, we show that certain classes of semigroups considered in Chapter 5 are of this type.

Theorem 7.4 *Let $S \subseteq M_n(K)$ be a finitely generated semigroup with permutational property. Then S has polynomial growth.*

Proof. It is known that a finitely generated PI - algebra has finite Gelfand-Kirillov dimension, [67], Corollary 10.7. Hence, the assertion follows from Theorem 5.33. Let us note that a purely combinatorial proof, not referring to PI - algebras, can be derived from Shirshov's height theorem, [87], Theorem 19.4. $\square$

Assume now that $S = \langle a_1, \dots, a_t \rangle \subseteq M_n(K)$ is an almost unipotent semigroup that is in the block upper triangular form of Proposition 5.9. As in Section 5.2, let e_j be the identity of the diagonal block Z_j, $j = 1, \dots, r$, and $a_1^{(i)}, \dots, a_r^{(i)}$ the diagonal blocks of a_i, that is $a_j^{(i)} = e_j a_i e_j$. Write $x_{kj}^{(i)} = e_k a_i e_j$, so that $a_i = \sum_{k,j=1, k \leq j}^r x_{kj}^{(i)}$. Also define the sets

$$C_{i,i} = e_i S e_i, \quad C_{i,i+1} = C_{i,i}^1 X_{i,i+1} C_{i+1,i+1}^1$$

where $X_{i,j} = \{x_{ij}^{(k)} \mid k = 1, \ldots, t\}$, and inductively, for $i < j - 1$,

$$C_{i,j} = C_{i,i+1}C_{i+1,j} \cup \ldots \cup C_{i,j-1}C_{j-1,j} \cup C_{i,i}^1 X_{i,j} C_{j,j}^1.$$

Let $D_{i,j}, i \leq j$, be the R-submodule of the ring $M_n(K)$ generated by $C_{i,j}$, where $R = \mathbf{Z}$ if $\mathrm{ch}(K) = 0$ and R is the prime subfield of K otherwise.

From Lemma 5.10 we know that, if $s = a_{i_1} \ldots a_{i_m}, i_j \in \{1, \ldots, t\}$, and $i < j$, then $e_i s e_j$ is a sum of at most $h_{j-i}(m)$ elements of the set $C_{i,j}$, where h_k is a polynomial in m of degree k, for $k = 1, \ldots, n - 1$.

Corollary 7.5 *Let $S \subseteq M_n(K)$ be a finitely generated almost unipotent semigroup. Then S has polynomial growth.*

Proof. If $s = a_{i_1} \ldots a_{i_m} = \sum e_i s e_j \in S$, then $e_i s e_j$ is an element of length $\leq h_{j-i}(m)$ in the generators $C_{i,j}$ of the (additive) abelian group $D_{i,j}$. Therefore the claim follows. $\square$

Note that, if $\mathrm{ch}(K) > 0$, then a finitely generated almost unipotent semigroup $S \subseteq M_n(K)$ is finite because it is periodic, Theorem 7.3. Each $D_{i,j}$ is then finite.

Our last class of examples of semigroups of polynomial growth requires preparatory results on uniform semigroups.

Lemma 7.6 *Assume that $J = \mathcal{M}(G, X, Y, P)$ is a completely 0-simple ideal of a semigroup S and U is a uniform subsemigroup of J such that J is the completely 0-simple closure of U in J and $Q = U \cup (S \setminus J)$ is a subsemigroup of S. Then*

1. *$\mathrm{GK}(S) \leq 2\sup(\mathrm{GK}(S/J), \mathrm{GK}(G))$, provided that at least one of the sets X, Y is finite,*

2. *$\mathrm{GK}(S) = \mathrm{GK}(Q) = \sup(\mathrm{GK}(S/J), \mathrm{GK}(G))$ if X, Y are finite.*

Proof. We consider the case where $r = |X| < \infty$ only, because the proof of assertion 1) in case $|Y| < \infty$ is symmetric. Since G is a group generated by a subsemigroup of Q, by Theorem 7.2 we have $\mathrm{GK}(G) \leq \mathrm{GK}(Q)$. Also, $Q/U = S/J$ is a homomorphic image of Q, hence $\mathrm{GK}(S/J) \leq \mathrm{GK}(Q)$. It remains to show that

$$\mathrm{GK}(S) \leq \lambda \sup(\mathrm{GK}(S/J), \mathrm{GK}(G)),$$

where $\lambda = 1$ if $|Y| < \infty$ and $\lambda = 2$ otherwise.

Let $e = e^2 \in J, e \neq \theta$. Choose nonzero elements $e_1 = e, e_2, \ldots, e_r$, such that $e_i \in J$ and $e_i \mathcal{L} e$ for all i, but e_i, e_j are not in the same $\mathcal{R}$-class of J if $i \neq j$. As in Proposition 3.16 (see also Section 4.2), we have a homomorphism

$$\phi : K_0[S] \longrightarrow M_r(K[G])$$

whose kernel N is the left annihilator of $K_0[J]$. Consider

$$\psi : K_0[S] \longrightarrow M_r(K[G]) \oplus K_0[S/J]$$

given by $\psi(x) = \phi(x) + \bar{x}$, where $\bar{x}$ is the image of x under the natural map $K_0[S] \longrightarrow K_0[S/J]$. Let $A = N \cap K_0[J]$. Then $A^2 = 0$ and $\ker(\psi) = A$.

Introduce a congruence $\sim$ on S by the rule

$$s \sim t \Longleftrightarrow s = t \quad \text{or} \quad s, t \in J \quad \text{and} \quad (s - t)J = 0.$$

(In other words, $(s - t) \in N$, so $S/\sim$ can be identified with $\psi(S)$.) Thus ψ factors through $K_0[S/\sim]$ and hence the square of the kernel of the map $K_0[S] \longrightarrow K_0[S/\sim]$ is zero. From [67], Corollary 5.10, it follows that $\mathrm{GK}(K_0[S]) \leq 2\ \mathrm{GK}(K_0[S/\sim])$. But

$$\mathrm{GK}(K_0[S/\sim]) \leq \sup(\mathrm{GK}(M_r(K[G])), \mathrm{GK}(S/J)).$$

Since we know that $\mathrm{GK}(K[G]) = \mathrm{GK}(M_r(K[G]))$, [67], Proposition 5.5, it follows that $\mathrm{GK}(S/\sim) \leq \mathrm{GK}(G)$ and

$$\mathrm{GK}(S/\sim) \leq \sup(\mathrm{GK}(G), \mathrm{GK}(S/J)).$$

This proves assertion 1).

Now, assume that $|Y|$ is finite. Let $s \in J$. If $t \sim s$, then $tx = sx$ for every $x \in J$ easily implies that $t\mathcal{R}s$. Moreover, every $\mathcal{H}$-class of J that is $\mathcal{R}$-related to s contains at most one element t of this type. Therefore, we get the following relation between the growth functions d_S and $d_{S/\sim}$ for S and $S/\sim$ subject to the generators $s_1, \ldots, s_m$ of S and $\phi(s_1), \ldots, \phi(s_m)$ of $S/\sim$, respectively

$$d_S(n) \leq d_{S/\sim}(n)|Y|.$$

It follows that $\mathrm{GK}(S) = \mathrm{GK}(S/\sim)$. So, the inequality displayed before implies that 2) holds. $\square$

Corollary 7.7 *Assume that U is a uniform subsemigroup of a completely 0-simple semigroup J with closure $\hat{U}$. Then $\mathrm{GK}(U) = \mathrm{GK}(\hat{U}) = \mathrm{GK}(T)$ for any intersection T of U with a maximal subgroup of $\hat{U}$.*

Proof. We may assume that $J = \hat{U}$. Let $J = \mathcal{M}(G, X, Y, P)$. Let H be the maximal subgroup of J containing T. Then $H \simeq G$ is the group generated by T. It is enough to show that $\mathrm{GK}(J) \leq \mathrm{GK}(T)$. So, choose any finitely generated subsemigroup I of J. Then I is contained in a completely 0-simple semigroup $J' = \mathcal{M}(G, X', Y', P')$, where $X' \subseteq X, Y' \subseteq Y$ are finite subsets. Therefore, replacing J by J', we may assume that the sets X, Y are finite. The assertion follows from Lemma 7.6. $\square$

Note that the group ring $C = K[H]$ is a subalgebra of $A = K_0[\hat{U}]$ and it is a 'corner' in A in the sense that $CAC \subseteq C$. It is known that $\mathrm{GK}(AC^2A) = \mathrm{GK}(C)$ in this case, cf. [1]. Since $\hat{U}H^2\hat{U} = \hat{U}$, in view of Theorem 7.1 we also come to $\mathrm{GK}(T) = \mathrm{GK}(H) = \mathrm{GK}(\hat{U})$.

The following result, relying on the lemma above, extends the ideas of Theorem 7.3 to certain classes of linear semigroups of particular interest.

Theorem 7.8 *Assume that the maximal subgroups of $M_n(K)$ intersecting a finitely generated semigroup $S \subseteq M_n(K)$ are contained in finitely many $\mathcal{R}$-classes of $M_n(K)$. Then the following conditions are equivalent*

1. *S has polynomial growth,*

2. *every nonempty intersection $S \cap G$ with a maximal subgroup G of $M_n(K)$ has finite Gelfand - Kirillov dimension,*

3. *S has no free noncommutative subsemigroups,*

4. *every group associated to S is finitely generated and almost nilpotent.*

Moreover, in this case $\mathrm{GK}(S)$ is bounded by a function $f(n, r)$, where $r = \max_T\{\mathrm{GK}(T)\}$ for T running over the set of all nonempty intersections $S \cap G$.

Proof. It is clear that 1) $\Rightarrow$ 2) and 2) $\Rightarrow$ 3), cf. Proposition 3.7. The implication 3) $\Rightarrow$ 4) follows Theorem 6.11 and from Proposition 3.16. Thus, assume that 4) holds. From Corollary 3.8 we know that S has an ideal chain $S = I_1 \supset \cdots \supset I_k$ such that, if $T_j = I_j/I_{j+1}$ for $j = 1, \ldots, k-1$, and $T_k = I_k$, then every T_i either is a uniform subsemigroup of a completely 0-simple semigroup $J_i \subseteq M_r/M_{r-1}$ for some r, or it is power nilpotent of index not exceeding $\binom{n}{q}$ for some $q \le n$, and $k < 2^n + n$. Moreover, in the former case the maximal subgroup of J_i is isomorphic to the group of quotients of every maximal nonzero cancellative subsemigroup of T_i (note that 4) implies that these semigroups have groups of quotients). The hypothesis implies that J_i has finitely many $\mathcal{R}$-classes. From Lemma 3.11 we know that $J_i \cup (S \setminus I_i)$ has a natural semigroup structure (as a Rees factor of a semigroup containing S) and contains S/I_{i+1} as a subsemigroup. By induction on k we show that $\mathrm{GK}(S) < \infty$.

If $k = 1$, then S is power nilpotent or uniform. In the former case 1) is obviously satisfied, while in the latter case 1) follows from Corollary 7.7. Assume now that $k > 1$. By the induction hypothesis $\mathrm{GK}(S/I_{k-1}) < \infty$ because it has a shorter ideal chain with the required properties. If S is power nilpotent, then the fact that $K_0[S/I_{k-1}] = K_0[S]/K_0[I_{k-1}]$ implies in view of [67], Corollary 5.10, that $\mathrm{GK}(S) < r \, \mathrm{GK}(S/I_{k-1})$, where r is the nilpotency index of I_{k-1}. Otherwise I_{k-1} is uniform, so $\mathrm{GK}(S) < \infty$ by Lemma 7.6.

The remaining assertion is a direct consequence of the proof. $\square$

For the basic facts on rings with Krull dimension, we refer to [81].

Corollary 7.9 *Assume that $S \subseteq M_n(K)$ is (Malcev) nilpotent or the algebra $K[S]$ has right Krull dimension. Then S has polynomial growth if and only if every nonempty intersection $S \cap G$, for a maximal subgroup G of $M_n(K)$, has finite Gelfand - Kirillov dimension.*

Proof. Let R be an $\mathcal{R}$-class of $M_n(K)$ such that $S \cap R$ is contained in a uniform component U of S. We know that there exists an ideal I of S such that U is an ideal of S/I. Then $K_0[R \cap S]$ is a right ideal of $K_0[S/I]$. If $K[S]$ has right Krull dimension, the same is true of $K_0[S/I]$. In particular, $K_0[S/I]$ has finite right Goldie dimension. It follows that U intersects finitely many $\mathcal{R}$-classes of $M_n(K)$. If S nilpotent, then the latter follows from Theorem 5.18. Thus, the assertion is a consequence of Theorem 7.8. $\square$

We conclude with a general observation on linear semigroups of polynomial growth. Recall that the unipotent radical of a linear group G is the largest normal subgroup of G consisting of unipotent matrices.

Proposition 7.10 *Assume that $S \subseteq M_n(K)$ is a finitely generated semigroup of polynomial growth. Then*

1. *S satisfies an identity,*

2. *there exists a finitely generated subring $R \subseteq K$ such that $\mathrm{cl}(S) \subseteq M_n(R)$,*

3. *groups associated to S are almost nilpotent and finitely generated modulo unipotent radical.*

Proof. 1) is a direct consequence of Theorem 6.11.

Let G be the subgroup of $M_n(K)$ generated by a nonempty intersection $S \cap D$ with a maximal subgroup D of $M_n(K)$. From Theorem 7.1 we know that $\mathrm{GK}(G) < \infty$ and G is almost nilpotent by Theorem 6.11. Let H be a nilpotent normal subgroup of G of finite index. By induction on the nilpotency class of H we show that G has a finitely generated subgroup N such that a power of every $g \in G$ lies in N. Let A be a finitely generated subgroup of $Z(H)$ of maximal torsion-free rank. If H is abelian, then put $N = A$. Now H, N have the same rank, so that H is torsion modulo N and the claim follows. Otherwise, by the induction hypothesis $H/Z(H)$ has a finitely generated subgroup $B/Z(H)$ with the desired property. Let N be the group generated by $b_1, \dots, b_t, A$, where $b_1, \dots, b_t, Z(H)$ generate B. It is clear that for every $h \in H$ there exists $k \geq 1$ with $h^k \in N$. Thus, the claim follows.

Since G is the group generated by $S \cap D$, it follows that $N = \mathrm{gp}(S \cap N)$, cf.[87], Lemma 7.5. Hence, we can find generators $u_1, \dots, u_k$ of the group N such that the inverses v_j of u_j in G lie in $S \cap D$. Then the semigroup $U = \langle S \cap D, u_1, \dots, u_k \rangle$ contains N. This easily implies that $G \subseteq U \subseteq M_n(R)$, where $R \subseteq K$ is the subring generated by the entries of (finitely many) generators of S, and the entries of $u_1, \dots, u_k$.

Now, in view of the remark following Corollary 3.11, $\mathrm{cl}(S)$ can be constructed from S by a finite chain of extensions, each obtained from the previously constructed $T \supseteq S$ by inverting the elements of $T \cap M$ for a group M associated to T. The above shows that we only need to invert finitely many elements in each step. Therefore 2) follows.

Let U be the unipotent radical of G. From [134], Theorem 3.6, we know that G has a normal subgroup F of finite index which is triangularizable in $M_n(\overline{K})$. So $u^{-1}Fu$ is triangular for some $u \in GK_n(\overline{K})$ and $u^{-1}Fu \subseteq M_n(R_1)$ for the subring R_1 of $\overline{K}$ generated by R, u and u^{-1}. Therefore $F/U \simeq (u^{-1}Fu)/(u^{-1}Uu)$ embeds into $(R_1^*)^m$ for the group of units R_1^* of R_1, where m is the rank of matrices in G. Since R_1 is finitely generated, R_1^* is a finitely generated group, cf. [58], Corollary 4.10. Therefore assertion 3) follows. $\square$

7.2   Linear growth and repetitivity

In view of Theorem 7.1 and of the recurrent behaviour of sequences of matrices (see Theorem 1.16 and the comment following it), one might expect that $GK(S)$ either is a natural number or it is infinite, for every $S \subseteq M_n(K)$. It seems that, at this stage, one cannot hope for a description of semigroups $S \subseteq M_n(K)$ of a given natural Gelfand - Kirillov dimension r. However, this is possible in the distinguished case $r = 1$, in which we say that S has linear growth. In this section we give such a criterion and compare it to a combinatorial property for S.

Let S be a semigroup, and let $s = s_1, \ldots, s_m$ be a sequence of elements of S. A k-factorization of s is a sequence $t = t_1, \ldots, t_k$, where the t_j's are the values of k consecutive factors of s, that is, $t_j = s_{i_j} s_{i_j+1} \cdots s_{i_{j+1}-1}$ for $j = 1, \ldots, k$ and some $1 \le i_1 < i_2 < \cdots < i_{k+1} \le m + 1$. We say that t is a power k-factorization if $t_1 = \cdots = t_k$. A semigroup S is said to be repetitive if and only if, for each finite subset X of S and every integer $k > 0$, there exists a positive integer $L = L(S, X, k)$ such that every sequence $s_1, \ldots, s_L$ of elements of X has a power k-factorization.

Repetitive semigroups were introduced by Justin in [51]. He then developed the theory in a series of papers. For example, a well-known corollary to Ramsey's theorem implies that every finite semigroup is repetitive (cf. [73, §4.1]).

Our first main result (Theorem 7.17) describes all repetitive linear semigroups, generalizing [55] and [63]. It allows us to construct examples of finitely generated repetitive linear semigroups of subexponential growth, and also of polynomial but not linear growth. Our second main result (Theorem 7.20) describes all semigroups $S \subseteq M_n(K)$ of Gelfand-Kirillov dimension one. Such semigroups turn out to be repetitive. Further, we show that if we impose an additional structural restriction on a semigroup $S \subseteq M_n(K)$, then repetitivity becomes equivalent to sublinear growth (Corollary 7.22).

Clearly, the class of repetitive semigroups is closed for subsemigroups and for homomorphic images. Also, a semigroup is repetitive if and only if all its finitely generated subsemigroups are repetitive.

Lemma 7.11 *Let I be an ideal of a semigroup S. Then S is repetitive if and only if S/I and I are repetitive.*

Proof. Choose a finite subset X of S. For any $k \geq 1$, let $m = L(S/I, X, k)$. Next, put $Y = X^m \cap I, n = L(I, Y, k)$ and $L = mn$. We claim that L satisfies the definition of $L(S, X, k)$. Consider a sequence $s = s_1, \ldots, s_L$ of elements of X and its factorization $u = u_1, \ldots, u_n$, where $u_i = s_{(i-1)m+1} \cdots s_{im}$ for $i = 1, \ldots, n$. If $u_1, \ldots, u_n \in I$, then u has a power k-factorization by the choice of n. Otherwise, $u_i \notin I$ for some i and therefore u_i has a power k-factorization by the choice of m. The latter is a power k-factorization of s as well. Hence S is repetitive. $\square$

Lemma 7.12 *Let U be a uniform subsemigroup of a completely 0-simple semigroup M. Then U is repetitive if and only if $U \cap G$ is repetitive for every maximal subgroup G of M.*

Proof. The necessity is clear. Thus, assume that every $U \cap G$ is repetitive. Since repetitivity of U can be verified on finitely generated subsemigroups of U, it is enough to consider the case where the sets of $\mathcal{R}$- and $\mathcal{L}$-classes of M are finite. Then $U = V_1 \cup \cdots \cup V_k$, where each V_i is the intersection of U with a principal left ideal of M. Moreover, each $V_i = W_{i1} \cup \cdots \cup W_{im}$, where W_{ij} is the intersection of V_i with a principal right ideal of M. Hence $W_{ij} \setminus \{\theta\}$ either is equal to some $U \cap G$ or $W_{ij}^2 = \theta$. So W_{ij} is repetitive. Therefore, it is enough to show that a semigroup S which is a finite union of its repetitive right (left) ideals $R_i, i = 1, \ldots, r$, is itself repetitive. By induction, it is enough to consider the case $r = 2$. By symmetry, we consider the right ideals only.

Let X be a finite subset of U. Put $X_i = X \cap R_i, i = 1, 2$. Let $m = L(R_2, X_2, k)$ and $Y_1 = X_1 \cup X_1 X^m$. Since Y_1 is contained in the repetitive right ideal R_1, the number $L(R_1, Y_1, k)$ can be defined and we put $n = m(L(R_1, Y_1, k) + 1)$. We claim that n can serve as $L(U, X, k)$. Let $s = s_1, \ldots, s_n$ be a sequence of elements of X. Let $s_{i_1}, \ldots, s_{i_l}, i_1 < \cdots < i_l$, be the subsequence of all elements of s that are in X_1. If s has m consecutive elements which all belong to X_2, then the factor formed by these elements has a power k-factorization by the choice of m. Otherwise, every factor of s of length m contains an element of X_1. Then $i_1 \leq m$ and $i_{j+1} \leq i_j + m$ for all $j = 1, \ldots, l - 1$. Hence $n < (l+1)m$ and by the choice of n we get $l > L(R_1, Y_1, k)$. Since R_1 is a right ideal, all elements $u_j = s_{i_j} \cdots s_{i_{j+1}-1}$ of the factorization $u = u_1 \cdots u_{l-1}$ belong to Y_1. Therefore s has a power k-factorization because $l - 1 \geq L(R_1, Y_1, k)$. The assertion follows. $\square$

Lemma 7.13 *Assume that H is a subgroup of finite index in a group G. If H is repetitive, then G is repetitive.*

Proof. Since H contains a normal subgroup of G of finite index, we may assume that H is a normal subgroup of G. Let X be a finite subset of G and $k \geq 1$. Consider an infinite word $x_1 x_2 \ldots$ with $x_i \in X$. From Corollary 1.15 it follows that there exists $p \geq 1$ such that for every $q \geq 1$ there exist $i_1 < i_2 < \cdots < i_{q+1}$ such that $i_{j+1} - i_j \leq p$ for $j = 1, \ldots, q$ and the consecutive factors $u_j = x_{i_j+1} \cdots x_{i_{j+1}}, j = 1, \ldots, q$, are in H. Now, let A be the set of all elements of G which are words of length at most p in the elements of X. Choose $q = L(H, A, k)$. By the choice of q we know that $u_1 \cdots u_q$ has a factor (as a word in u_i) which is a k-power. Hence $x_1 \cdots x_{i_{q+1}}$ has such a factor. The assertion follows from Lemma 1.13. $\square$

The next two fundamental lemmas are due to Justin [53], [54].

Lemma 7.14 *If a commutative semigroup S is repetitive, then S does not contain free commutative subsemigroups of rank two.*

Proof. Let $X = \langle x, y \rangle^1$ be the free monoid on x, y. Let A be the commutative free monoid on x, y and $\alpha : X \longrightarrow A$ the natural map. The idea is to construct arbitrarily long words w in x, y such that $\alpha(w) \in A$ has no power 5-factorization.

For a word $x_1 \cdots x_n \in X$ we denote by $w[i,j]$ the factor $x_i \cdots x_j$, where $i < j$. Consider the homomorphism $\phi : X \longrightarrow A$ given by

$$\phi(x) = xxxxy, \ \phi(y) = xyyyy.$$

Let $g \in X$ and $f = \phi(g)$. Suppose that $\alpha(f)$ contains a 5-power, say, $f_1 f_2 f_3 f_4 f_5$ with $f_u = f[i_u + 1, i_{u+1}]$ are such that $\alpha(f_1) = \cdots = \alpha(f_5)$. Let $i_u = 5r_u + j_u$, where $0 \le j_u < 5$, and let z_u be the $r_u + 1$-th letter of g, for $u = 1, \ldots, 6$. Write $S_u = \phi(z_u)$ and $W_u = S_u[1, j_u]$ (with $W_u = 1$ if $j_u = 0$). Consider the homomorphism $\eta : X \longrightarrow \mathbf{Z}_3$ given by $\eta(x) = 1, \eta(y) = 2$. Define $d_u = \eta(W_u)$.

Now $i_{u+1} - i_u = j_{u+1} - j_u \pmod 5$ for $i = 1, 2, 3, 4, 5$, and $i_{u+1} - i_u$ is equal to the length of f_u. Since all f_u have equal length (as they are equal modulo α), we come to

$$j_{u+2} - j_{u+1} = j_{u+1} - j_u \pmod 5 \text{ for } u = 1, 2, 3, 4.$$

Suppose one of $j_{u+1} - j_u, u = 1, 2, 3, 4, 5$, is nonzero. Then it is easy to see that $j_1 = j_6$, a contradiction. Therefore $j_u = j$ for some j.

Since all $\alpha(f_u)$ are equal, also all $\eta(f_u)$ must be equal. Therefore $\eta(f[1, i_u]) + \eta(f_u) = \eta(f[1, i_{u+1}])$ implies that $\eta(f[1, i_u]) - \eta(f[1, i_{u+1}]) \in \mathbf{Z}_3$ does not depend on u. Since $\eta\phi(x) = 0$ and $\eta\phi(y) = 0$, it follows that $\eta(f[1, i_u]) = \eta(w_u) = d_u$. This leads to

$$d_{u+2} - d_{u+1} = d_{u+1} - d_u \pmod 3 \text{ for } u = 1, 2, 3, 4.$$

Since $\eta(xx) \ne \eta(xy)$ and $\eta(xxx) \ne \eta(xyy)$, it follows that either all z_u are equal or $j \ne 2, 3$. So, either $j = 0, j = 1, j = 4$, or all z_u are equal. In all these cases, if $u \in \{1, \ldots, 6\}$, we put $i'_u = i_u + t$ with $t = -j$ if $j \ne 4$, or $t = 1$ if $j = 4$, and we define $f'_u = f[i'_u + 1, i'_{u+1}]$. It is easy to see that in every case $f'_1 f'_2 f'_3 f'_4 f'_5$ is a 5-power modulo α. Moreover, since $i'_u = 0 \pmod 5$, we have $f'_u = \phi(g_u)$, where g_u are consecutive factors of g. But $\langle \alpha(x), \alpha(y) \rangle \subseteq A$ is a free commutative subsemigroup, so $\alpha\phi(w)$ determines uniquely $\alpha(w)$ for any $w \in X^1$. Hence, the factor $g_1 g_2 g_3 g_4 g_5$ of g is a 5-power modulo α.

Now, put $h_1 = x$ and $h_{n+1} = \phi(h_n)$ for $n > 1$. We have shown that every $\alpha(h_n) \in A$ has no power 5-factorization. The assertion follows. $\square$

The next result includes the fact that the infinite cyclic group is repetitive, which is a generalization of the van der Waerden theorem on arithmetic progressions (see [73]).

Lemma 7.15 *Every almost cyclic group is repetitive.*

Proof. By Lemma 7.13 it is enough to consider the case of the infinite cyclic group $\mathbf{Z}$. Let $T \subseteq \mathbf{Z}$ be a finite subset and k any natural number. Put $m = \sup(T)$ and $m' = \inf(T)$. Let X be a free semigroup of rank $|T|$ and $\phi : X \longrightarrow \mathbf{Z}$ a homomorphism which is one-to-one on the generators of X. Assume first that T consists of positive integers. Let Y be the free semigroup on $y_1, \ldots, y_m$. Define a homomorphism $\psi : X \longrightarrow Y$ by

$$\psi(x) = y_t \cdots y_1 \text{ if } x \in X \text{ and } \phi(x) = t.$$

It is clear that $\phi(f)$ is the length of $\psi(f)$, for every $f \in X$. Let $w \in X$, with $\psi(w) = z_1 \cdots z_t$, $z_j \in Y, 1 \leq j \leq t$. For $1 \leq j \leq t$ let w_j denote the shortest initial factor of w such that $\phi(w_j) \geq j$. If $z_j = y_p$, we have

$$\phi(w_j) = j + p - 1.$$

Applying van der Waerden theorem ([73]) to the word $\psi(w) \in Y$ we see that there exists an integer $V = V(m, k + 1) \geq 1$ such that for $t \geq V$ there exist $j_1 < \ldots < j_{k+1}$, some $r \geq 1$ and some y_p such that

$$
\begin{aligned}
j_{s+1} - j_s &= r & 1 \leq s \leq k \\
z_{j_s} &= y_p & 1 \leq s \leq k+1
\end{aligned}
$$

Next, we define factors h_s of w by

$$w_{j_{s+1}} = w_{j_s} h_s \text{ for } 1 \leq s \leq k.$$

From the displayed equalities it follows that

$$\phi(h_s) = (j_{s+1} + p - 1) - (j_s + p - 1) = r.$$

Hence, w has k consecutive factors h_s whose images under ϕ are equal if $t \geq V$. Since $t \geq m'|w|$, the latter holds if the length of w exceeds $\frac{V}{m'} + 1$.

Now, consider the general case, where $A \subseteq \mathbf{Z}$ is any finite subset. Consider an infinite word S in the generators of X. By Lemma 1.13 it is enough to show that it has k consecutive factors with equal images under ϕ. Write $S = f_i S_i$, where f_i is the initial factor of S of length i, for $i \geq 1$. Assume first that $\phi(f_i)$ are bounded. Then there are infinitely many indices $i_1 < i_2 < \cdots$ such that $\phi(f_{i_j}) = \phi(f_{i_{j+1}})$ for all $j \geq 1$. If the words h_j are defined by $f_{i_j} h_j = f_{i_{j+1}}, j = 1, 2, \ldots$, then we have $\phi(h_j) = 0$. Since h_j are consecutive factors of S, we are finished. Therefore, suppose that $\phi(f_i)$ are not bounded. Then S can be factorized as

$$S = g_1 g_2 \cdots$$

where $g_j \in X$ is such that $1 \leq \phi(g_j) \leq m = \sup(A)$. In fact, it is enough to put $S_1 = S$ and $S_j = g_j S_{j+1}$ where g_j is the shortest initial factor of S_j such that $\phi(g_j) \geq 0$. Now, let Z be the free semigroup on the set $\{z_1, z_2, \ldots\}$ and $\eta : Z \longrightarrow X$ be defined by $\eta(z_i) = g_i$ for $i \geq 1$. Then the homomorphism $\psi = \phi \eta : Z \longrightarrow \mathbf{Z}$ satisfies $\psi(z_i) \geq 1$. Therefore, by the first part of the proof, the infinite word $z_1 z_2 z_3 \cdots$ has k consecutive factors $v_1, \ldots, v_k$ with equal images under ψ. Then $h_i = \eta(v_i), i = 1, \ldots, k$, are consecutive factors of S with equal images under ϕ. This completes the proof of the lemma. $\square$

Recall that a semigroup S is divided by a semigroup T if T is a homomorphic image of a subsemigroup of S.

Lemma 7.16 *Let G be an almost nilpotent group, and let S be a finitely generated subsemigroup of G. If S is not divided by a free commutative semigroup of rank 2, then the group of quotients SS^{-1} is almost cyclic.*

Proof. Since SS^{-1} is also almost nilpotent, we may assume that $G = SS^{-1}$, so G is finitely generated. Let N be a nilpotent normal subgroup of finite index in G. Then [87], Lemma 7.5, implies that $(S \cap N)(S \cap N)^{-1} = N$, because $[G : N] < \infty$. Also, N is finitely generated. Denote by n the nilpotency class of N. We proceed by induction on n.

If $n = 1$, then N is an abelian group. By the hypothesis, S has no free commutative subsemigroups of rank two. Therefore $S \cap N$ is of torsion-free rank ≤ 1. By [87], Proposition 23.1, $\mathrm{rk}(N) = \mathrm{rk}(S \cap N)$. A finitely generated abelian group of torsion-free rank ≤ 1 is almost cyclic. Hence G is almost cyclic.

Suppose that $n > 1$. Consider $N/Z(N)$, where $Z(N)$ is the center of N. Since the nilpotency class of $N/Z(N)$ is less than n, the factor $N/Z(N)$ is almost cyclic. Choose a normal subgroup A of N such that $Z(N) \subseteq A$, N/A is finite and $A/Z(N)$ is cyclic. In particular, the index of A in G is finite. There exists an element x in A such that A is generated by x and $Z(N)$. Hence A is abelian. As above it follows that A is almost cyclic. Therefore G is almost cyclic. $\square$

Now we can state the first main theorem of this section, [62].

Theorem 7.17 *The following conditions are equivalent for a linear semigroup S*

1. *S is repetitive,*

2. *S is not divided by a free commutative semigroup of rank 2,*

3. *all the associated groups are locally almost cyclic.*

Proof. Let $S \subseteq M_n(K)$ be a semigroup which is not divided by a free commutative semigroup of rank 2. Take any maximal subgroup D of $M_n(K)$ intersecting S, and consider the associated linear group $G = \mathrm{gp}(D \cap S) \subseteq D$. Certainly S does not contain free noncommutative subsemigroups of rank two. If we look at any finitely generated cancellative subsemigroup T of G, then it follows from the generalised Tits alternative, Theorem 6.11, that the group TT^{-1} is almost nilpotent. Hence it is almost cyclic by Lemma 7.16. Since every subgroup H of G generated by a finite number of elements $h_1 = s_1 t_1^{-1}, \ldots, h_k = s_k t_k^{-1}$, where $s_1, t_1, \ldots, s_k, t_k \in D \cap S$, is contained in the group of quotients of the cancellative subsemigroup of S generated by $s_1, t_1, \ldots, s_k, t_k \in S$, we see that G is locally almost cyclic.

Suppose now that S is a linear semigroup such that all associated groups of S are locally almost cyclic. By Corollary 3.8 S possesses a finite ideal chain such that each factor is nilpotent or is contained in a completely 0-simple semigroup H with maximal subgroup isomorphic to a group associated to S. Clearly, all nilpotent factors are repetitive. So, consider a factor of the latter type. Every almost cyclic group is repetitive by Lemma 7.15. Therefore all associated linear groups of S are repetitive. From Lemma 7.12 it follows that the corresponding semigroup H is repetitive, too. Since the class of repetitive semigroups is closed for ideal extensions by Lemma 7.11, we see that S is repetitive.

Finally, if S is repetitive, then S is not divided by the free commutative semigroup of rank 2 by Lemma 7.14. This completes the proof of the theorem. $\square$

Note that the 'if' part of Theorem 7.17 can be also proved via combinatorial methods developed in [42].

There exist repetitive linear semigroups whose growth is not even polynomial. Namely, the semigroup $S \subseteq M_3(\mathbf{Z})$, considered in Section 7.1, generated by the matrices

$$e = \begin{pmatrix} 1 & 0 & 0 \\ 0 & 0 & 0 \\ 0 & 0 & 1 \end{pmatrix} \quad \text{and} \quad g = \begin{pmatrix} 1 & 1 & 1 \\ 0 & 2 & 1 \\ 0 & 0 & 1 \end{pmatrix}$$

is of this type. In fact, S is repetitive by Theorem 7.17.

On the other hand, there exist repetitive semigroups $S \subseteq M_n(K)$ with polynomial but nonlinear growth.

Example Denote by S the subsemigroup generated in $M_2(\mathbf{Z})$ by the elements

$$a = \begin{pmatrix} \alpha & 0 \\ 0 & 0 \end{pmatrix}, \quad b = \begin{pmatrix} 0 & 0 \\ 0 & \beta \end{pmatrix} \quad e = \begin{pmatrix} 0 & 1 \\ 0 & 0 \end{pmatrix},$$

where $\alpha, \beta \in \mathbf{Z}$ are relatively prime. It is readily verified that the elements $a^n e b^m$, for $m, n = 1, 2, \dots$, are all different. Therefore $\mathrm{GK}(S) \geq 2$. Since $S = \{a^n e b^m \mid m, n \geq 0\} \cup \{a^n, b^n \mid n \geq 1\}$, it is easy to see that in fact $\mathrm{GK}(S) = 2$. Note that S is contained in a union of four $\mathcal{H}$-classes of $M_2(\mathbf{Q})$. All the associated groups are cyclic infinite, so that S is repetitive.

The next two examples show that it is impossible to replace 'locally almost cyclic' by 'almost cyclic' in Theorem 7.17, even if the whole semigroup is finitely generated.

Example There exist finitely generated repetitive linear semigroups whose associated groups are not almost cyclic. Namely, denote by S the subsemigroup generated in $M_3(\mathbf{Z})$ by

$$e = \begin{pmatrix} 1 & 0 & 0 \\ 0 & 0 & 0 \\ 0 & 0 & 1 \end{pmatrix} \quad \text{and} \quad g = \begin{pmatrix} 1 & 1 & 1 \\ 0 & 1/2 & 1 \\ 0 & 0 & 1 \end{pmatrix}.$$

Then $g^n = \begin{pmatrix} 1 & a_n & b_n \\ 0 & 1/2^n & a_n \\ 0 & 0 & 1 \end{pmatrix}$, where for $n \geq 0$

$$
\begin{aligned}
a_n &= a_{n-1} + 1/2^{n-1} \\
&= 1 + 1/2 + \cdots + 1/2^{n-1} \\
&= 2 - 1/2^{n-1}
\end{aligned}
$$

and

$$
\begin{aligned}
b_n &= b_{n-1} + a_{n-1} + 1 \\
&= a_{n-1} + a_{n-2} + \cdots + a_2 + a_1 + b_1 + n - 1 \\
&= (2 - 1/2^{n-2}) + (2 - 1/2^{n-3}) + \cdots + (2 - 1) + 1 + (n - 1) \\
&= 2(n - 1) + n - (1 + 1/2 + \cdots + 1/2^{n-2}) \\
&= 3n - 2 - (2 - 1/2^{n-2}) = 3n - 4 + 1/2^{n-2}.
\end{aligned}
$$

Denote by D the maximal subgroup of $M_3(\mathbf{Q})$ containing e. Put $G = (D \cap S)(D \cap S)^{-1}$. Then

$$
eg^n e \in D \cap S \subseteq \left\{ \begin{pmatrix} 1 & 0 & * \\ 0 & 0 & 0 \\ 0 & 0 & 1 \end{pmatrix} \right\}.
$$

Clearly, G is a torsion-free abelian group.

Consider the associated linear groups of S. One of them is the cyclic group generated by g. It contains all elements of rank three. Further, for any n, there exists k such that kb_n is an integer. Therefore

$$
(eg^n e)^k = \begin{pmatrix} 1 & 0 & kb_n \\ 0 & 0 & 0 \\ 0 & 0 & 1 \end{pmatrix}
$$

belongs to the cyclic group generated by $\begin{pmatrix} 1 & 0 & 1 \\ 0 & 0 & 0 \\ 0 & 0 & 1 \end{pmatrix}$. It follows that the set of all matrices of rank two in S forms a uniform component of S with associated group G being abelian and a periodic extension of a cyclic group. Thus S is repetitive by Theorem 7.17.

If G were finitely generated, then it would be cyclic, because it is a torsion-free abelian group, which is a periodic extension of a cyclic group. But $\{3n - 4 + 1/2^{n-2} \mid n \geq 1\} \subseteq \mathbf{Q}$ is not contained in any set of the form $a\mathbf{Z}$, for $a \in \mathbf{Q}$. Hence G is not cyclic, and so it is not finitely generated.

A similar example can also be given in the case of positive characteristic.

Example Let p be a prime, $\mathbf{F}_p$ a finite field of p elements, and let S be the subsemigroup generated in $M_3(\mathbf{F}_p(x))$ by

$$e = \begin{pmatrix} 1 & 0 & 0 \\ 0 & 0 & 0 \\ 0 & 0 & 1 \end{pmatrix} \text{ and } g = \begin{pmatrix} 1 & 1 & 0 \\ 0 & x & 1 \\ 0 & 0 & 1 \end{pmatrix}.$$

As above we can verify that, up to isomorphism, the associated groups of S are the cyclic group generated by g and a p-group G consisting of matrices of the form

$$\begin{pmatrix} 1 & 0 & * \\ 0 & 0 & 0 \\ 0 & 0 & 1 \end{pmatrix}.$$

Therefore S is repetitive. However, it is easy to see that G is not finite, hence not almost cyclic.

Lemma 7.18 *Let G be a linear group, and let $G = TT^{-1}$ for a semigroup T. Then the following are equivalent:*

1. *G is repetitive,*

2. *$GK(G) \leq 1$,*

3. *for every $x, y \in T$ there exist integers i, j such that $(i, j) \neq (0, 0)$ and $x^i = y^j$.*

Proof. The equivalence of 1) and 2) follows from Theorem 7.17 and Theorem 7.1.

If G is a group with $GK(G) \leq 1$, then for every $x, y \in G$ there exist $i, j, k, \ell \geq 1$ with $x^i y^j = x^k y^\ell$ and $(i, j) \neq (k, \ell)$. Therefore 3) is a consequence of 2).

Assume that 3) holds. To prove 1) it is enough to consider the case where G is finitely generated. Condition 3) clearly implies that

T is not divided by the free commutative semigroup of rank 2. From Theorem 6.11 it follows that G is almost nilpotent. By Lemma 7.16 G is almost cyclic. Therefore Lemma 7.15 implies that G is repetitive. $\square$

We will use a special property of semigroups of linear growth, [67], the proof of Lemma 2.4, or [52]. If $S = \langle s_1, \ldots, s_k \rangle$, then let $\alpha : X \longrightarrow S$ be the homomorphism of the free semigroup on $x_1, \ldots, x_k$ extending the map $s_i \mapsto x_i, i = 1, \ldots, k$. We equip X with the order $\prec$ defined by

$v \prec w$ if $|v| < |w|$ or $|v| = |w|$ and v precedes w lexicographically.

For each $s \in S$, $\alpha^{-1}(s)$ contains a unique word v which is minimal with respect to $\prec$. Then $\alpha(v)$, viewed as a word in $s_1, \ldots, s_k$, is called the normal form of s.

Proposition 7.19 *The following conditions are equivalent for a finitely generated semigroup* $S = \langle s_1, \ldots, s_k \rangle$

1. *S has linear growth,*

2. *there exists a finite set $F \subseteq S^1$ such that $S \subseteq A$, where $A = \{uz^\ell v \mid u, v, z \in F, z \neq 1, \ell \geq 1\}$. Moreover, if $s = s_{i_1} \cdots s_{i_m} \in S$ is the normal form of s, then some $uz^\ell v$ is a factorization of $s_{i_1} \cdots s_{i_m}$, for u, v, z and ℓ as above. In particular, $m \geq l$.*

Now we are ready to describe all semigroups $S \subseteq M_n(K)$ with linear growth. This involves a simple extension of condition 3) of Lemma 7.18. The result comes from [62].

Theorem 7.20 *Let $S \subseteq M_n(K)$ be a semigroup. Then the following conditions are equivalent:*

1. *$\mathrm{GK}(S) \leq 1$,*

2. *for every $x, y, z \in S$, there exist positive integers i, j, k, ℓ such that $x^i y z^j = x^k y z^\ell$ and $(i, j) \neq (k, \ell)$.*

If, in addition, S is finitely generated, then the groups associated to S are almost cyclic.

Proof. Since the set of words $\{x^i y z^j \mid i, j \geq 0\}$ in the alphabet x, y, z has $n(n+1)/2$ words of length $\leq n$, it is clear that 2) is a consequence of 1).

Now assume that S satisfies condition 2). It is enough to consider the case where S is generated by a finite number of elements, say $s_1, \ldots, s_k$. We can assume that S is a semigroup containing the zero matrix, since otherwise we could adjoin it to S. We know that S has an ideal chain $0 = I_0 \subseteq I_1 \subseteq \cdots \subseteq I_\ell = S$, where each I_{j+1}/I_j is a subsemigroup of a principal factor M_i/M_{i-1} of $M_n(K)$ and it is either nilpotent or isomorphic to a uniform component of S. By induction on m we shall show that $\mathrm{GK}(S/I_{\ell-m}) \leq 1$ for $m = 0, 1, \ldots, \ell$. This will prove 1).

The case $m = 0$ is clear. Suppose that $\mathrm{GK}(S/I_{j+1}) \leq 1$ for some $j < \ell$.

Case 1. I_{j+1}/I_j can be identified with a uniform component U of S. As in the proof of Theorem 7.8 S/I_j can be considered as a subsemigroup of a disjoint union $\widehat{S} = (S \setminus I_{j+1}) \cup \widehat{U}$, where $\widehat{U}$ is a completely 0-simple ideal of $\widehat{S}$ and U intersects all nonzero $\mathcal{H}$-classes of $\widehat{U}$.

Given that $\mathrm{GK}(S/I_{j+1}) \leq 1$, from Proposition 7.19 it follows that there exists a finite set $F \subseteq S^1$ such that $S \setminus I_{j+1} \subseteq A$, where $A = \{u z^\ell v \mid u, z, v \in F, \ell \geq 1\}$. Moreover, if $s = s_{i_1} \cdots s_{i_m} \in S \setminus I_{j+1}$, then $s = u z^\ell v$ is a word of length $\leq m$ in $s_1, \ldots, s_k$.

Suppose that $\widehat{U} = M(G, X, Y, P)$ for infinite sets X, Y. Take any element $s \in U$, say $s = s_{i_1} \cdots s_{i_m} \neq \theta$. Denote by b the shortest terminal factor of s that lies in U. Then $s = ab$ and $b = s_{i_t} c$ for some t and some $c \in (S \setminus I_{j+1})^1$. Hence $b \in s_{i_t} A$. Therefore all possible b belong to the set

$$A' = \bigcup_{i=1}^{k} s_i A = \{ u'(z')^\ell v' \mid u', z', v' \in F', \ell \geq 1 \}$$

for a finite set $F' \subseteq S^1$. Moreover $b = s_{i_t} u z^\ell v$ with $u, z, v \in F$ and $\ell \leq m$. Note that s and b generate the same left ideals in $\widehat{U}$. In particular, there exists N such that the elements of U of length at most m lie in at most Nm $\mathcal{L}$-classes of $\widehat{U}$.

Similarly, the shortest initial factor e of s that is in $\widehat{U}$ comes from a set A'' of the same type (defined by a finite set $F'' \subseteq S^1$), and s, e generate the same right ideals in $\widehat{U}$.

This means that each of the sets X, Y is a union of finitely many subsets, each corresponding to one of the sets $\{ u z^\ell v \mid \ell \geq 1 \}$ as u, z, v

range over F', F'' respectively. The above shows also that every $\mathcal{H}$-class of $\hat{U}$ is $\mathcal{L}$-related to an element of

$$B = \{uz^{\ell}v \mid u, z, v \in E, \ell \geq 1\}$$

for the finite set $E = F \cup F' \cup F'' \subseteq S^1$, and is $\mathcal{R}$-related to an element of this set.

From Lemma 3.24 we know that there exists a finite subset Q of U such that for every $t \in U$ we have $qtq' \neq \theta$ for some $q, q' \in Q$. Clearly, $t\mathcal{R}tq'$ in $\hat{U}$ in this case. Since E and Q are finite, but the set X of $\mathcal{R}$-classes of $\hat{U}$ is assumed to be infinite, it follows that there exists $a \in Q$ such that $\{uz^{\ell}v \mid \ell \geq 1\}a$ intersects infinitely many $\mathcal{R}$-classes of $\hat{U}$ for some fixed $u, z, v \in E$. Similarly, since Y is infinite, $b\{cx^{\ell}d \mid \ell \geq 1\}$ intersects infinitely many $\mathcal{L}$-classes of $\hat{U}$ for some $b \in Q$ and some $c, d, x \in E$. Since there exists $f \in U$ such that $afb \neq \theta$ in $\hat{U}$ (because U intersects all nonzero $\mathcal{H}$-classes of $\hat{U}$), it follows that

$$\{uz^{\ell}vafbcx^{k}d \mid k, \ell \geq 1\}$$

intersects infinitely many $\mathcal{L}$-classes and infinitely many $\mathcal{R}$-classes of $\hat{U}$. The same must be true of the set $\{z^{\ell}yx^{k} \mid k, \ell \geq 1\}$, where $y = vafbc$. Note that $z^{\ell}yx^{k}\mathcal{R}z^{\ell}y$ and $z^{\ell}yx^{k}\mathcal{L}yx^{k}$ in $\hat{U}$ whenever $z^{\ell}yx^{k}$ is nonzero. If $z^{\ell}yx^{k} = \theta$ for some ℓ, k, then $z^{\ell'}yx^{k'} = \theta$ for all $\ell' \geq \ell, k' \geq k$, which contradicts that $\{z^{\ell'}yx^{k'} \mid \ell', k' \geq 0\}$ intersects infinitely many $\mathcal{R}$- and $\mathcal{L}$-classes of $\hat{U}$. Hence $z^{\ell}yx^{k} \neq \theta$ for every ℓ, k. The cyclic semigroup generated by z acts by left multiplication on the set of $\mathcal{R}$-classes of $\hat{U}$. Hence $z^{\ell}y\mathcal{R}z^{p}y$ if and only if $p = \ell$. Similarly, $yx^{k}\mathcal{L}yx^{r}$ if and only if $k = r$. Therefore all elements $z^{\ell}yx^{k}$, for $k, \ell = 1, 2, \ldots$, are different. This contradicts condition 2), proving that either $|X| < \infty$ or $|Y| < \infty$.

Lemma 7.18 and Proposition 3.16 show in view of Theorem 7.17 that G is almost cyclic.

Our next aim is to prove that G must be finite if one of the sets X and Y is infinite. For example, assume $|X| = \infty$. Suppose also that G is infinite. Then there exists a nonperiodic element $x \in U \cap H$ for a maximal subgroup H of $\hat{U}$. Again, we see that there exist $u, z, v \in S/I_j$, $a \in U$, such that $\{uz^{\ell}va \mid \ell \geq 1\}$ intersects infinitely many $\mathcal{R}$-classes of $\hat{U}$. Then the set $\{z^{\ell}va \mid \ell \geq 1\}$ also has this property. Choose $f \in U$ so that $afx \neq \theta$ in $\hat{U}$. As above, it follows that $z^{\ell}vafx^{k}, z^{p}vafx^{r}$ are not in the same $\mathcal{R}$-class of $\hat{U}$ if $p \neq \ell$. Therefore all elements $z^{\ell}vafx^{k}$, for $k, \ell \geq 1$, are different. This contradicts condition 2) and proves the claim.

Assume first that $\hat{U} = M(G, X, Y, P)$, where $|G| < \infty$ and $|X| < \infty$ (the case where $|Y| < \infty$ goes similarly). Then $U = \hat{U}$, because $H \cap S = H$ for any maximal subgroup H of $\hat{U}$ (since $H \cap S$ generates H). Let T be any $\mathcal{L}$-class of $\hat{U}$. Then $|T| \leq |X| \cdot |G| < \infty$.

As noted above, the number of $\mathcal{L}$-classes of $\hat{U}$ containing elements of length $\leq m$ in $s_1, \ldots, s_k$ is at most Nm. Therefore there are at most $|T|Nm$ elements of U that are words of length $\leq m$. Also, by the induction hypothesis, there is a linear bound on the number of elements of length $\leq m$ in $S \setminus I_{j+1}$. It follows that $\mathrm{GK}(S/I_j) \leq 1$, as desired.

Consider finally the second case where $\hat{U} = M(G, X, Y, P)$ and X, Y are finite. By Theorem 7.17 and Lemma 7.18, since G is almost cyclic, and from the induction hypothesis and from Lemma 7.6 it follows that

$$\mathrm{GK}(S) \leq \mathrm{GK}(\hat{S}) = \sup(\mathrm{GK}(S/I_{j+1}), \mathrm{GK}(G)) \leq 1,$$

proving the inductive claim again.

Case 2. We consider the case that $N = S_{j+1}/S_j$ is nilpotent, say $N^t = \theta$ where we may assume that $t \geq 2$. Consider any finite subset X of S^1. By 2),

$$(\exists K(X) \geq 1)(\forall x, y, z, w \in X)(\exists i, j, k, \ell)$$

$$1 \leq i, j, k, \ell \leq K(X), (i, j) \neq (k, \ell) \text{ and } x^i y w z^j = x^k y w z^\ell.$$

Define

$$X' = X \cup \{uz^p vw, uvz^p w \mid u, z, v, w \in X, p = 1, \ldots, K(X)\},$$

which is also finite. Let m be a positive integer. Using the technique of Case 1 (which relies on Proposition 7.19) there is a finite set $F \subseteq S^1$ such that if w is a word in the generators of length $\leq m$, which, when evaluated in S/S_j, does not lie in N, or lies in N but has no proper subword which evaluates to an element of N, then w lies in

$$A_1 = \{uz^\ell v \mid u, z, v \in F, 1 \leq \ell \leq m\}.$$

Put $F_1 = F$ and, for $i = 2, \ldots, t - 1$,

$$F_i = (F_{i-1})' \text{ and } A_i = \{uz^\ell v \mid u, z, v \in F_i, 1 \leq \ell \leq m\}$$

Note that $A_i \subseteq A_{i+1}$ for $i = 1, \ldots, t-2$.

Let $s = s_{i_1} \cdots s_{i_{m'}}$, where $m' \le m$. We prove that $s \in A_{t-1}$. If $s \in S \backslash S_{j+1}$ then $s \in A_1 \subseteq A_{t-1}$ and we are done. Suppose that $s \in N$. Clearly s can be written in the form $s = x_1 y_1$ where x_1 is the shortest initial factor of s which lies in N. If $y_1 = 1$ then $s = x_1$. Consider the case $y_1 \ne 1$. If $y_1 \notin N$ then put $x_2 = y_1$. If $y_1 \in N$ then $y_1 = x_2 y_2$ where x_2 is the shortest initial factor of y_1 which lies in N. Continuing in this way we obtain a decomposition

$$s = x_1 \cdots x_r \text{ for } 1 \le r < t$$

where, for $i = 1, \ldots, r$, no proper initial segment of x_i lies in N. Hence $x_1, \ldots, x_r \in A_1$. We prove by induction on $i = 1, \ldots, r$ that $x_1 \cdots x_i \in A_i$. The induction begins since $x_1 \in A_1$. Let $i > 1$. Suppose, as inductive hypothesis, that $x_1 \cdots x_{i-1} \in A_{i-1}$. Also $x_i \in A_1 \subseteq A_{i-1}$. Hence, for some $u_1, u_2, z_1, z_2, v_1, v_2 \in F_{i-1}$, and some ℓ_1, ℓ_2 such that $1 \le \ell_1, \ell_2 \le m$,

$$x_1 \cdots x_{i-1} = u_1 z_1^{\ell_1} v_1 \text{ and } x_i = u_2 z_2^{\ell_2} v_2.$$

Thus

$$x_1 \cdots x_i = u_1 z_1^{\ell_1} v_1 u_2 z_2^{\ell_2} v_2.$$

We may suppose that ℓ_1 and ℓ_2 have been chosen so that (ℓ_1, ℓ_2) is minimal with respect to the lexicographic order such that the previous equation holds for some $u_1, z_1, v_1, u_2, z_2, v_2 \in F_{i-1}$. But

$$z_1^i v_1 u_2 z_2^j = z_1^k v_1 u_2 z_2^\ell$$

for some i, j, k, ℓ such that $1 \le i, j, k, \ell \le K(F_{i-1})$ and $(i, j) < (k, \ell)$ in the lexicographic order. If $\ell_1, \ell_2 > K(F_{i-1})$ then

$$x_1 \cdots x_i = u_1 z_1^{\ell_1 - k + k} v_1 u_2 z_2^{\ell_2 - \ell + \ell} v_2 = u_1 z_1^{\ell_1 - (k-i)} v_1 u_2 z_2^{\ell_2 - (\ell - j)} v_2$$

which is a contradiction, since

$$(\ell_1 - (k - i), \ell_2 - (\ell - j)) < (\ell_1, \ell_2)$$

with respect to the lexicographic order. Thus $\ell_1 \le K(F_{i-1})$ or $\ell_2 \le K(F_{i-1})$. If $\ell_1 \le K(F_{i-1})$, then $u_1 z_1^{\ell_1} v_1 u_2, z_2, v_2 \in (F_{i-1})' = F_i$, so $x_1 \cdots x_i \in A_i$. If $\ell_2 \le K(F_{i-1})$, then $u_1, z_1, v_1 u_2 z_2^{\ell_2} v_2 \in (F_{i-1})' = F_i$, so again $x_1 \cdots x_i \in A_i$. It follows by induction that $x_1 \cdots x_i \in A_i$ for $i = 1, \ldots, r$. In particular, $s \in A_r \subseteq A_{t-1}$.

It follows now that S/S_j has at most $|A_{t-1}| \leq m|F_{t-1}|^3$ elements that are words in the generators $s_1, \ldots, s_k$ of length at most m. But $S/S_j = N \cup S\backslash S_{j+1}$ and $\mathrm{GK}(S/S_{j+1}) \leq 1$ by an inductive hypothesis. Therefore $\mathrm{GK}(S/S_j) \leq 1$, which completes the inductive step.

Thus 1) holds.

Finally, note that in the course of the proof we have shown that the associated groups are almost cyclic whenever S is finitely generated and satisfies 2). $\square$

We note that the assertion of Theorem 7.20 is not valid for arbitrary semigroups. For example, consider any finitely generated periodic group which is not almost nilpotent.

The following corollaries compare the class of repetitive semigroups with the class of semigroups of Gelfand-Kirillov dimension at most one.

Corollary 7.21 *Every semigroup $S \subseteq M_n(K)$ of Gelfand - Kirillov dimension one is repetitive.*

Proof. The associated linear groups have Gelfand - Kirillov dimension at most one by Theorem 7.2. Hence they are locally almost cyclic by Theorem 7.1. Therefore Theorem 7.17 implies that S is repetitive. $\square$

Recall that a linear semigroup $S \subseteq M_n(K)$ is the union of its uniform components if and only if, for every $a \in S$, there exist $x, y \in S$ such that the rank of xay is equal to the rank of a, and xay lies in a maximal subgroup of $M_n(K)$. For example, regular semigroups are of this type.

The proof of Theorem 7.20 shows that uniform components U of a finitely generated semigroup $S \subseteq M_n(K)$ of polynomial growth can be of two types only. Either U is a completely 0-simple semigroup over a finite group and U has finitely many $\mathcal{R}$-classes or finitely many $\mathcal{L}$-classes, or U intersects finitely many $\mathcal{H}$-classes of $M_n(K)$ and the associated group is infinite. So, for S equal to the union of its uniform components, this can be rephrased as:

($*$) S is contained in the union of a finite number of $\mathcal{R}$- and $\mathcal{L}$-classes of $M_n(K)$ and if $S \cap H$ is infinite for a maximal subgroup H of $M_n(K)$, then S intersects finitely many $\mathcal{H}$-classes of $M_n(K)$ that are $\mathcal{R}$- or $\mathcal{L}$-related to H.

Corollary 7.22 *Assume that $S \subseteq M_n(K)$ is a finitely generated semigroup that is the union of its uniform components. Then the following conditions are equivalent*

1. $GK(S) \leq 1$,

2. *S is repetitive and S satisfies condition* $(*)$,

3. *the groups associated to S are almost cyclic and S satisfies condition* $(*)$.

Proof. The proof of Theorem 7.20 actually shows that 1) and 3) are equivalent. The equivalence of 2) and 3) follows from Theorem 7.17 and Proposition 3.16. $\square$

Note that in the second example after Theorem 7.17 S is repetitive and intersects finitely many $\mathcal{H}$-classes of $M_2(\mathbf{Q})$, but $GK(S) > 1$. Thus Corollary 7.22 does not extend to arbitrary semigroups of matrices.

7.3   Almost semisimple semigroups

The first aim of this section is to show that the growth function of a finitely generated linear semigroup $S \subseteq M_n(K)$ is controlled by its behaviour on finitely many cancellative subsemigroups of S. From Section 7.1 we know that, if the growth of S is polynomial, then every cancellative subsemigroup T of S has a group of quotients $G \subseteq M_n(K)$ that is almost nilpotent and of finite rank. We prove that the latter condition, strengthened by the hypothesis that the unipotent radical of every such G is finite, is sufficient for S to have polynomial growth. Moreover, the degree of growth of S is then bounded by a polynomial $f(n, d)$ in n and the maximal degree d of growth of cancellative $T \subseteq S$.

Let $S = \langle s_1, \ldots, s_r \rangle$ be a semigroup. By $|s|$ mean the minimal length of $s \in S$ in the generators $s_1, \ldots, s_r$. If $A \subseteq S$ is a subset, then by $f_A(m)$ we denote the cardinality of the set of elements of S that can be written as words of length at most m in $s_1, \ldots, s_r$. In particular, $d_S(m) = f_S(m)$ is the growth function corresponding to this set of generators.

Let J be an ideal of a finitely generated semigroup $S = \langle s_1, \ldots, s_r \rangle$ and $J \subseteq I$ for a completely 0-simple semigroup I. Let $h(m)$ denote the number of $\mathcal{H}$-classes of I that contain elements of S that are words of length $\leq m$ in $s_1, \dots, s_r$. We call $h(m)$ the growth function for the number of $\mathcal{H}$-classes of I with respect to J and the generators $s_1, \dots, s_r$.

Lemma 7.23 *Let I be a completely 0-simple ideal of a semigroup U and $S = \langle s_1, \dots, s_r \rangle$ a subsemigroup of U. Let $J = S \cap I$ and suppose that S/J has polynomial growth of degree d. Then the growth function of the number of $\mathcal{H}$-classes of I with respect to J and the generators $s_1, \dots, s_r$ is bounded by a polynomial of degree not exceeding $4d$.*

Proof. Let

$$D_n = \{w = s_{i_1} \cdots s_{i_k} \in J \,|\, k \leq n, \text{ proper subwords of } w \text{ are not in } J\}.$$

Then $|D_n| \leq r(d_{S/J}(n))$ since every $w \in D_n$ is of the form $w = s_i z$, where $i \in \{1, \dots, r\}$ and $z \notin J$.

Let $0 \neq w = s_{i_1} \cdots s_{i_k} \in J$. Then $w = w_1 v$, where $w_1 = s_{i_1} \cdots s_{i_t}, v = s_{i_{t+1}} \cdots s_{i_k}$ and w_1 is the shortest initial factor of w that lies in J. Write $w = x_1 y_1$, where y_1 is the shortest terminal factor of w_1 (treated as the word $s_{i_1} \cdots s_{i_t}$) that lies in J. Note that v, x_1 can be empty words. Then $w = x_1 y_1 v$, so that $w \mathcal{R} x_1 y_1$ in I. Moreover, $x_1 \notin J$ and $y_1 \in D_n$. Hence, such an element w can lie in at most $d_{S/J}(n)|D_n|$ $\mathcal{R}$-classes of I. A symmetric argument shows that w can lie in at most $d_{S/J}(n)|D_n|$ $\mathcal{L}$-classes of I. Hence, $h(n) \leq d_{S/J}(n)^2 |D_n|^2 \leq r^2 d_{S/J}(n)^4$. The assertion follows. $\square$

Corollary 7.24 *Let I be a completely 0-simple ideal of a semigroup U and $S = \langle s_1, \dots, s_r \rangle \subseteq U, J = S \cap I$. If S/J has polynomial growth and for every $\mathcal{H}$-class H in I we have $f_{S \cap H}(m) \leq cm^d$ for some $c, d \geq 0$, independent of m and H, then S has polynomial growth.*

Proof. $f_J(m) \leq cm^d h(m)$ has polynomial growth by Lemma 7.23. Hence, the assertion follows from the fact that $d(m) \leq f_J(m) + d_{S/J}(m)$. $\square$

Our first aim is to show that the growth of a linear semigroup S is in some sense determined by the cancellative subsemigroups of S.

Theorem 7.25 *Let $S = \langle s_1, \dots, s_r \rangle \subseteq M_n(K)$ be a semigroup. For each uniform factor I_i/I_{i+1} of a structural chain $S = I_0 \supset I_1 \supset \cdots \supset I_t$ of S choose a cancellative subsemigroup T that is of the form $S \cap H$ for a maximal subgroup of $M_n(K)$. Then the following conditions are equivalent*

1. *S has polynomial growth,*

2. *$f_T(m)$ has polynomial growth for every cancellative subsemigroup T of S which is of the form $T = S \cap H$ for a maximal subgroup H of $M_n(K)$,*

3. *$f_{T_i}(m)$ has polynomial growth for every T_i.*

Moreover, in this case, $\mathrm{GK}(S) \leq g(n, d)$, where g is a function of n and the maximal degree d of polynomials bounding the functions $f_{T_i}(m)$.

Proof. Proof. The implications 1) $\Rightarrow$ 2) and 2) $\Rightarrow$ 3) are obvious. Thus, assume that 3) holds. We will inductively show that $\mathrm{GK}(S/I_i) < \infty$.

If S/I_1 is nilpotent, then it is finite and $\mathrm{GK}(S/I_1) = 0$. If S/I_1 is uniform, then from Corollary 7.7 it follows that $\mathrm{GK}(S/I_1) = \mathrm{GK}(T_1)$.

Suppose we know that $\mathrm{GK}(S/I_i) < \infty$. Again consider two cases. If I_i/I_{i+1} is nilpotent, then $(I_i/I_{i+1})^k = 0$, where $k \leq \binom{n}{j}$, j denoting the common rank of matrices in $I_i \setminus I_{i+1}$. Hence $\mathrm{GK}(S/I_{i+1}) \leq k\, \mathrm{GK}(S/I_i) < \infty$ by [67], Corollary 5.10. Assume now that I_i/I_{i+1} is uniform. From Lemma 3.24 we know that for every $\mathcal{H}$-class H in $M_n(K)$ such that $T = H \cap (I_i \setminus I_{i+1}) \neq \emptyset$ there exist $x, y \in I_i \setminus I_{i+1}$ (chosen from a finite set Z) such that $0 \neq xTy \subseteq T_i$. Since $z \mapsto xzy, z \in T$, is a one-to-one mapping, it follows that $f_T(m) \leq f_{T_i}(m + N)$, where $N = 2\{\max |w| \,|\, w \in Z\}$. Hence $f_T(m)$ is bounded by a polynomial independent of the choice of H. Corollary 7.24 implies that $\mathrm{GK}(S/I_{i+1}) < \infty$. This establishes 1).

Existence of the function g follows from the above proof and the fact that every structural chain must satisfy $t \leq 2^{n+1}$. $\square$

As noted in examples in Section 7.1, there exist finitely generated semigroups $S \subseteq M_n(K)$ whose all maximal cancellative subsemigroups have finite Gelfand - Kirillov dimension, but the growth of S is not polynomially bounded.

From Proposition 7.10 we know that, whenever a finitely generated $S \subseteq M_n(K)$ has polynomial growth, then every group associated to S is almost nilpotent and is finitely generated modulo its unipotent radical. Our main goal is to show that the problem (as that in the examples mentioned above) can only come from the unipotent radical of these groups. We will prove that for a class of $S \subseteq M_n(K)$, the polynomiality of the growth of S can be decided by measuring the Gelfand - Kirillov dimension of all (in fact, of finitely many) such H. We will need some preparatory lemmas.

Lemma 7.26 Let $s_1, \ldots, s_r \in M_n(K), e = \begin{pmatrix} 1 & 0 \\ 0 & 0 \end{pmatrix} \in M_n(K)$ be the idempotent of rank one, and let $0 \neq es_{i_1} \cdots s_{i_m} e = \begin{pmatrix} x & 0 \\ 0 & 0 \end{pmatrix}$ for some $i_j \in \{1, \ldots, r\}, m \geq 1, x \in K$. Then

1. If A is the set of entries of the matrices $s_1, \ldots s_r$, then x is a sum of n^{m-1} elements of the form $a_1 \cdots a_m, a_i \in A$.

2. Let E be a subfield of K with $[K : E] = k < \infty$, and $\phi : M_n(K) \longrightarrow M_{nk}(E)$ the embedding coming from the regular representation $K \longrightarrow M_k(E)$. If B is the set of entries of the matrices $\phi(s_1), \ldots, \phi(s_r)$ and if $x \in E$, then x is a sum of $(nk)^{m-1}$ elements of the form $b_1 \cdots b_m, b_i \in B$.

3. Assume that $E \subseteq K$ is a Galois extension and $N_{K/E}(x) = 1$. Let $\overline{\sigma}$ be the natural extension of $\sigma \in G(K/E)$ to an automorphism of $M_n(K)$. Let C be the set of entries of all matrices $\overline{\sigma}(s_j), \sigma \in G(K/E), j = 1, \ldots, r$. Then x^{-1} is a sum of $n^{(m-1)l}$ elements of the form $c_1 \cdots c_{ml}$, where $c_i \in C$ and $|G(K/E)| = l + 1$.

4. Let v be a discrete valuation of rank one on K. Then $v(x) \geq mv(a)$ for $a \in A$ with minimal valuation and, if 3) holds, then $|v(x)| \leq mN$ for some N dependent on A, C and l only.

Proof. 1) is clear. 2) follows from 1) and the fact that x can be viewed as a scalar $k \times k$ submatrix of $M_{nk}(E)$.

3) By the hypothesis $x^{-1} = \sigma_1(x) \cdots \sigma_l(x)$, for $\sigma_i \in G(K/E)$. Thus, x^{-1} is the nonzero entry of the product of $e\overline{\sigma}_j(s_{i_1}) \cdots \overline{\sigma}_j(s_{i_m})e, j = 1, \ldots, l$. So, the assertion follows as in 1).

4) By 1) we have $v(x) \geq \min\{v(a_1 \cdots a_m)\} \geq mv(a)$. But 3) implies that $v(x^{-1}) \geq mlv(c)$ for some $c \in C$. Hence 4) follows. $\square$

Lemma 7.27 *Let* $X = (x_{ij}), i = 1, \ldots, n, j = 1, \ldots, t$, *be a matrix of rank* t *with real coefficients. Then there exists* $a \in \mathbf{R}$ *such that, if*

$$|a_1 x_{11} + \cdots + a_t x_{1t}| \leq M$$

$$\cdots$$

$$|a_1 x_{n1} + \cdots + a_t x_{nt}| \leq M$$

for some $a_1, \ldots, a_t, M \in \mathbf{R}$, *then* $|a_j| \leq aM$ *for* $j = 1, \ldots, t$.

Proof. If $t = 1$, then $|a_1 x_{k1}| \leq M$ for every k and $x_{i1} \neq 0$ for some i because X has rank 1. Hence, we can put $a = |x_{i1}|^{-1} M$.

Assume that $t > 1$. Let for example x_{11} be a nonzero element of the first column of X. For $j = 2, \ldots, n$ we have

$$\begin{aligned}
|a_2(x_{j2} &- x_{12} x_{j1} x_{11}^{-1}) + \cdots + a_t(x_{jt} - x_{1t} x_{j1} x_{11}^{-1})| \\
&\leq |a_1 x_{j1} + \cdots + a_t x_{jt}| + |a_1 x_{11} x_{j1} x_{11}^{-1} + \cdots + a_t x_{1t} x_{j1} x_{11}^{-1}| \\
&\leq M + M|x_{j1} x_{11}^{-1}|.
\end{aligned}$$

This yields a system of inequalities with an $(n-1) \times (t-1)$ matrix Y that has rank $t - 1$. (We performed elementary row operations on X eliminating the nonzero entries below x_{11}). This allows us to complete the proof by induction on t. $\square$

Lemma 7.28 *Let* $L \supseteq D = F(t_1, \ldots, t_u) \supseteq F$ *be field extensions such that* $[L : D] = p < \infty$ *and* $t_1, \ldots, t_u$ *is a transcendence basis of* D *over* F. *Let* $S = \langle s_1, \ldots, s_r \rangle \subseteq M_n(L)$, *and* $\sigma \in \mathrm{Aut}(F)$. *Assume that* $e \in M_n(L)$ *is a diagonal idempotent of rank 1 and* $e s_{i_1} \cdots s_{i_m} e$ *has the only nonzero entry* $x \in F$. *There exists a field isomorphism* $\tau : L \longrightarrow L'$ *such that the only nonzero entry of* $e \bar{\tau}(s_{i_1}) \cdots \bar{\tau}(s_{i_m}) e \in M_n(L')$ *is equal to* $\sigma(x)$, *where* $\bar{\tau} : M_n(L) \subseteq M_n(L')$ *is the natural extension of* τ.

Proof. Extend σ to D, sending each t_i to t_i, and then to $\sigma' : M_p(D) \longrightarrow M_p(D)$. Let L be embedded into $M_p(D)$ via the regular representation. Put $L' = \sigma'(L)$ and extend $\tau = \sigma'_{|L}$ to $\bar{\tau} : M_n(L) \longrightarrow M_n(L')$. All these mappings agree with σ when restricted to F. Hence, if I denotes the identity matrix, then

$$e \bar{\tau}(s_{i_1}) \cdots \bar{\tau}(s_{i_m}) e = \bar{\tau}(e s_{i_1} \cdots s_{i_m} e) = \bar{\tau}(e(xI)e) = e \bar{\tau}(xI)e = e \sigma(x)e,$$

has the nonzero entry equal to $\sigma(x)$. $\square$

Proposition 7.29 *Let $S = \langle s_1, \ldots, s_r \rangle \subseteq M(K)$ be a finitely generated semigroup. Let $e \in M_n(K)$ be a diagonal idempotent of rank one and $z_1, \ldots, z_t$ free generators of a subgroup H of K^*. Suppose that A is a subset of S such that for each $a = s_{j_1} \cdots s_{j_m} \in A$ with $0 \neq s = eae, j_k \in \{1, \ldots, r\}, m \geq 1$, the nonzero entry of the matrix s is of the form $x = z_1^{i_1} \cdots z_t^{i_t}$ for some $i_j \in \mathbf{Z}$. Then there exists a constant M (independent of a, m and $j_1, \ldots, j_m$) such that $|i_j| \leq Mm$ for $j = 1, \ldots, t$.*

Proof. We shall proceed by induction on t. However, the case $t = 1$ will be dealt with later. First, we show how to proceed with the induction step.

Let L be a finitely generated subfield of K such that $S \subseteq M_n(L)$ and let $K_0 \subseteq K_0(t_1, \ldots, t_s) \subseteq L$ for the prime subfield K_0 of K and a transcendence basis $t_1, \ldots, t_s$ of L over K_0. It is known that this can be chosen so that $E = K_0(t_1, \ldots, t_s) \subseteq L$ is a separable field extension, cf.[135], Theorems II.13.30 and II.13.31, so that extending L we can assume that $E \subseteq L$ is a finite Galois extension. Let $N = N_{L/E}$ be the corresponding norm. Then for $G = G(L/E)$ we have

$$\prod_{\sigma \in G} \sigma(x) = N(x) = N(z_1^{i_1}) \cdots N(z_t^{i_t})$$

and clearly $N(x), N(z_i) \in E$. The only nonzero entry of the matrix $e\overline{\sigma}(s_{j_1}) \cdots \overline{\sigma}(s_{j_m})e$ is equal to $\sigma(x)$, where $\overline{\sigma}$ is the natural extension of σ to an automorphism of $M_n(L)$. We consider two cases.

Case 1. $N_{L/E}(x)$ is not a root of unity.

Let for example $N(z_1), \ldots, N(z_p), 1 \leq p \leq t$, be a maximal subset of $\{N(z_i) | i = 1, \ldots, t\}$ that freely generates a subgroup of L^*. If $p < t$, then there exists $k \in \mathbf{N}$ such that $N(z_j)^k = N(z_1)^{a_{1j}} \cdots N(z_p)^{a_{pj}}$ for $j = p + 1, \ldots, t$ and some $a_{ij} \in \mathbf{Z}$. Hence, $N(x)^k = N(z_1)^{b_1} \cdots N(z_p)^{b_p}$, where $b_q = ki_q + i_{p+1}a_{q,p+1} + \cdots + i_t a_{q,t}$ for $q = 1, \ldots, p$. Moreover, the matrix in $eM_n(L)e$ with the only nonzero entry equal to $N(x)^k$ is a word of length $(m + 2)k|G|$ in the generators of the semigroup $S' = \langle e, \overline{\sigma}(s_i) | \sigma \in G, i = 1, \ldots, r \rangle \subseteq M_n(L)$. If $G = \{\sigma_1, \ldots, \sigma_{|G|}\}$, then define $A' = \{a' = (e\overline{\sigma}_1(a)e\overline{\sigma}_2(a)e \cdots e\overline{\sigma}_{|G|}(a)e)^k | a \in A\}$. It is clear that $A' \subseteq S'$ satisfy the hypotheses of the proposition with respect to the free generators $N(z_1), \ldots, N(z_p)$ of a subgroup of L^*. Since $p < t$ and $N(x)^k$ is the only nonzero entry of the matrix a', we have performed an induction step. Hence, the induction hypothesis shows that $|b_q| \leq$

$Ml(a') \leq M(m+2)k|G|, q = 1, \ldots, p$, for a constant M, where $l(a')$ is the minimal length of a' in the generators of S'. Note that M is independent of $(m+2)k|G|$, and so of m because k depends on $z_1, \ldots, z_t$ only, and M is independent of the choice of $a \in A$ and $j_1, \ldots, j_m$. Thus, $|b_q| \leq M'm$ for $q = 1, \ldots p$, where $M' = 3Mk|G|$. Now

$$x^k z_1^{-b_1} = y_2^{i_2} \cdots y_t^{i_t},$$

where $y_j = z_j^k$ for $j = 2, \ldots, p$, and $y_j = z_j^k z_1^{-a_{1j}}$ for $j = p+1, \ldots, t$. But $y_2, \ldots, y_t$ are free generators of a subgroup of L^*. Hence the above equality fulfills the hypotheses of the proposition with respect to the semigroup $S'' = \langle S, u, v \rangle$, where $u, v \in M_n(L)$ are such that the nonzero entries of $u = eue$, $v = eve$ are z_1, z_1^{-1} respectively, and the subset $A'' = \{(eae)^k v^{b_1} | a \in A, b_1 > 0\} \cup \{(eae)^k u^{b_1} | a \in A, b_1 \geq 0\}$. Consequently, the induction hypothesis implies in view of $|b_1| \leq M'm$ that

$$|i_j| \leq M''((m+4)k + |b_1|) \leq (5k + M')M''m \text{ for } j = 2, \ldots, t$$

for some M'' (independent of $a \in A, m, j_1, \ldots, j_m$) because $x^k z_1^{-b_1}$ is the nonzero entry of ewe for a word w of length $(m+4)k + |b_1|$ in S''. In view of the bound $|b_1| \leq M'm$, this completes the inductive argument (establishing a linear bound on all $|i_j|$).

Assume now that $p = t$. Then we replace the elements $x, z_1, \ldots, z_t$ by $N(x), N(z_1), \ldots, N(z_t) \in E$. Thus, passing to the semigroup S' and its subset $\{e\bar{\sigma}_1(a)e\bar{\sigma}_2(a)e \cdots e\bar{\sigma}_{|G|}(a)e | a \in A\}$ we can assume that $x, z_1, \ldots, z_t \in E$. If $K_0 = E$, and $ch(K) = p > 0$, all $z_i \in \mathbf{F}_p$ have finite order, a contradiction. Otherwise, we treat E as the field of quotients of the polynomial ring $R = \mathbf{F}_p[t_1, \ldots, t_s]$ or of $R = \mathbf{Z}[t_1, \ldots, t_s]$. Let q be a prime element of R and δ the q-adic valuation on E. Write $c_i = \delta(z_i)$ for $i = 1, \ldots, t$. From Lemma 7.26 2) it follows that x is a sum of $(nk)^{m-1}$ elements of the form $b_1 \cdots b_m$ with b_i chosen from a finite set $B \subseteq E$ independent of x. Writing $b_i = f_i g^{-1}$ for some $f_i, g \in R$, we see that $x = fg^{-m}$ where f is the sum of all $f_1 \cdots f_m$. If q is a polynomial of a nonzero degree, then it follows that $\delta(x)$ does not exceed the maximal degree of $f_1 \cdots f_m$. If q is a constant, then $ch(K) = 0$ and $\delta(x)$ does not exceed the absolute value of any nonzero coefficient in the polynomial f. Therefore $\delta(x) \leq (nk)^{m-1} h^m$ for some $h \in \mathbf{R}$ independent of x. Hence, we always have $|c_1 i_1 + \cdots + c_t i_t| \leq Nm$ for some N (dependent on $s_1, \ldots, s_r$ only).

Suppose $c_1 = \cdots = c_t = 0$ for every such q. Then $z_1, \ldots, z_t$ lie in the group of units of R, cf.[135], Corollary VI.10.3. Since the latter is finite, this is a contradiction. Thus, for some such δ one of the c_i, say c_1, is nonzero. Now, $x = z_1^{k_1}(z_2 z_1^{-c_2})^{i_2} \cdots (z_t z_1^{-c_t})^{i_t}$, where $k_1 = c_1 i_1 + c_2 i_2 + \cdots + c_t i_t$, and $y_i = z_i z_1^{-c_i}, i = 2, \ldots, t$, are free generators of a subgroup of L^*. Then

$$x z_1^{-k_1} = y_2^{i_2} \cdots y_t^{i_t}$$

and the left side is the nonzero entry of ewe for a word w of length $\leq m + Nm$ in $s_1, \ldots, s_r, u, v$, the latter two defined as above. Applying the induction hypothesis to the semigroup $\langle S, u, v \rangle$ and its subset consisting of these elements w, we see that $|i_j| \leq M(N+1)m$ for $j = 2, \ldots, t$, for some M (independent of m). Since $|c_1 i_1 + \cdots + c_t i_t| \leq Nm$, and $c_1 \neq 0$, we also get a desired linear bound on $|i_1|$. This completes the inductive argument in Case 1.

Case 2. $N_{L/E}(x)$ is a root of unity.

As above, passing to $x^k = (z_1^k)^{i_1} \cdots (z_t^k)^{i_t}$ for some $k \geq 1$, we can assume that $N_{L/E}(x) = 1$. Then Lemma 7.26 4) shows that $|v(x)| \leq Nm$ for some N (independent of m) and every valuation with value group $\mathbf{Z}$ of L. If for some such v one of $v(z_1), \ldots, v(z_t)$ is nonzero, then we can decrease t as in Case 1, performing the inductive step. Therefore, we need to consider only the case where $v(z_1) = \cdots = v(z_t) = 0$ for every such v. It is well known that this implies that each z_i, z_i^{-1} is integral over $C = \mathbf{Z}$ or $C = \mathbf{F}_p$, depending on the characteristic of K. (In fact, if $a \in L$ is not integral, then $v(a) \neq 0$ for the $a^{-1}C[a^{-1}]$-adic valuation v on $C[a^{-1}]$. Then v can be extended to a valuation on L that is again discrete of rank one, cf.[135], pp.51,52 and p.85). If $C = \mathbf{F}_p$, we have $\langle z_i \rangle \subseteq \mathbf{F}_p[z_i]$ is finite, contradicting the assumption on z_i. Thus, assume that $\mathrm{ch}(K) = 0$. Let F be a finite Galois extension of $\mathbf{Q}$ containing $z_1, \ldots, z_t$. Extending L we can assume that $F \subseteq L$. We know that the integral closure B of $\mathbf{Z}$ in F contains the group H generated by $z_1, \ldots, z_t$. Let $u = u_1 + u_2$, where $\sigma_i \in G(F/\mathbf{Q}), i = 1, \ldots, u_1$, are such that $\sigma_i(F) \subseteq \mathbf{R}$ and $\sigma_i, \sigma_i' \in G(F/\mathbf{Q}), i = u_1 + 1, \ldots, u$, are the conjugate pairs of the remaining automorphisms of F. It is known that there exists $k \geq 1$ such that the $u \times t$ matrix

$$(\ln |\sigma_i(z_j^k)|)_{i,j}$$

has rank t, cf.[83], 47.2. Using the presentation $x^k = (z_1^k)^{i_1} \cdots (z_t^k)^{i_t}$ and the fact that $z_1^k, \ldots, z_t^k$ freely generate a subgroup of L^*, as before, we can reduce our problem to the case where $k = 1$. Now, for every $\sigma = \sigma_i \in G(F/\mathbf{Q})$, we have $|\sigma(x)| = |\sigma(z_1)^{i_1} \cdots \sigma(z_t)^{i_t}|$. Moreover $|\sigma(x)| \leq a^m$ and $|\sigma(x)^{-1}| \leq b^m$ for some $a, b \in \mathbf{R}$ independent of m, by Lemma 7.26 1),3) and Lemma 7.28. Therefore $|\ln|\sigma(x)|| \leq Mm$ for some M independent of m. Hence

$$|i_1 \ln|\sigma(z_1)| + \cdots + i_t \ln|\sigma(z_t)|| \leq Mm.$$

Lemma 7.27 yields $|i_j| \leq Nm$ for $j = 1, \ldots, t$, and some N independent of m. This completes the inductive argument in Case 2.

It remains to check the validity of the assertion in case $t = 1$, that is $x = z_1^{i_1}$. However, it is clear from the above reasoning that it takes care of this case, too. This completes the proof of the proposition. $\square$

We are now in a position to prove the first main result of this section.

Theorem 7.30 Let $S = \langle s_1, \ldots, s_r \rangle \subseteq M_n(K)$ be a finitely generated semigroup such that every nonempty intersection $T = S \cap F$ with a maximal subgroup F of $M_n(K)$ has a group of quotients $G \subseteq F$ that is almost nilpotent and finitely generated modulo its unipotent radical U. Then the number of cosets in G/U that contain elements of T that are words of length $\leq m$ in $s_1, \ldots, s_r$ is bounded by a polynomial in m, depending only on the torsion-free rank of G/U and on the rank of matrices in T.

Proof. From [134], Theorem 3.6, we know that the group G of quotients of T has a nilpotent normal subgroup D of finite index that is triangularizable in $f M_n(\overline{K})f = M_{\mathrm{rank}(f)}(\overline{K})$, where $f = f^2 \in D$. Extending K we can assume that f is diagonal and D is triangular in $M_n(K)$. Note that U contains the unipotent radical D_u of D because the latter is a normal subgroup of G. Clearly, D_u is the kernel of the diagonal projection $D \longrightarrow diag(D)$. Then $diag(D)$ has a subgroup B of finite index with all rank one diagonal projections being torsion-free subgroups of K^* (the group $D/(D \cap U) \simeq diag(D) \subseteq (K^*)^{\mathrm{rank}(f)}$ is finitely generated since it has finite index in the finitely generated group G/U). Thus, there is a finite set $Z \subseteq G$ such that for every $g \in G$ there exists $z \in Z$ with $zg \in D$ and $diag(zg) \in B$. Let

$e \leq f$ be a diagonal idempotent of rank one. From Proposition 7.29 it follows that for each nonzero $a = e z s_{i_1} \cdots s_{i_k} e, k \leq m, z \in Z$, such that $diag(z s_{i_1} \cdots s_{i_k}) \in B$, the power of each generator of the projection eBe of B appearing in the presentation of a has absolute value $\leq N(m+1)$, where N is independent of m and $i_1, \ldots, i_k$. Hence, the set $\{ezse | z \in Z, diag(zs) \in B,$ length of $s \leq m\}$ has at most $(2N(m+1)+1)^t$ elements, where t is the rank of eBe. Since this argument works for every diagonal of rank one $e \leq f$, we get a polynomial bound on $q_m = |\{diag(z s_{i_1} \cdots s_{i_k}) \in B | z \in Z, k \leq m\}|$. Now, the number of cosets in G/U that contain elements of T that are words of length $\leq m$ in $s_1, \ldots, s_r$ does not exceed $q_m |Z|$. This proves the theorem. $\square$

For any semigroup S we define the rank of S by the formula $rk(S) = \sup\{t | S$ has a free commutative subsemigroup on t free generators$\}$. It is clear that, if $S \subseteq M_n(K)$, then $rk(S) = rk(T)$ for a cancellative subsemigroup T of S contained in a maximal subgroup of $M_n(K)$, see Proposition 3.7. If T has a group of quotients G which has a finite normal subgroup A such that G/A is almost abelian, then $rk(T) = rk(G)$ and the latter is equal to the Gelfand-Kirillov dimension of G, [87], Proposition 23.2. In particular, this applies to the case described in Theorem 7.31. (Note that groups associated to a semigroup S satisfying conditions of this theorem are in fact almost abelian and finitely generated, cf. Lemma 5.25 and Proposition 7.10).

The above result leads naturally to a condition under which we are able to prove the finiteness of $GK(S)$. Namely, assume that there exists $q \in \mathbf{N}$ such that $|\phi_T^{-1}\phi_T(t)| \leq q$ for every $t \in T$, where $\phi_T : T \longrightarrow G/U$ is the natural homomorphism. Suppose that $a_1, \ldots, a_r \in G$ are distinct elements that lie in the same coset of U. Since G is the group of quotients of $T, a_i = w_i v^{-1}$ for some $w_i, v \in T$. Then $w_i = a_i v \in T$ are in the same coset of U, so that $r \leq q$. It follows that the above condition is equivalent to saying that U is a finite group. (So, U must be trivial if $ch(K) = 0$.) In this case, Theorem 7.30 yields a polynomial bound on $f_T(m)$.

Theorem 7.31 *Let $S \subseteq M_n(K)$ be a finitely generated semigroup of finite rank r such that every group G associated to S is almost nilpotent, (equivalently, S has finite rank and satisfies an identity). If each G has finite unipotent radical, then S has polynomial growth of degree bounded by $f(n, r)$, where f is a function of n and r.*

Proof. We know that G has a normal subgroup D of finite index with finite unipotent radical U, and D/U embeds into $(L^*)^k$ for a finitely generated subfield $L \subseteq K$ and some $k \leq n$. Since $\mathrm{rk}(G) = \mathrm{rk}(D/U)$, [87], Proposition 23.2, it follows that D/U is finitely generated. In view of Theorem 7.25 we need to show that $f_T(m)$ is polynomially bounded for every cancellative subsemigroup of S of the form $T = S \cap G$. This is a direct consequence of Theorem 7.30. The existence of the function $f(n,r)$ follows from the proof. $\square$

Remark 1) From the remark preceding Theorem 7.31 and the proof it follows that $S \subseteq M_n(K)$ has polynomial growth if and only if the associated groups $G = TT^{-1}$ are almost nilpotent and for each $t \in T$ that is a word of length $\leq m$ in the generators of S we have $|\phi_T^{-1}\phi_T(t)| \leq q(m)$ for a fixed polynomial q. In fact, in this case $f_T(m)$ also is polynomially bounded.

2) The hypotheses of the theorem amount to say that each G is an almost diagonalizable group (so, our $S \subseteq M_n(K)$ can be called 'almost semisimple') which is also finitely generated.

If K is of characteristic zero, then the hypotheses of the theorem can be reformulated, due to the following observation.

Proposition 7.32 *Assume that* $\mathrm{ch}(K) = 0$. *Then* $S \subseteq M_n(K)$ *is almost semisimple if and only if the groups associated to S are almost nilpotent and a power of every element $s \in S$ is diagonalizable.*

Proof. The necessity is clear. So, suppose that D is a nilpotent subgroup of finite index in a group G associated to S. If D_u, D_s denote the sets of unipotent and diagonalizable elements of D respectively, then D_u, D_s are subgroups and $\langle D_u, D_s \rangle = D_u \times D_s$, [134], Theorem 7.11. If a power of every $s \in S$ is diagonalizable, then $S \cap D \subseteq D_s$ because $\mathrm{ch}(K) = 0$. Hence $D = (S \cap D)(S \cap D)^{-1} \subseteq D_s$, so D is diagonalizable. $\square$

We conclude with the case of 2×2 matrices, in which the description can be simplified.

Corollary 7.33 *Let* $S \subseteq M_2(K)$ *be a finitely generated semigroup. Then the following conditions are equivalent*

1. S *has polynomial growth,*

2. S *satisfies an identity and has finite rank,*

3. *each cancellative subsemigroup of S embeds into a finitely generated linear group of polynomial growth,*

4. $S \cap GL_2(K)$ *has an almost nilpotent group of quotients, if non-empty, and every nonempty intersection $S \cap H$ with a maximal subgroup $H \neq GL_2(K)$ of $M_2(K)$ lies in a finitely generated subgroup of H.*

Proof. Clearly, $S \cap GL_n(K)$ is finitely generated (if nonempty) whenever so is S. If 1) holds, then groups associated to S are finitely generated modulo unipotent radical by Proposition 7.10. So all such groups are finitely generated. Hence 3) follows via Proposition 3.7 and Theorem 7.1.

A nilpotent subgroup of $GL_2(K)$ is almost abelian, cf.[134], Chapter 7. Hence 2) is a consequence of 3), cf. Lemma 5.3.

A subgroup of finite rank of the multiplicative group of a finitely generated field is finitely generated by Lemma 1.7. Hence, 4) follows from 2).

The implication 4) $\Rightarrow$ 1) is a consequence of the proof of Theorem 7.31 because the unipotent radical of every $S \cap H$ is trivial. $\square$

We conclude this chapter with some remarks concerning the general case. From Proposition 7.10 we know that the groups associated to a linear semigroup S are almost nilpotent and finitely generated modulo the unipotent radical, whenever S has polynomial growth. The basic idea of the approach relies on the fact that every almost nilpotent linear group is isomorphic to a subdirect product of an almost unipotent linear group and an almost diagonalizable linear group. The interaction between the 'group elements' z of S (that is, elements $z \in S \cap D$ for a maximal subgroup D of $M_n(K)$) with the 'unipotent parts' of the uniform components U of S that lie below z decides whether S is of polynomial growth. It was shown in the course of the proof of Theorem 7.30 that the problem does not come from the 'semisimple part' of the group associated to U.

We introduce a formalism that allows us to study S via its corresponding 'decompositions'. Let D be a maximal subgroup of $M_n(K)$

such that $S \cap D$ generates an almost nilpotent group G and lies in a uniform component U of S. Let $T \subseteq M_n(K)$ be the semigroup generated by $S \cup G$. Denote by S_U the Rees factor of S with respect to the ideal $S_j \setminus U$, where j is the rank of matrices in U and $S_j = \{x \in S \mid \text{rank}(x) \leq j\}$. From the structure theorem and from Lemma 3.11, it follows that the layers of T are of the form $T_j / T_{j-1} = \widehat{U} \cup U_2' \cup \ldots \cup U_t' \cup N'$, where $\widehat{U}$ is the completely 0-simple closure of U in M_j/M_{j-1}, $\widehat{U} = U_1', U_2', \ldots, U_t'$ are the uniform components of T and N' is the nilradical of T_j/T_{j-1}. Moreover, the semigroup $S_{\widehat{U}} = T/(T_j \setminus \widehat{U})$ is a disjoint union of $S \setminus S_j = T \setminus T_j$ and $\widehat{U}$, so it contains S_U.

T modulo T_{j-1} and N' is a subdirect product of $T/(T_j \setminus U_i')$, $i = 1, \ldots, t$. So, by [67], Corollary 5.10, $\text{GK}(S) < \infty$ if and only if each of $S/(S_j \setminus U_i)$ has polynomial growth. So, we consider S_U only. Since S is finitely generated, it is easy to see that $T \subseteq M_n(L)$ for a finitely generated subfield L of K, see Proposition 7.10.

Let $\sim_{Ann(\widehat{U})}$ denote the congruence

$$\{(x, y) \in S_{\widehat{U}} \times S_{\widehat{U}} \mid x, y \in \widehat{U}, \widehat{U}(x - y)\widehat{U} = 0\}.$$

The construction of Lemma 3.25 allows us to find a linear representation ξ of T such that $\xi(T \setminus S_{\widehat{U}}) = 0$ and ξ restricted to $S_{\widehat{U}}$ is determined by $\sim_{Ann(\widehat{U})}$. So, the kernel of the map $K_0[T] \longrightarrow K_0[\xi(T)]$ is nilpotent, hence again by [67], Corollary 5.10, we may assume that $S_{\widehat{U}}$ is linear and all matrices of the least nonzero rank in $S_{\widehat{U}}$ are in $\widehat{U}$.

Lemma 7.34 *Assume that G is an almost nilpotent group associated to a semigroup $S \subseteq M_n(K)$. Then there exist normal subgroups G_u, G_s of G such that G_u is unipotent, G_s is diagonalizable, and G has an almost diagonalizable rational linear representation $G^{(s)}$ and an almost unipotent rational linear representation $G^{(u)}$ with $G^{(s)} \simeq G/G_u$, $G^{(u)} \simeq G/G_s$.*

Proof. We can assume that K is algebraically closed. It is well known that the Zariski closure F of G also is almost nilpotent, so that the connected component F^c of F must be nilpotent (use [134], Lemma 5.3, Theorem 5.11 and Exercise 7.2). Let $G_u = G \cap F_u$ where F_u is the unipotent radical of F^c, and let G_s be the maximal closed diagonalizable normal subgroup of G, cf. [134], Theorem 14.22. Then the assertion follows from [134], Theorem 6.4, Lemma 6.5 and Lemma 6.6. $\square$

So G is a subdirect product of an almost diagonalizable linear group $G^{(s)}$ and an almost unipotent linear group $G^{(u)}$. The map $G \longrightarrow G^{(u)}$ has a natural extension to

$$\hat{U} = \mathcal{M}(G, X, Y, P) \longrightarrow \hat{U}^{(u)} = \mathcal{M}(G/G_s, X, Y, P_u).$$

It is easy to see that we get an extension to a homomorphism $S_{\hat{U}} \longrightarrow S_{\hat{U}}^{(u)} = (S_{\hat{U}} \setminus \hat{U}) \cup \hat{U}^{(u)}$, which is identity on $S_{\hat{U}} \setminus \hat{U}$. Similarly, one constructs $\hat{U}^{(s)} = \mathcal{M}(G/G_u, X, Y, P_s)$ and a homomorphism $S_{\hat{U}} \longrightarrow S_{\hat{U}}^{(s)} = (S \setminus \hat{U}) \cup \hat{U}^{(s)}$. We approach S_U via the embedding $S_U \longrightarrow S_U^{(s)} \times S_U^{(u)}$, where $S_U^{(s)}, S_U^{(u)}$ are the images of S_U under the respective homomorphisms of $S_{\hat{U}}$.

Let ϕ_s be the linear representation of $S_{\hat{U}}$ extending $G \longrightarrow G/G_u$ and one-to-one on $S_{\hat{U}} \setminus \hat{U}$, which arises from Lemma 3.26. Let $\varphi : S_{\hat{U}} \longrightarrow \phi_s(S_{\hat{U}}) \times S_{\hat{U}}^{(u)}$ be the resulting map. If $\varphi(x) = \varphi(y)$ for $x, y \in \hat{U}$, then $x \mathcal{H} y$ in $\hat{U}$ and consequently $x = y$, because the kernel of $\phi_{s|G}$ is equal to G_u and G_u trivially intersects G_s. So, ϕ_s factors through $S_{\hat{U}}^{(s)}$ and $S_{\hat{U}}$ embeds into $\phi_s(S_{\hat{U}}) \times S_{\hat{U}}^{(u)}$. Now, $\phi_s(S_U)$ is linear, and it is easy to see that $\phi_s(U)$ is an ideal which is a uniform component whose associated group is also rationally isomorphic to $G^{(s)}$.

Trying to prove that $\mathrm{GK}(S) < \infty$ by induction on the number of uniform components, we may assume that $\mathrm{GK}(S_U/U) < \infty$. Since $S_U/U \simeq S_U^{(s)}/U^{(s)} \simeq \phi_s(S_U)/\phi_s(U)$ and the group associated to $\phi_s(U)$ is almost diagonalizable, Theorem 7.25 and Theorem 7.30 imply that $\mathrm{GK}(\phi_s(S_U)) < \infty$. So, it remains to see when $S_U^{(u)}$ has polynomial growth, because this is now equivalent to $\mathrm{GK}(S_U) < \infty$, see [67], Proposition 3.12 and Proposition 3.13.

It was claimed in [92], [93], that roughly speaking, a finitely generated linear semigroup S has polynomial growth if and only if the groups associated to S are finitely generated and almost nilpotent and for every uniform component U of S the semigroup $S_U^{(u)}$ modulo $\sim_{Ann(U^{(u)})}$ is almost unipotent (and that the latter condition can be dropped if $\mathrm{ch}(K) > 0$). An equivalent condition was also claimed in terms of linear recurrencies $az^m a, m = 1, 2, \dots$, for $a, z \in S$. However, there is a gap in the proof, resulting from the fact that in general $S_U^{(u)}$ does not allow a reduction to the linear case, see the example after Proposition 3.26.

In case $S_U^{(u)}$ can be replaced by a linear semigroup $\phi_u(S_U)$, as $S_U^{(s)}$ was replaced by $\phi_s(S_U)$ above, one shows that the problem reduces to the case of subsemigroups of the form $\langle a, z \rangle$, where $a \in U, z \in S_U$, and $\text{rank}(z^2) = \text{rank}(z) > \text{rank}(a)$. So, it is not surprising that the recurrent properties of the sequence $az^m a, m = 1, 2, \ldots$, should be decisive, because the group H associated to $\langle a, z \rangle$ and containing a is generated by nonzero elements of this type and by powers of a, (see also Theorem 7.25).

We briefly discuss the case where $\text{ch}(K) > 0$, since the conjectured result seems quite readable in this case. Namely, one expects $S_U^{(u)}$ to be finite modulo $\sim_{Ann(U^{(u)})}$ (for this, it is enough to check that the factor semigroup is periodic). This in turn implies that the completely 0-simple semigroup $U^{(u)} / \sim_{Ann(U^{(u)})}$ is finitely generated (cf. [87], Lemma 2.1), so it has finitely many $\mathcal{H}$-classes. Then its maximal subgroup $G^{(u)}$ is finitely generated, whence G is finitely generated as we assume that G is finitely generated modulo its unipotent radical. Consequently, the unipotent radical of G is finite. This is exactly what is desired, in view of Theorem 7.31.

Let a, z be as above. Assume that $a \in G$. Fix some $p, q \in \{1, \ldots, n\}$ and consider the sequence $u_m = u_m^{(pq)}$ of (p, q)-th entries of $az^m a, m = 1, 2, \ldots$. Matrices $az^m a$ satisfy a linear recurrence coming from the characteristic polynomial of z. So

$$u_m = \sum_{i=1}^{n} P_i(m)\lambda_i^m$$

for some polynomials P_i and $\lambda_i \in \overline{K}$, see Section 1.4. Extending the methods of [93] one can show that the group $H \subseteq G$ associated to $\langle a, z \rangle$ and containing a is finitely generated. So, in view of Theorem 7.31, we see that $\text{GK}(\langle a, z \rangle) < \infty$ exactly when H is finitely generated (note that $\langle a, z \rangle$ has only one uniform component consisting of matrices of ranks equal to $\text{rank}(a)$). However, for lack of linear methods in $S_U^{(u)}$, we are unable to extend this to S_U.

One might expect that the polynomiality of the growth of S can always be verified on 2-generated subsemigroups $\langle a, z \rangle$ of our special type, provided we know that the associated linear groups G are almost nilpotent and finitely generated modulo the unipotent radical. It seems also that the growth functions of linear semigroups have a rather regular behaviour. In particular, one might expect that the Gelfand-Kirillov dimension of $S \subseteq M_n(K)$ is an integer whenever it is finite.

Chapter 8

Monoids of Lie type

In this chapter we discuss an exceptional class of finite monoids. They are finite analogues of linear algebraic monoids, in a way that finite groups of Lie type are analogues of algebraic groups. Their theory was created as an abstraction of the theory of linear algebraic monoids and developed in a series of papers by Putcha and Renner. The full linear monoid $M_n(\mathbf{F}_q)$, discussed in Chapter 2, is the basic example. Actually, among the topics presented before, only this and the material of Chapter 4 are relevant for our aims here.

After giving the necessary background on groups of Lie type, we state the definition and the main properties of such monoids. A universal object in the class of monoids built on a given group of Lie type, as $M_n(\mathbf{F}_q)$ is built on $GL_n(\mathbf{F}_q)$, is then described. Exceptional properties relating irreducible representations of monoids of Lie type are presented in the last section. Throughout this chapter F will stand for a field.

8.1 Definition and examples

Let G be a finite group. We say that G has a Tits system (or admits a BN-pair) if there are subgroups B, N of G which generate G and such that $T = B \cap N$ is a normal subgroup of N and $W = N/T$ (called the Weyl group) has a generating set S (referred to as a set of Coxeter generators for G) consisting of order two elements and satisfying

$$sBa \subseteq BsaB \cup BaB \quad \text{for all} \quad a \in W, s \in S$$
$$sBs \neq B \quad \text{for all} \quad s \in S$$

Here, abusing notation, we write Ba for the coset $Bn_a, n_a \in N$, such that $a = Tn_a$, with a similar convention for the cosets aB.

If $a \in W$, then the length $l(a)$ is defined as the minimal length of a in the generators from S. We put $l(1) = 0$. It turns out that W has a unique element a_0 of maximal length. Then a_0 is of order two and $a_0 S a_0 = S$. Define $B^- = a_0 B a_0$. If $I \subseteq S$, then by W_I we denote the subgroup of W generated by I. Here $W_\emptyset = \{1\}$. Then $P_I = BW_I B$ and $P_I^- = B^- W_I B^-$ are subgroups of G which are their own normalizers. Moreover, these are the only subgroups containing B, B^- respectively. The axioms imply that $G = \bigcup_{a \in W} BaB$, a disjoint union. This is called the Bruhat decomposition of G. Since $B^- = a_0 B a_0$, we also have $G = \bigcup_{a \in W} B^- a B$, a disjoint union. The conjugates of B are called the Borel subgroups, the conjugates of P_I are the parabolic subgroups of G. If $x \in G$, then $x^{-1} P_I x, x^{-1} P_I^- x$ are called opposite parabolic subgroups. Hence, if P, P^- are opposite parabolic subgroups, then $x^{-1} P^- x, x \in P$, is the set of all opposites of P and there exists $g \in G$ such that $g^{-1} P g = P_I, g^{-1} P^- g = P_I^-$ for some $I \subseteq S$. If $I, K \subseteq S$, then $P_I \cap P_K = P_{I \cap K}$.

Speaking about a group of Lie type, we will assume that G admits a split BN-pair satisfying some commutator relations, see [13], page 61. In other words, G satisfies some extra conditions, which in particular are satisfied by any group H_σ of constants of a connected reductive group $H \subseteq GL_n(\overline{\mathbf{F}}_q)$ with respect to the Frobenius map σ defined by $\sigma((a_{ij})) = (a_{ij}^q)$. We note that groups of this type have played a crucial role in the classification of finite simple groups.

Then there is a normal subgroup U of B such that $B = UT, U \cap T = \{1\}$. So $B^- = U^- T$, where $U^- = a_0 U a_0$. Moreover, if $L_I = P_I \cap P_I^-$ for $I \subseteq S$, then there exist normal subgroups U_I in P_I and U_I^- in P_I^- such that $U_I \cap P_I = \{1\}, U_I^- \cap P_I^- = \{1\}$, and $P_I = L_I U_I, P_I^- = L_I U_I^-$. This is called the Levi decomposition of P_I, P_I^-. Also U_I, U_I^- are called the unipotent radicals of P_I, P_I^- and L_I is called a Levi factor of P_I. More generally, if $P = x^{-1} P_I x, P^- = x^{-1} P_I^- x$ are any opposite parabolic subgroups of G, and $L = P \cap P^-$, then $P = LR_u(P), P^- = LR_u(P^-)$, where $R_u(P) = x^{-1} U_I x, R_u(P^-) = x^{-1} U_I^- x$. The unipotent radical $R_u(P)$ does not depend on the choice of P^-, while any two Levi factors of P are conjugate by an element of $R_u(P)$.

U is generated by certain subgroups $X_\alpha, \alpha \in \Phi^+$, called the root subgroups. In fact, U is the product of all X_α in any order. Similarly, U^- is the product of some root subgroups $X_\alpha, \alpha \in \Phi^- = -\Phi^+$. Moreover, T

normalizes the root subgroups $X_\alpha, \alpha \in \Phi = \Phi^+ \cup \Phi^-$, and W permutes the subgroups $X_\alpha, \alpha \in \Phi$. If $I \subseteq S$, then L_I is generated by T and the X_α's such that $X_\alpha, X_{-\alpha}$ are both contained in P_I. Then U_I is the product, in any order, of the $X_\alpha, \alpha \in \Phi^+$, such that $X_{-\alpha} \not\subseteq P_I$. Similarly, U_I^- is the product, in any order, of the $X_{-\alpha}$ with $X_\alpha \subseteq U_I$. Each L_I is a group of Lie type with a BN-pair $B \cap L_I, N \cap L_I$. Moreover $(B \cap L_I)U_I = B$ and $(B^- \cap L_I)U_I^- = B^-$. All this can be found in [13].

If $s \in S, a \in W$, then $l(as) = l(a) \pm 1$. Moreover, we have the following multiplication rule for the components of the Bruhat decomposition

$$BsB \cdot BaB = \begin{cases} BsaB & \text{if } l(sa) = l(a) + 1 \\ BsaB \cup BaB & \text{if } l(sa) = l(a) - 1 \end{cases} \qquad (8.1)$$

Applying the map $g \mapsto g^{-1}$ we also get $(BaB)(BsB) = BasB \cup BaB$ or $(BaB)(BsB) = BasB$.

Example Let $G = GL_n(\mathbf{F}_q)$. Then we define B as the group of upper triangular matrices and T as the group of diagonal matrices. Next, W consists of permutation matrices and $N = \mathbf{F}_q^* W$. Then $S = \{s_1, \dots, s_{n-1}\}$, where s_i is the simple reflection interchanging $i, i+1$. Now $a_0 = E_{n,1} + E_{n-1,2} + \cdots + E_{1,n}$ with $E_{i,j}$ standing for the matrix with 1 in the (i,j) position and zeros elsewhere. Then B^- is the group of lower triangular matrices, and $U \subseteq B, U^- \subseteq B^-$ are the subgroups of all unipotent matrices. If $I \subseteq S$, then P_I, P_I^- consist of the corresponding block upper, respectively lower, triangular matrices. Now L_I is the group of block diagonal matrices. Φ^+ is in one-to-one correspondence with ordered pairs (i,j) with $1 \le i < j \le n$. Φ^- is in one-to-one correspondence with the pairs (i,j) with $1 \le j < i \le n$. The root subgroups are of the form $X_{i,j} = \{I + \alpha E_{i,j} \mid \alpha \in \mathbf{F}_q\}$.

We will need the following technical result, which is a special case of standard general facts on the intersections of parabolic subgroups, see [13], Proposition 2.8.6, Theorem 2.8.7, Proposition 2.8.9. It follows via the root subgroup calculus and it is stated in the form needed later. Note that $a^{-1}L_I a, a^{-1}U_I a, a^{-1}U_I^- a$ make sense for $a \in W, I \subseteq S$, because T normalizes L_I, U_I and U_I^-.

Lemma 8.1 Let G be a group of Lie type, $a \in W$ and $I, K \subseteq S$. Then

1. $P_I \cap a^{-1}U_I^- a = (L_I \cap a^{-1}U_I^- a)(U_I \cap a^{-1}U_I^- a)$,

2. $aP_I a^{-1} \cap L_I$ is a parabolic subgroup of L_I with unipotent radical $aU_I a^{-1} \cap L_I$,

3. $P_K \cap P_I^- = L_{I \cap K}(L_K \cap U_I^-)(U_K \cap P_I^-)$ and $U_{I \cap K}^- = U_K^-(L_K \cap U_I^-)$,

4. if $K \subseteq I$, then $U_I \subseteq U_K$, $P_K = (P_K \cap L_I)(P_K \cap U_I) = (P_K \cap L_I)U_I$ and $P_K \cap L_I$ is a parabolic subgroup of L_I with Levi decomposition $P_K \cap L_I = (U_K \cap L_I)L_K$.

Let M be a finite monoid with group of units G. If $X \subseteq M$, by $E(X)$ we denote the set of idempotents of X. Write $E = E(M)$. If $e \in E$, then we define

$$P(e) = \{x \in G \mid xe = exe\}, \quad P^-(e) = \{x \in G \mid ex = exe\},$$

$$U(e) = \{x \in G \mid xe = e\}, \quad U^-(e) = \{x \in G \mid ex = e\},$$

$$L(e) = \{x \in G \mid xe = ex\}.$$

It is clear that these are subgroups of G. (For $P(e), P^-(e)$ we use the finiteness of G here.) M is called a monoid of Lie type on the group G if M is a finite regular monoid with zero such that the following two conditions are satisfied

1. the $\mathcal{J}$-classes of M are of the form GaG for $a \in M$,

2. for all $e \in E$, $P(e), P^-(e)$ are opposite parabolic subgroups of G and $R_u(P(e)) \subseteq U(e)$, $R_u(P^-(e)) \subseteq U^-(e)$.

Our definition differs from that originally given in [109], but as it will be seen in Corollary 8.6, the two definitions are equivalent.

Lemma 8.2 *Let M be a monoid of Lie type on a group G. If $e \in E(M)$, then the $\mathcal{H}$-class of e in M is equal to $P(e)eP^-(e) = eL(e) = L(e)e$ and $L(e) = P(e) \cap P^-(e)$ is the common Levi factor of $P(e), P^-(e)$. Moreover, the $\mathcal{R}$- and $\mathcal{L}$-classes of e in M are eG, Ge respectively. In particular, M is unit regular, that is, $M = E(M)G = GE(M)$.*

Proof. If $xey\mathcal{H}e$ for some $x, y \in G$, then $xe\mathcal{H}e$ and $ey\mathcal{H}e$ because the principal factor of e in M is completely 0-simple. Therefore $exe = xe$ and $eye = ey$, so we get $x \in P(e), y \in P^-(e)$. It is clear that $L(e) = P(e) \cap P^-(e)$. Since $R_u(P(e))e = e$ and $eR_u(P^-(e)) = e$, it follows that

$xey \in L(e)eL(e)$. But $le = ele = el$ for every $l \in L(e)$ implies that $L(e)e = eL(e)$. Let $J = GeG$. Now, $eG = (eL(e))G = (eGe \cap J)G = eJ \cap J$, and similarly $Ge = Je \cap J$. The remaining assertion is now clear. □

It is easy to see that $M_n(\mathbf{F}_q)$ is a monoid of Lie type on the group $GL_n(\mathbf{F}_q)$, see Chapter 2. Other important examples arise from the following more general construction.

Example Let $M \subseteq M_n(\overline{\mathbf{F}}_q)$ be a Zariski closed connected submonoid with zero such that the group of units G of M is a reductive group. That is, $R_u(G) = \{1\}$. Assume that a surjective morphism $\sigma : M \longrightarrow M$ is given. If $M_\sigma = \{x \in M \,|\, \sigma(x) = x\}$ is a finite monoid, we call it a finite reductive monoid. It is known that M_σ is a monoid of Lie type, cf. [111]. For example, if σ is the Frobenius map $\sigma((a_{ij})) = (a_{ij}^q)$ and $M = M_n(\overline{\mathbf{F}}_q)$, then $M_\sigma = M_n(\mathbf{F}_q)$. An excellent introduction to the theory of monoids of this type can be found in [130].

Examples of another type will be given in the next two sections.

8.2 Universal monoids of Lie type

The aim of this section is to introduce, for a given group of Lie type G, a monoid $\mathcal{M} = \mathcal{M}_{(G)}$ on G which has certain universal property. Namely, the group G with a principal factor of any other monoid of Lie type on G form a monoid which is a homomorphic image of the corresponding monoid arising from $\mathcal{M}$. Therefore, the local properties of M can often be deduced by studying $\mathcal{M}$. Moreover, the intersection properties of structural subgroups of G are beautifully reflected in $\mathcal{M}$, and consequently the representation theory of $\mathcal{M}$ can be set up in terms of G, as we will see in the next section.

Suppose that $uv = z$ for some $u \in U_I^-, v \in U_I, z \in L_I$. Then $u = zv^{-1} \in U_I^- \cap P_I = U_I^- \cap L_I = \{1\}$, so also $z = 1 = v$. Since $P_I^- P_I = U_I^- L_I U_I$ and L_I normalizes U_I, this easily implies that we have a natural map $P_I^- P_I \longrightarrow L_I$, defined by $ulv \mapsto l$ for $u \in U_I^-, l \in L_I, v \in U_I$. We extend this map to a map $\eta_I : G \longrightarrow L_I \cup \{\theta\}$ defining $\eta_I(x) = \theta$ for $x \notin P_I^- P_I$. Since L_I normalizes U_I, U_I^-, it follows also that

$$\eta_I(ax) = \eta_I(a)\eta_I(x), \quad \eta_I(xb) = \eta_I(x)\eta_I(b) \qquad (8.2)$$

for $a \in P_I^-, b \in P_I, x \in G$. Hence, η_I restricted to P_I, P_I^- is a homomorphism and

$$\eta_I(ax) = a\eta_I(x), \quad \eta_I(xa) = \eta_I(x)a \tag{8.3}$$

for $a \in L_I, x \in G$.

The following result comes from [111].

Theorem 8.3 *Let G be a group of Lie type with Coxeter generators S. We start with idempotents $e_I, I \subseteq S$. Let $\mathcal{M} = \bigcup_{I \subseteq S} J_I \cup \{\theta\}$, a disjoint union, where $J_I = Ge_I G/ \equiv$, with relation $\equiv$ given by*

$$xe_I y \equiv x_1 e_I y_1 \text{ if } x_1^{-1}x \in P_I, \; y_1 y^{-1} \in P_I^- \text{ and } \eta_I(x_1^{-1}x) = \eta_I(y_1 y^{-1})$$

for $x, x_1, y, y_1 \in G$. If $a = xe_I y \in J_I, b = se_K t \in J_K$, then define

$$ab = \begin{cases} xle_{I \cap K}mt & \text{if } ys \in U_I^- lmU_K, \; l \in L_I, \; m \in L_K, \\ \theta & \text{if } ys \notin P_I^- P_K \end{cases}$$

Then $\mathcal{M}$ is a monoid of Lie type on G.

Proof. Since η_I is a homomorphism when restricted to P_I or P_I^-, it is clear that $\equiv$ is an equivalence relation. First, we check that the multiplication on $\mathcal{M}$ is well-defined. Let $a = xe_I y, a_1 = x_1 e_I y_1 \in J_I$ and $b = se_K t, b_1 = s_1 e_K t_1 \in J_K$ be such that $a_1 \equiv a, b_1 \equiv b$. Then

$$x_1^{-1}x \in P_I, \; y_1 y^{-1} \in P_I^-, \; \eta_I(x_1^{-1}x) = \eta_I(y_1 y^{-1}),$$
$$s_1^{-1}s \in P_K, \; t_1 t^{-1} \in P_K^-, \; \eta_K(s_1^{-1}s) = \eta_K(t_1 t^{-1}).$$

Now $y_1 s_1 = (y_1 y^{-1})(ys)(s^{-1}s_1)$ and it follows that $ys \in P_I^- P_K$ if and only if $y_1 s_1 \in P_I^- P_K$. Let $ys, y_1 s_1 \in P_I^- P_K = U_I^- L_I L_K U_K$. Then there exist $l, l_1 \in L_I, m, m_1 \in L_K$ such that $ys \in U_I^- lmU_K$ and $y_1 s_1 \in U_I^- l_1 m_1 U_K$. Hence $ab = xle_{I \cap K}mt, a_1 b_1 = x_1 l_1 e_{I \cap K}m_1 t_1$. Since $x_1^{-1}x \in P_I$, we get $x_1^{-1}x \in U_I \eta_I(x_1^{-1}x)$. Since U_I is a normal subgroup of P_I,

$$l_1^{-1}x_1^{-1}xl \in U_I l_1^{-1}\eta_I(x_1^{-1}x)l \tag{8.4}$$
$$= U_I l_1^{-1}\eta_I(y_1 y^{-1})l = U_I \eta_I(l_1^{-1}y_1 y^{-1}l)$$

because $l, l_1 \in L_I$ (see (8.3)). Next

$$\begin{aligned} z &= l_1^{-1}y_1 y^{-1}l = l_1^{-1}(y_1 s_1)s_1^{-1}s(ys)^{-1}l \\ &\in l_1^{-1}(U_I^- l_1 m_1 U_K)s_1^{-1}s(U_K m^{-1}l^{-1}U_I^-)l \\ &= U_I^- m_1 U_K s_1^{-1}sU_K m^{-1}U_I^- = U_I^- m_1 s_1^{-1}sm^{-1}U_K U_I^- \end{aligned}$$

since $m, s_1^{-1}s \in P_K$. So, for some $u_1, u_2 \in U_I^-$ and $v \in U_K$ we have

$$z = u_1 m_1 s_1^{-1} s m^{-1} v u_2.$$

Let $z_1 = m_1 s_1^{-1} s m^{-1} v$. Now $z \in P_I^-$, because $y_1 y^{-1} \in P_I^-$ and $l, l_1 \in L_I$. Hence $z_1 = u_1 z u_2$ implies that $z_1 \in P_I^-$ and $\eta_I(z) = \eta_I(z_1)$. Since $s_1^{-1}s \in P_K$ and $m \in L_K$, we have $z_1 = m_1 \eta_K(s_1^{-1}s)m^{-1}w$ for some $w \in U_K$. Let $z_2 = m_1 \eta_K(s_1^{-1}s)m^{-1}$. Then $z_2 \in L_K$ and

$$z_1 = z_2 w \in P_K \cap P_I^- = L_{I \cap K}(L_K \cap U_I^-)(U_K \cap P_I^-)$$

by Lemma 8.1. Since $L_{I \cap K} \subseteq L_K$ and $P_K = L_K U_K$ is a semidirect product, it follows that $z_2 \in L_{I \cap K}(L_K \cap U_I^-), w \in U_K \cap P_I^- = (U_K \cap L_I)(U_K \cap U_I^-)$. Hence, in view of (8.2),

$$\eta_I(z) = \eta_I(z_1) = \eta_I(z_2 w) = \eta_I(z_2)\eta_I(w),$$
$$\eta_I(z_2) \in L_{I \cap K}, \quad \eta_I(w) \in U_K \cap L_I.$$

In particular, $\eta_I(z) \in P_I \cap P_K = P_{I \cap K}$. So, by (8.4)

$$l_1^{-1}x_1^{-1}xl \in U_I \eta_I(z) \subseteq U_I P_{I \cap K} \subseteq P_{I \cap K}. \tag{8.5}$$

Moreover, since $U_I \cup U_K \subseteq U_I \cap U_K$ and $\eta_I(w) \in U_K$,

$$\eta_{I \cap K}(l_1^{-1}x_1^{-1}xl) = \eta_{I \cap K}(\eta_I(z)) = \eta_{I \cap K}(\eta_I(z_2)\eta_I(w)) = \eta_{I \cap K}(\eta_I(z_2)),$$

which in view of $U_I^- \subseteq U_{I \cap K}^-$ is equal to

$$\eta_{I \cap K}(z_2) = \eta_{I \cap K}(m_1 \eta_K(s_1^{-1}s)m^{-1}). \tag{8.6}$$

Similarly to (8.5) one shows that $m_1 t_1 t^{-1} m^{-1} \in P_{I \cap K}^-$. Now, since $t_1 t^{-1} \in P_K^-$ and $m \in L_K$, we get $m_1 t_1 t^{-1} m^{-1} \in m_1 \eta_K(t_1 t^{-1})m^{-1}U_K^-$. Since $U_K^- \subseteq U_{I \cap K}^-$ and $\eta_I(s_1^{-1}s) = \eta_I(t_1 t^{-1})$, in view of (8.6) we have

$$\eta_{I \cap K}(m_1 t_1 t^{-1} m^{-1}) = \eta_{I \cap K}(m_1 \eta_K(t_1 t^{-1})m^{-1})$$
$$= \eta_{I \cap K}(m_1 \eta_K(s_1 s^{-1})m^{-1}) = \eta_{I \cap K}(l_1^{-1}x_1^{-1}xl).$$

It follows that the multiplication on $\mathcal{M}$ is indeed well-defined.

Next, we prove associativity. Let $a = xe_I y \in J_I, b = se_K t \in J_K, c = ue_N v \in J_N$, for some $I, K, N \subseteq S$ and $x, y, s, t, u, v \in G$. Put $H =$

$I \cap K \cap N$. Suppose that $(ab)c \neq \theta$. Then $ys \in U_I^- l_1 l_2 U_K$ for some $l_1 \in L_I, l_2 \in L_K$, and $ab = x l_1 e_{I \cap K} l_2 t$. Since $(ab)c \neq \theta$,

$$l_2 t u \in U_{I \cap K}^- l_3 l_4 U_N$$

for some $l_3 \in L_{I \cap K}, l_4 \in L_N$. So $(ab)c = x l_1 l_3 e_H l_4 v$.

Now $U_{I \cap K}^- = U_K^-(U_I^- \cap L_K)$ by Lemma 8.1. Hence

$$tu \in l_2^{-1} U_{I \cap K}^- l_3 l_4 U_N = l_2^{-1} U_K^-(U_I^- \cap L_K) l_3 l_4 U_N = U_K^- l_2^{-1}(U_I^- \cap L_K) l_3 l_4 U_N.$$

So, for some $z \in U_I^- \cap L_K$,

$$tu \in U_K^- l_2^{-1} z l_3 l_4 U_N$$

and

$$bc = s l_2^{-1} z l_3 e_{K \cap N} l_4 v. \tag{8.7}$$

Now

$$(ys) l_2^{-1} z l_3 \in (U_I^- l_1 l_2 U_K) l_2^{-1} z l_3 = U_I^- l_1 U_K z l_3 = U_I^- l_1 z l_3 U_K$$

(because $z, l_3 \in L_K$), which is equal to

$$U_I^- l_1 z l_1^{-1} l_1 l_3 U_K = U_I^- l_1 l_3 U_K,$$

the latter since $z \in U_I^-, l_1 \in L_I$. Since $U_K \subseteq U_{K \cap N}$ and $l_1 l_3 \in L_I$, in view of (8.7) we see that

$$a(bc) = x l_1 l_3 e_H l_4 v = (ab)c.$$

Similarly, $a(bc) \neq \theta$ implies that $a(bc) = (ab)c$. Therefore $\mathcal{M}$ is a monoid with group of units $G = J_S$.

It remains to show that $\mathcal{M}$ satisfies the conditions defining monoids of Lie type. Every $\mathcal{J}$-class J of $\mathcal{M}$ contains some $Ge_I G, I \subseteq S$. It is also clear that $\mathcal{M} e_I \mathcal{M} \setminus Ge_I G$ consists of non-generators of the ideal $\mathcal{M} e_I \mathcal{M}$. Therefore $J = Ge_I G$. Since $\mathcal{M}$ is finite and every $\mathcal{J}$-class contains an idempotent, $\mathcal{M}$ is regular.

Let $x \in P(e_I)$. Then $xe_I = e_I xe_I$. So $e_I xe_I \neq \theta, x \in P_I^- P_I$. Thus $xe_I = e_I xe_I = \eta_I(x)e_I$. Hence $x^{-1}\eta_I(x) \in P_I$. Since $\eta_I(x) \in L_I$, we have $x \in P_I$. Therefore $P(e_I) \subseteq P_I$. Clearly $L_I \subseteq P(e_I)$. If $u \in U_I$, then

$\eta_I(u) = 1$. So $ue_I = e_I$ and $U_I \subseteq P(e_I)$. Hence $P_I = L_I U_I \subseteq P(e_I)$. So $P(e_I) = P_I$. Similarly $P^-(e_I) = P_I^-$ and $e_I U_I^- = \{e_I\}$.

We claim that every idempotent $f \in Ge_I G$ is conjugate to e_I. So, let $f = xe_I y$. Then $xe_I y xe_I y = xe_I y$ implies that $e_I y xe_I = e_I$. Hence $yx \in P_I^- P_I, \eta_I(yx) = 1$. Then $yx \in U_I^- U_I$. So $u^{-1}y = vx^{-1}$ for some $u \in U_I^-, v \in U_I$ and

$$ f = xe_I y = xv^{-1} e_I u^{-1} y = (xv^{-1}) e_I (vx^{-1}), $$

so indeed $f = g^{-1} e_I g$ for some $g \in G$.

It follows easily that $P(f) = g^{-1} P(e_I) g, P^-(f) = g^{-1} P^-(e_I) g$ and $U(f) = g^{-1} U(e_I) g, U^-(f) = g^{-1} U(e_I) g$. Therefore, the defining conditions checked above for the idempotent e_I also hold for f. This completes the proof of the theorem. $\square$

Remark We have seen in the proof of the theorem that, for every $I \subseteq S$, one has $P(e_I) = P_I$ and $U_I \subseteq U(e_I)$ in the monoid $\mathcal{M}$. If $xe_I = e_I$ for some $x \in G$, then $x \in P_I$ and $\eta_I(x) = 1$. Hence $x \in U_I$. Therefore we get $U(e_I) = U_I$. Similarly $P^-(e_I) = P_I^-, U^-(e_I) = U_I^-$. Also $e_I x = xe_I$ if and only if $x \in P_I \cap P_I^- = L_I$. Hence $L(e_I) = L_I$ and it is isomorphic to the maximal subgroup eL_I of $\mathcal{M}$.

Choose any subset $I \subseteq S$. Consider $J_I^0 = J_I \cup \{\theta\} \subseteq \mathcal{M}$. Then J_I^0 is a subsemigroup of $\mathcal{M}$ and it is isomorphic to a principal factor of $\mathcal{M}$. So, it must be completely 0-simple. We will describe a Rees matrix presentation of J_I^0. By the above remark, every maximal subgroup of J_I^0 is isomorphic to $e_I J_I e_I \setminus \{\theta\} \simeq L_I$. Write G/P_I for a complete set of representatives $a_1, \ldots, a_t$ of cosets $gP_I, g \in G$. Similarly, let $G/P_I^- = \{b_1, \ldots, b_t\}$ be the representatives of all $P_I^- g, g \in G$. The sandwich matrix is defined by $Q_I = (\eta_I(b_j a_i))$. In other words, we get $J_I^0 \simeq \mathcal{M}(L_I, G/P_I, G/P_I^-, Q_I)$. Here, for $xe_I y \in J_I$ we write $x = a_i l_1 u_1, y = u_2 l_2 b_j$ for some i, j and some $l_1, l_2 \in L_I, u_1 \in U_I, u_2 \in U_I^-$. So, $xe_I y$ is identified with $(l_1 l_2, a_i, b_j)$. This can be compared with Proposition 2.5. Note that $\{J_I \mid I \subseteq S\}$ is the set of nonzero $\mathcal{J}$-classes of $\mathcal{M}$ and that $J_I^0 J_K^0 = J_{I \cap K}^0$ for any $I, K \subseteq S$. Now, let $\mathcal{M}_I = G \cup J_I \cup \{\theta\}$. Then $\mathcal{M}_I$ is itself a monoid of Lie type on G.

Lemma 8.4 *If $I \subseteq S$, then $e_I \mathcal{M} e_I$ is isomorphic to the universal monoid of Lie type on the group L_I.*

Proof. It is easy to see that $e_I(Ge_KG)e_I = L_{Ie_{I\cap K}L_I} \cup \{\theta\}$ for every $K \subseteq S$. So $e_I\mathcal{M}e_I = \bigcup_{K \subseteq I} L_{Ie_K}L_I \cup \{\theta\}$. Let $K \subseteq I$. As noted above, $P(e_K) = P_K$ and $L(e_K) = L_K$ in $\mathcal{M}$. Hence, by Lemma 8.1 $\{x \in L_I \mid xe_K = e_Kxe_K\} = P_K \cap L_I$ is a parabolic subgroup of L_I with unipotent radical $U_K \cap L_I = \{x \in L_I \mid xe_K = e_K\}$. Since a symmetric assertion holds for P_K^-, U_K^-, the result follows easily. $\square$

For an $\mathcal{J}$-class of a monoid M of Lie type on G the set $J \cup \{\theta\}$, with multiplication $a \cdot b = ab$ if $ab \in J$ and $a \cdot b = \theta$ otherwise, becomes a completely 0-simple semigroup isomorphic to a principal factor of M. Then $M(J) = G \cup J \cup \{\theta\}$, with obvious multiplication, becomes a monoid of Lie type. This is called a local monoid of M. The universal property of $\mathcal{M}$ has now the following form.

Corollary 8.5 *Let M be a monoid of Lie type on the group G. Assume that $J = GeG$ is an $\mathcal{J}$-class of M with $e \in E(M)$ and $P(e)$ conjugate to P_I for some $I \subseteq S$. Then there exists an onto homomorphism $\phi : \mathcal{M}_I \longrightarrow M(J)$ such that $f = \phi(e_I)$ satisfies $P(f) = P_I, P^-(f) = P_I^-$ and ϕ is the identity map on G.*

Proof. We know that $g^{-1}P(e)g = P_I, g^{-1}P^-(e)g = P_I^-$ for some $g \in G$. Let $f = g^{-1}eg$. Then $P(f) = P_I$ and $P^-(f) = P_I^-$, and hence $U_If = f = fU_I^-$. Assume that $xe_Iy = se_It$ in Ge_IG for some $x, y, s, t \in G$. Then $s^{-1}x \in P_I, ty^{-1} \in P_I^-$ and $\eta_I(s^{-1}x) = \eta_I(ty^{-1})$. So $s^{-1}x \in lU_I, ty^{-1} \in U_I^-l$ for some $l \in L_I$. Since $U_If = f = fU_I^-$ in M, we get $s^{-1}xf = lf$ and $fty^{-1} = fl$. Now $xfy = sft$ because $lf = fl$ by Lemma 8.2. Hence, $xe_Iy \mapsto xfy$ determines a map $\mathcal{M}_I \longrightarrow M(J)$, which is identity on G. To check that this is a homomorphism, note that $xfysft = xf\eta_I(ys)ft$ for every $x, y, s, t \in G$ because $U_If = f = fU_I^-$. $\square$

As an application, we derive an important technical result concerning the set $E(M)$ of idempotents of any monoid of Lie type M. We will use the fact that $P_IU_I^-U_I = P_IP_I^-P_I = G$ for $I \subseteq S$. For this, it is enough to know that $BB^-B = G$. By induction on the length of $a \in W$ we first check that $BaB \cap B^- \neq \emptyset$. This is clear if $l(a) = 0$ since $B \cap B^- = T$. Let $a = sc$, where $c \in W, s \in S$ are such that $l(a) > l(c)$. Then $BsBBcB = BaB$ by (8.1). Since $BcB \cap B^- \neq \emptyset \neq BsB \cap B^-$ by the induction hypothesis, we get $BaB \cap B^- \neq \emptyset$, as desired. From the Bruhat decomposition it now follows that $G = BB^-B$.

Corollary 8.6 *Let M be a monoid of Lie type on the group G. Then*

1. *if $e, f \in E(M)$, then $e \mathcal{J} f$ if and only if $g^{-1} f g = e$ for some $g \in G$,*

2. *if $e, f \in E(M)$ and $e \mathcal{J} f$, then there exist $e_1, e_2 \in E(M)$ such that $e \mathcal{R} e_1, e_1 \mathcal{L} e_2, f \mathcal{R} e_2$,*

Proof. Because of Corollary 8.5 it is enough to check that the assertions hold for the universal monoid $\mathcal{M}$ on G. For 1) this was done in the proof of Theorem 8.3. To prove 2), conjugating, we may assume that $e = e_I$ for some $I \subseteq S$. Moreover, $f = x^{-1} e_I x$ for some $x \in G$. Note that $x \in G = P_I P_I^- U_I$. Hence, we may choose $y \in P_I^- P_I x \cap U_I$. Then there exists $z \in P_I x$ such that $y \in P_I^- z$. Now $y \in U_I$ implies that $e_1 = ey$ is an idempotent. Clearly $e_1 \mathcal{R} e$. Moreover $zx^{-1} = q \in P_I$ implies that the idempotent $e_2 = z^{-1} e z = x^{-1} q^{-1} e z$ satisfies $e_2 \mathcal{R} f$. Similarly, $y z^{-1} \in P_I^-$ implies that $e_1 \mathcal{L} e_2$. $\square$

Finally, we state without proof an analogue of Theorem 2.7 in our more general context. As in the case of $GL_n(K) \subseteq M_n(K)$, it is a basis for developing a combinatorial approach to any monoid M of Lie type with group of units G. By $\mathcal{U} = \mathcal{U}(M)$ we denote the set of $\mathcal{J}$-classes of M. We have seen above that for every $J_1, J_2 \in \mathcal{U}(\mathcal{M})$ there exists $J \in \mathcal{U}(\mathcal{M})$ such that $J_1 J_2 \subseteq J \cup \{\theta\}$. Then we write $J = J_1 \wedge J_2$. In this way $\mathcal{U}(\mathcal{M})$ becomes a lattice. From the multiplication rule in $\mathcal{M}$ it follows easily that $e_K e_I = e_{K \cap I}$ for $K, I \subseteq S$. So, this lattice can be represented by $\{e_I \mid I \subseteq S\} \cup \{\theta\}$.

Let J be a $\mathcal{J}$-class of M. Then $J = GeG$ for some $e \in E(M)$ and $P(e), P^-(e)$ are opposite parabolic subgroups of G. So there exists $x \in G$ such that $x^{-1} P(e) x = P_I, x^{-1} P^-(e) x = P_I^-$ for some $I \subseteq S$. Then $e_J = x^{-1} e x \in J$ is an idempotent such that $P(e_J) = P_I$ and $P^-(e_J) = P_I^-$. We put $\Lambda(M) = \{e_J \mid J \in \mathcal{U}(M)\}$ and $E_0 = \bigcup_{n \in N} n^{-1} \Lambda(M) n$.

Theorem 8.7 *Assume that M is a monoid of Lie type. Define $R = \langle N, \Lambda(M) \rangle \subseteq M$. Then*

1. *$\Lambda(M)$ is a cross section lattice of idempotents of M, that is, for all $J_1, J_2 \in \mathcal{U}(M)$ we have $e_{J_1} e_{J_2} = e_{J_2} e_{J_1} = e_{J_3}$ for some $J_3 \in \mathcal{U}(M)$,*

2. *$R = N E_0$ is an inverse monoid with group of units N and $E(R) = E_0$,*

3. $\equiv$ *defined by* $ex \equiv e'y$ *if* $e = e', xy^{-1} \in T$, *is a congruence on* R,

4. $R' = R/\equiv$ *is an inverse monoid with group of units* W *and* $E(R') \simeq E_0$,

5. $BuB = BvB$ *if and only if* $u \equiv v$, *for* $u, v \in R$, *and* $M = \bigcup_{u \in R} BuB$, $R' = \bigcup_{e \in \Lambda(M)} We'W$, *where* e' *denotes the image of* e *in* R'.

8.3 Connections with group representations

Foundations of the representation theory of monoids of Lie type are presented in this section. First, we show that the complex algebra $\mathbf{C}[M]$ is semisimple. This extends Theorem 2.34. Secondly, we prove that all irreducible modular representations of M restrict to irreducible representations of the group G. These are exceptional properties within the class of all finite regular monoids. The results come from [98] and [111].

We shall consider representations $\phi : M \longrightarrow M_n(F)$, over a field F, such that $\phi(\theta) = 0$. Therefore, irreducible representations of M over F correspond to irreducible $F_0[M]$-modules. If V is an F-space, then $GL(V)$ denotes the group of F-automorphisms of V. We first explain the role of subgroups of the type $P(e)$ for the representation theory of the group G.

Lemma 8.8 *Let* M *be a finite monoid with zero and group of units* G. *Let* $e \in E(M)$. *Suppose* $J = GeG$ *is a* $\mathcal{J}$-*class of* M *such that* $J^0 = J \cup \{\theta\}$ *is an ideal of* M. *Let* V *be an irreducible* $F_0[M]$-*module, for a field* F, *such that* $eV \neq 0$. *Let* $P = \{a \in G \mid ae = eae\}$ *and* $\phi : P \longrightarrow GL(eV), \psi : G \longrightarrow GL(V)$ *denote the associated representations. Suppose the sandwich matrix* Q *of* J^0 *(under a Rees presentation of* J^0*) is* $t \times t$ *and invertible. Then* ψ *is equivalent to the induced representation* ϕ_P^G. *In particular* $\dim(V) = |G/P| \dim(eV)$.

Proof. From Section 4.2 we know that invertibility of Q implies that $\dim(V) = t \cdot \dim(eV)$. Let $g_1 = 1, g_2, \ldots, g_k$ be the left coset representatives of P in G. Suppose that $g_i e \mathcal{R} g_j e$. Then $g_i^{-1} g_j e \mathcal{R} e$, so $g_i^{-1} g_j \in P$.

Hence $i = j$. It follows that $k \leq t$. Let $x \in G$. Then $xP = g_i P$ for some i. Since $Pe = ePe$, we get $xeV = xPeV = g_i PeV = g_i eV$. So $W = \sum_i g_i eV$ is invariant under J^0. Since J^0 is an ideal of M, and V is an irreducible $\mathbf{C}_0[M]$-module, $W = V$. Hence $\dim(V) \leq k \cdot \dim(eV)$, so that $k = t$ and $V = g_1 eV \oplus \cdots \oplus g_t eV$.

Let $g \in G$. For every i there exists j such that $gg_i P = g_j P$. So $g_j^{-1} gg_i \in P$ and $g_j^{-1} gg_i(eV) = eV$. This means that V is the $F[G]$-module induced from the $F[P]$-module eV, see [17], §12D. The assertion follows. $\square$

The first main result of this section, originally proved in [98], reads as follows.

Theorem 8.9 *Let M be a finite monoid of Lie type. Then $\mathbf{C}[M]$ is a semisimple algebra.*

According to the results of Sections 4.2 and 4.3, every irreducible complex representation ϕ of M comes from a representation of a maximal subgroup H of M. We need to show that every ϕ is completely reducible. The proof is based on an induction of Harish-Chandra type, the starting point being the case of so called cuspidal representations of H. The crucial step is to show that certain system of linear equations is triangularizable in the cuspidal case. We start with a series of preparatory lemmas.

Lemma 8.10 *Let $Z = \mathcal{M}(H, t, t, Q)$ be a finite completely 0-simple semigroup. Assume that for every two nonzero idempotents $e, f \in Z$ there exists an automorphism $\sigma \in \mathrm{Aut}(Z)$ such that $\sigma(e) = f$. If $x_1, \ldots, x_t \in F[H]$, for a field F, are such that $(x_1, \ldots, x_t)Q = (1, 0, \ldots, 0)$, then Q is invertible in $M_t(F[H])$.*

Proof. Let e be an idempotent lying in the $\mathcal{R}$-class of Z corresponding to the first row of $Z = \mathcal{M}(H, t, t, Q)$. The hypothesis on Q implies that there exists an element $x \in F_0[Z]$ such that $\mathrm{supp}(x) \subseteq eZ$ and for every $y \in F_0[Z]$

$$xy = \begin{cases} y & \text{if } \mathrm{supp}(y) \subseteq eZ \\ 0 & \text{if } \mathrm{supp}(y) \cap eZ = \emptyset \end{cases}$$

(x corresponds to the matrix whose first row is $(x_1, \ldots, x_t)$, the other rows being zero, under the identification $F_0[Z] \simeq \mathcal{M}(F[H], t, t, Q)$).

In other words, x is a left identity of the algebra $F_0[eZ]$ such that it annihilates on the left the remaining $\mathcal{R}$-classes of Z. Choose nonzero idempotents $e = e_1, \ldots, e_t$ from all distinct nonzero $\mathcal{R}$-classes of Z. By the hypothesis, there exist automorphisms σ_i of Z such that $\sigma_i(e) = e_i, i = 1, \ldots, t$. Since every σ_i permutes the nonzero $\mathcal{R}$-classes of Z, it is clear that every $\sigma_i(x)$ is a left identity of $F_0[e_i Z]$, which annihilates $e_j Z, j \neq i$, on the left. Here σ_i are treated as the induced automorphisms of $F_0[Z]$. It follows easily that $\sum_{i=1}^t \sigma_i(x)$ is a left identity of $F_0[Z]$. Thus, Q is left invertible, and since $F_0[Z]$ is finite dimensional over F, we see that Q is invertible in $M_t(F[H])$. $\square$

Recall that, if $Q = (q_{ij})$ is a matrix over H^0, for a group H, and ϕ is a representation of H, then we let $\phi(Q) = (\phi(q_{ij}))$.

Corollary 8.11 *Assume that for every irreducible complex representation ϕ of a maximal subgroup H of a finite completely 0-simple semigroup $Z = \mathcal{M}(H, t, t, Q)$ we have $(A_1, \ldots, A_t)\phi(Q) = (1, 0, \ldots, 0)$ for some matrices $A_1, \ldots, A_t \in M_t(\mathbf{C}[H])$. Assume also that for every nonzero $e, f \in E(Z)$ there exists $\sigma \in \text{Aut}(Z)$ such that $\sigma(e) = f$. Then Q is invertible in $M_t(\mathbf{C}[H])$ and $\mathbf{C}[Z]$ is a semisimple algebra.*

Proof. The algebra $\mathbf{C}[H]$ is semisimple by Maschke's theorem. Since the assumption allows us to solve the equation

$$(x_1, \ldots, x_t)Q = (1, 0, \ldots, 0)$$

in every projection onto a simple component of this algebra, the hypotheses of the foregoing lemma are satisfied. The result follows. $\square$

We will need another technical result on groups of Lie type.

Lemma 8.12 *Let $I \subseteq S$, $a_1, a_2 \in W$ be such that a_1 is of minimal length in the coset $W_I a_1$. If $l(a_1) \geq l(a_2)$ and $a_1 \neq a_2$, then $a_1 B \cap P_I^- P_I a_2 = \emptyset$.*

Proof. Since $B \cap L_I, N \cap L_I$ form a BN-pair for L_I, by the Bruhat decomposition for L_I we have

$$P_I^- P_I = U_I^- L_I U_I = U_I^-(B^- \cap L_I)W_I(B \cap L_I)U_I = B^- W_I B.$$

For any $a \in W, s \in S$, by (8.1) we get

$$B^- aBs = a_0 Ba_0 aBs \subseteq a_0 Ba_0 aB \cup a_0 Ba_0 asB = B^- aB \cup B^- asB.$$

Let $a_2 = s_1 \cdots s_t$, where $s_i \in S$ and $l(a_2) = t$. Then it follows that

$$P_I^- P_I a_2 = B^- W_I Bs_1 \cdots s_t \subseteq \bigcup_{i_1 < \cdots < i_r} B^- W_I s_{i_1} \cdots s_{i_r} B.$$

Suppose that $a_1 B \cap P_I^- P_I a_2 \neq \emptyset$. By the Bruhat decomposition the union $G = \bigcup_{a \in W} B^- aB$ is disjoint. Hence $a_1 \in W_I s_{i_1} \cdots s_{i_r}$ for some $i_1 < \cdots < i_r$. Since a_1 is of minimal length in $W_I a_1$ and $l(a_1) \geq l(a_2) = t$, we must have $r = t$ and $a_1 = a_2$. This contradiction completes the proof. $\square$

A representation ϕ of $L = L_I$ is called cuspidal if for every parabolic subgroup P of L with unipotent radical V we have $\sum_{v \in V} \phi(v) = 0$. The study of all irreducible complex representations of L reduces, in some sense, to the case of cuspidal representations. Namely, if ϕ is an irreducible representation of L, then there exist $K \subseteq I, K \neq I$, and an irreducible cuspidal representation π of L_K such that, if $\bar{\pi}$ is the trivial extension of π to $P_K \cap L$ (that is, $\bar{\pi}(u) = 1$ for $u \in U_K \cap L$), then ϕ is a component of the induced representation $\psi = \bar{\pi}_{P_K \cap L}^L$, see [13], Chapter 9. This often allows an inductive approach to representations of G. The following lemma, being the starting point of such an induction, is the first step in the proof of Theorem 8.9.

Lemma 8.13 *Let $\phi : L_I \longrightarrow GL_n(\mathbf{C})$ be an irreducible cuspidal representation. Then there exist $A_1, \ldots, A_t \in M_n(\mathbf{C})$ such that*

$$(A_1, \ldots, A_t)\phi(Q_I) = (I_n, 0, \ldots, 0),$$

where Q_I is the $t \times t$ sandwich matrix corresponding to the principal factor $J_I^0 = Ge_I G \cup \{\theta\}$ of $\mathcal{M}$ and I_n is the identity $t \times t$ matrix.

Proof. Write $L = L_I, P = P_I, P^- = P_I^-$. We start by carefully choosing the coset representatives of G/P and G/P^-. We know that any coset of W_I in W has a unique element of minimal length, [13], Proposition 2.3.3. Let these right coset representatives be $a_1 = 1, a_2, \ldots, a_s$. Then, by the Bruhat decomposition,

$$G = B^- WB = \bigcup_i B^- W_I a_i B = \bigcup_i P^- a_i B. \qquad (8.8)$$

Thus $G = \bigcup_i P^- a_i B$. We claim that this union is disjoint. Suppose that $P^- a_i B = P^- a_j B$ for some i, j. Assume for example that $l(a_i) \geq l(a_j)$. Since $a_i B$ intersects $P^- a_j$, we see by Lemma 8.12 that $a_i = a_j$, as desired. Also by the the Bruhat decomposition

$$G = BWB = \bigcup_i Ba_i^{-1}W_I B = \bigcup_i Ba_i^{-1}P. \tag{8.9}$$

Thus $G = \bigcup_i Ba_i^{-1}P$. We claim that this union is also disjoint. Suppose that $Ba_i^{-1}P = Ba_j^{-1}P$. Then $Pa_i B = Pa_j B$. Again, if $l(a_i) \geq l(a_j)$, then by Lemma 8.12 we get $a_i = a_j$.

For $i = 1, \ldots, s$, let

$$V_i = U \cap a_i^{-1}U_I a_i, \quad Y_i = U \cap a_i^{-1}U_I^- a_i, \quad Z_i = U \cap a_i^{-1}La_i.$$

Since U is the product of the positive root subgroups in any order, we see that $U = Y_i Z_i V_i$. Hence

$$P^- a_i B = P^- a_i U = P^- a_i V_i, \quad Ba_i^{-1}P = Ua_i^{-1}P = Y_i a_i^{-1}P \tag{8.10}$$

for $i = 1, \ldots, s$, because a_i normalizes T and $U_I^-, L \subseteq P^-, U_I, L \subseteq P$. Let $a_i = Tn_i, n_i \in N$. Suppose that $v_1, v_2 \in V_i$ are such that $P^- n_i v_1 = P^- n_i v_2$. So $n_i v_1 v_2^{-1} n_i^{-1} \in P^-$. But it is easy to see that $n_i v_1 v_2^{-1} n_i^{-1} \in U_I$. Since $P^- \cap U_I = L \cap U_I = \{1\}$, we get $v_1 v_2^{-1} = 1$ and $v_1 = v_2$. Hence, by (8.8),(8.10), $n_i v, v \in V_i, i = 1, \ldots, s$, is a set of distinct right coset representatives for P^- in G. Suppose $y_1, y_2 \in Y_i$ are such that $y_1 n_i^{-1}P = y_2 n_i^{-1}P$. Then $n_i y_1^{-1} y_2 n_i^{-1} \in P$. But $n_i y_1 y_2^{-1} n_i^{-1} \in U_I^-$. Hence, similarly, $y_1 y_2^{-1} = 1$ and $y_1 = y_2$. Thus, by (8.9),(8.10), $yn_i^{-1}, y \in Y_i, i = 1, \ldots, s$, is a set of distinct left coset representatives of P in G. Therefore, as noted after Theorem 8.3, the entries of the matrix Q_I correspond to the pairs $(n_i v, yn_j^{-1})$, where $v \in V_i, y \in Y_j, i, j = 1, \ldots, s$. If $x \in G$, for simplicity let $\overline{x} = \eta_I(x)$. Then the entry in the position $(n_i v, yn_j^{-1})$ is $(\overline{n_i v y n_j^{-1}})$.

Let $N_W(L)$ be the normalizer of L in W. We next arrange the a's so that $a_1 = 1, \ldots, a_r \in N_W(L)$ and $a_k \notin N_W(L)$ for $k = r + 1, \ldots, s$. Further, let $l(a_1) \leq l(a_2) \leq \cdots \leq l(a_r)$. Since $U_I^- \subseteq U^-$, we have

$$Y_1 = U \cap U_I^- = \{1\}.$$

Thus, we are trying to find $A_{n_i v} \in M_n(\mathbf{C})$ for $v \in V_i, i = 1, \ldots, s$, such that

$$\sum_{n_i v} A_{n_i v} \phi(\overline{n_i v}) = I_n \tag{8.11}$$

and

$$\sum_{n_i v} A_{n_i v} \phi(\overline{n_i v y n_j^{-1}}) = 0 \text{ for } y \in Y_j, \; j = 2, \ldots, s. \tag{8.12}$$

Let $1 \le j \le r$. Then $a_i \in N_W(L)$. Thus, for any root subgroup X_α of U_I, $a_i X_\alpha a_i^{-1}$ is either contained in U_I or U_I^-. Since U_I is the product of its root subgroups in any order, it follows that

$$U_I = Y_i V_i \text{ for } i = 1, \ldots, r. \tag{8.13}$$

Now, for $y \in Y_i, v \in V_i, x \in G$ we have by (8.2)

$$\overline{n_i y v x} = \overline{(n_i y n_i^{-1})(n_i v x)} = \overline{n_i y n_i^{-1}} \, \overline{n_i v x} = \overline{n_i v x}. \tag{8.14}$$

From (8.13) and (8.14) it follows that for all $x \in G$

$$\sum_{u \in U_I} \phi(\overline{n_i u x}) = |Y_i| \sum_{v \in V_i} \phi(\overline{n_i v x}), \quad i = 1, \ldots, r. \tag{8.15}$$

Now let $j \ge r + 1$. By Lemma 8.1 $R = a_j P a_j^{-1} \cap L$ is a parabolic subgroup of L with unipotent radical $V = a_j U_I a_j^{-1} \cap L$. Suppose $R = L$. Then $a_j^{-1} L a_j \subseteq P$. Since $j \ge r + 1$, $a_j \notin N_W(L)$. Since L is generated by T and root subgroups, there is a root subgroup $X_\alpha \subseteq L$ such that $a_j^{-1} X_\alpha a_j \not\subseteq L$. So $a_j^{-1} X_\alpha a_j \subseteq U_I$. Hence $X_{-\alpha} \subseteq L, a_j^{-1} X_{-\alpha} a_j \subseteq U_I^-$, contradicting the assumption $a_j^{-1} L a_j \subseteq P$. We have shown that $R \ne L$. Since ϕ is cuspidal, it follows that

$$\sum_{a \in V} \phi(a) = 0. \tag{8.16}$$

Let $a \in V, i \le r$. Since $n_j^{-1} a n_j \in U_I$, we have by (8.3)

$$\sum_{u \in U_I} \phi(\overline{n_i u n_j^{-1}}) \phi(a) = \sum_{u \in U_I} \phi(\overline{n_i u n_j^{-1} a})$$
$$= \sum_{u \in U_I} \phi(\overline{n_i (u n_j^{-1} a n_j) n_j^{-1}}) = \sum_{u \in U_I} \phi(\overline{n_i u n_j^{-1}}).$$

Hence, considering such equalities for all $a \in V$, by (8.16) we get

$$\sum_{u \in U_I} \phi(\overline{n_i u n_j^{-1}}) = 0. \tag{8.17}$$

Now let $w \in U$. Then $w = lv$ for some $l \in L, v \in U_I$, because $U \subseteq P$. Since $n_i \in N_G(L)$, we have by (8.17), (8.3)

$$\sum_{u \in U_I} \phi(\overline{n_i u w n_j^{-1}}) = \sum_{u \in U_I} \phi(\overline{n_i u l v n_j^{-1}})$$

$$= \phi(n_i l n_i^{-1}) \sum_{u \in U_I} \phi(\overline{n_i(l^{-1}ul)v n_j^{-1}})$$

$$= \phi(n_i l n_i^{-1}) \sum_{u \in U_I} \phi(\overline{n_i u n_j^{-1}}) = 0.$$

Thus, for $w \in U$,

$$\sum_{u \in U_I} \phi(\overline{n_i u w n_j^{-1}}) = 0, \quad i \leq r, \; j \geq r+1. \tag{8.18}$$

From general considerations it follows that, if a solution to (8.11), (8.12) exists, it must be such that $A_{n_i v} = A_{n_i}$ for $v \in V_i, i = 1, \dots, r$. So at this time let $A_{n_i v} = A_{n_i}$ for $v \in V_i, i = 1, \dots, r$. We further let $A_{n_i v} = 0$ for $i \geq r+1, v \in V_i$. Then, by (8.15) applied to $x = y n_j^{-1}$ and (8.18) applied to $w = y$, we see that (8.12) is automatically valid for $j \geq r+1, y \in Y_j$. Hence (8.11), (8.12) now become by (8.15)

$$\sum_{i=1}^{r} \frac{1}{|Y_i|} A_{n_i} \sum_{u \in U_I} \phi(\overline{n_i u}) = I_n, \tag{8.19}$$

$$\sum_{i=1}^{r} \frac{1}{|Y_i|} A_{n_i} \sum_{u \in U_I} \phi(\overline{n_i u y n_j^{-1}}) = 0, \quad y \in Y_j, j = 2, \dots, r. \tag{8.20}$$

Now for $j \leq r, a_j \in N_W(L)$. Hence $a_j^{-1} U_I^- a_j \cap L = \{1\}$. So, by Lemma 8.1

$$Y_j = U \cap a_j^{-1} U_I^- a_j \subseteq P \cap a_j^{-1} U_I^- a_j = (L \cap a_j^{-1} U_I^- a_j)(U_I \cap a_j^{-1} U_I^- a_j) \subseteq U_I.$$

Thus (8.20) simplifies to

$$\sum_{i=1}^{r} \frac{1}{|Y_i|} A_{n_i} \sum_{u \in U_I} \phi(\overline{n_i u n_j^{-1}}) = 0, \quad j = 2, \dots, r. \tag{8.21}$$

Similarly, if $i \leq r$, then $a_i U_I a_i^{-1} \cap L = \{1\}$. Hence $a_i U_I a_i^{-1} \subseteq U_I^- U_I$. Thus

$$\overline{n_i u n_i^{-1}} = 1 \quad \text{for } u \in U_I, \; i = 1, \dots, r. \tag{8.22}$$

Furthermore, by Lemma 8.12

$$\overline{n_i u n_j^{-1}} = 0 \text{ for } 1 \leq j < i \leq r,\, u \in U_I. \tag{8.23}$$

By (8.22), (8.23) we see that (8.19), (8.21) finally become

$$|U_I| A_{n_1} = I_n \tag{8.24}$$

$$\frac{|U_I|}{|Y_j|} A_{n_j} + \sum_{i=1}^{j-1} \frac{1}{|Y_i|} A_{n_i} \sum_{u \in U_I} \phi(\overline{n_i u n_j^{-1}}) = 0 \text{ for } j = 2, \ldots, r. \tag{8.25}$$

This is a triangular system of equations in $A_{n_1}, \ldots, A_{n_r}$, which can now be solved inductively. This completes the proof of the lemma. □

Proof of Theorem 8.9. From Proposition 4.13 and Maschke's theorem it follows that the assertion is equivalent to the invertibility of the sandwich matrices of all principal factors of M. So, by Corollary 8.5, it is enough to prove that $\mathbf{C}[\mathcal{M}_I]$ is semisimple, where $\mathcal{M}_I = GUGe_IGU\{\theta\}$, for any $I \subseteq S, I \neq S$. We prove this by induction on $|I|$. If $|I| = 0$, then every irreducible representation of $L_I = B$ is cuspidal. Hence, the assertion follows from Lemma 8.13 and Corollary 8.11. So, assume that $|I| > 0$. Write $e = e_I$. Let Q_I be the sandwich matrix of $J_I^0 = GeGU\{\theta\}$ and $L = L_I \simeq eL$. By Corollary 8.11 and Lemma 8.13 it is enough to show that $\phi(Q_I)$ is invertible for any noncuspidal irreducible representation of L. Let ϕ be such a representation. As explained before Lemma 8.13, there exist $K \subseteq I, K \neq I$, and an irreducible cuspidal representation π of L_K such that if $\overline{\pi}$ is the trivial extension of π to $Z = P_K \cap L$ (that is, $\overline{\pi}(u) = 1$ for $u \in U_K \cap L$), then ϕ is a component of $\psi = \overline{\pi}_Z^L$. It therefore suffices to prove that $\psi(Q_I)$ is invertible. Let $P = P_I, P^- = P_I^-$. Since $K \subseteq I$, we have

$$P_K \subseteq P, \quad P_K^- \subseteq P^-, \quad L_K \subseteq L, \quad U_I^- \subseteq U_K^-, \quad U_I \subseteq U_K.$$

Let $f = e_K$ and $\mathcal{M}_K = G \cup GfG \cup \{\theta\}$. Let $\mathcal{M}_{K,I} = \mathcal{M}_K \cup \mathcal{M}_I$ with the G's identified and with the zeros identified. So

$$\mathcal{M}_{K,I} = G \cup GeG \cup GfG \cup \{\theta\}$$

can be considered as a submonoid of a Rees quotient of $\mathcal{M}$. Namely, for $a = xey \in GeG, b = sft \in GfG$ we have

$$ab = \begin{cases} x\eta_I(ys)ft & \text{if } ys \in P^-P \\ \theta & \text{otherwise} \end{cases}$$

and

$$ba = \begin{cases} sf\eta_I(tx)y & \text{if } tx \in P^-P \\ \theta & \text{otherwise} \end{cases}.$$

Clearly $\mathcal{M}_{K,I}$ is a monoid of Lie type. By Lemma 8.4 $e\mathcal{M}_{K,I}e$ is a monoid of Lie type on the group L, in fact $e\mathcal{M}e \simeq (\mathcal{M}_{(L)})_K$. Since $f < e$, $f \in e\mathcal{M}_{K,I}e$. Since $I \neq S$, $\mathbf{C}[e\mathcal{M}_{K,I}e]$ is semisimple by the induction hypothesis. Now $GfG \cup \{\theta\}$ is the unique minimal nonzero ideal of $\mathcal{M}_{K,I}$. Hence, by Theorem 4.26 and Theorem 4.28, there exists an irreducible $\mathbf{C}_0[\mathcal{M}_{K,I}]$-module V such that the $\mathbf{C}[L_K]$-module fV corresponds to the representation π. Now eV is an irreducible $\mathbf{C}_0[e\mathcal{M}_{K,I}e]$-module. By Lemma 8.8 the $\mathbf{C}[L]$-module eV corresponds to the representation ψ and

$$\dim(eV) = \dim(fV) \cdot \frac{|L|}{|P_K \cap L|} \tag{8.26}$$

Also by the induction hypothesis, the sandwich matrix of $GfG \cup \{\theta\}$ is invertible. Hence, by Lemma 8.8

$$\dim(V) = \dim(fV) \cdot \frac{|G|}{|P_K|} \tag{8.27}$$

Since $P_K \subseteq P_I = P$ and $U_I \subseteq U_K \subseteq P_K$, by Lemma 8.1 we have $P_K = (P_K \cap L)(P_K \cap U_I) = (P_K \cap L)U_I$ and

$$|P_K| = |(P_K \cap L)U_I| = |P_K \cap L| \cdot |U_I|.$$

Hence, by (8.26), (8.27)

$$\dim(V) = \dim(eV) \cdot \frac{|G|}{|L||U_I|} = \dim(eV) \cdot \frac{|G|}{|P|}. \tag{8.28}$$

Since $(GeG)^2 \subseteq GeG \cup \{\theta\}$, V is also a $\mathbf{C}_0[J_I^0]$-module. We claim that it is completely reducible. Let A be the image in $\text{End}(V)$ of the algebra $\mathbf{C}_0[J_I^0]$, corresponding to this module. Since J_K^0 is an ideal of

$\mathcal{M}_{K,I}$ which does not annihilate V, it follows that V is irreducible over $\mathbf{C}_0[J_K^0]$. Hence $\text{End}(V)$ is the the image of $\mathbf{C}_0[J_K^0]$. By Theorem 4.26 and Lemma 4.18 the radical of the algebra $\mathbf{C}_0[J_I^0]$, viewed as a Munn algebra, consists of elements x such that $Q_I \circ x \circ Q_I = 0$. In other words $(GeG)x(GeG) = 0$ in $\mathbf{C}_0[J_I^0]$, hence also in $\mathbf{C}_0[\mathcal{M}_{K,I}]$. Since

$$(GeG)(GfG) = (GfG)(GeG) = GfG \cup \{\theta\},$$

we see that $(GfG)x(GfG) = 0$. Hence, the image of x in $\text{End}(V)$ must be zero. This means that the radical of $\mathbf{C}_0[J_I^0]$ maps to zero, so A is a semisimple algebra. Now, from $(GeG)(GfG) = GfG \cup \{\theta\}$ it follows that A contains the identity of $\text{End}(V)$. This shows that V is indeed a completely reducible $\mathbf{C}_0[J_I^0]$-module. Applying the remark after Theorem 4.28 to the irreducible components of this module, we come to $\dim(V) = \text{rank}(\psi(Q_I))$. The sandwich matrix Q_I of $J_I^0 = GeG \cup \{\theta\}$ is $t \times t$, where $t = |G|/|P|$. Thus $\psi(Q_I)$, as a matrix over $\mathbf{C}$, is $t' \times t'$, where $t' = t \cdot \dim(eV)$. Hence (8.28) implies that $\psi(Q_I)$ is invertible. This completes the proof of the theorem. $\square$

As mentioned before, irreducible representations of G are classified, via cuspidal representations, according to which parabolic subgroup they come from, [13], Chapter 9. Irreducible representations of $\mathcal{M}$ are classified according to which principal factor of $\mathcal{M}$ they come from, see Section 4.3. The following result provides a connection.

Corollary 8.14 *Let M be a monoid of Lie type on the group G and $e \in E(M)$. Let π be an irreducible complex representation of the group $L(e)$ containing $U(e) \cap U^-(e)$ in its kernel. Let $\overline{\pi}$ be the trivial extension of π to $P(e)$ (that is, $\overline{\pi}(R_u(P(e))) = 1$) and ϕ the representation induced from $P(e)$ to G. Then ϕ extends to an irreducible representation of M. Moreover, every complex representation of G extending to an irreducible representation of M arises in this way.*

Proof. Let $J = GeG \cup \{\theta\}$ be the corresponding principal factor of M. The sandwich matrix of J is invertible by Theorem 8.9. If L is the maximal subgroup of J containing e, then $L = L(e)e = eL(e)$ by Lemma 8.2. So, if π satisfies the conditions of the corollary, then it factors through L, because $U(e) \cap U^-(e) = \{x \in L(e) \mid ex = e\}$. Hence, it determines an irreducible representation of J. Via the natural map

$M \longrightarrow J$, this leads to a representation μ of M. From Corollary 8.8 it follows that $\mu_G = \phi$, as desired. The second assertion is also an easy consequence of Theorem 8.9 and Corollary 8.8, applied to an idempotent $e \in M$ which is minimal with respect to having nonzero image under the considered representation of M. $\square$

An alternative proof of Theorem 8.9 was obtained in [120]. We note that a classification of irreducible complex representations of M can be given, as in Theorem 2.35, see also Section 4.3. However, the explicit formulas for the identities of the algebras of the principal factors of M are not known.

Next, we discuss irreducible modular representations of $\mathcal{M}$. We say that a prime number p is the natural characteristic of the group G if U is a p-group. For example, this is the case if $G = H_\sigma$ for a Frobenius map σ on a linear reductive group $H \subseteq GL_n(\overline{\mathbf{F}}_p)$. By a modular representation of a monoid M of Lie type on G we then mean a representation $M \longrightarrow M_n(\overline{\mathbf{F}}_p)$. The sandwich matrices of the principal factors of $\mathcal{M}$ are not invertible in the corresponding matrix rings $M_n(\overline{\mathbf{F}}_p[L_I])$. So, we do not get an analogue of Theorem 8.9 in the modular case. However, a very strong link, of another type, with the representation theory of G is a consequence of Theorem 8.16 below. In what follows $F = \overline{\mathbf{F}}_p$.

We will need some background on modular representations of groups of Lie type. First recall that irreducible modular representations of G are in one-to-one correspondence with ordered pairs (I, χ), where $I \subseteq S$ and $\chi : L_I \longrightarrow F^*$ is any homomorphism. If $\phi : G \longrightarrow GL_n(F)$ is an irreducible representation, then there exists a unique 1-dimensional subspace Y of the column space $V = F^n$ which is stabilized by B. Then $I \subseteq S$ is such that $P_I = \{x \in G \,|\, \phi(x)Y = Y\}$. Moreover Y is a 1-dimensional $F[L_I]$-module, yielding $\chi : L_I \longrightarrow F^*$, [16]. Let $V^{U_I} = \{v \in V \,|\, \phi(U_I)v = v\}$. Then

$$V = V^{U_I} \oplus (1 - \phi(U_I^-))V$$

and both summands are $F[L_I]$-submodules of V. Here $1 - \phi(U_I^-)$ stands for the F-subspace spanned by all matrices $1 - \phi(u), u \in U_I^-$. In particular, $eV = V^{U_I} = Y$ and $(1 - e)V = (1 - \phi(U_I^-))V$ for a unique idempotent $e \in M_n(F)$ of rank one, [12]. It is also known that every irreducible representation of G is the restriction of a rational representation $\overline{\phi}$ of the corresponding reductive group $H \subseteq GL_n(F)$ [131].

So $G = H_\sigma$ for a Frobenius map σ. Moreover, there exists a BN-pair N_H, B_H for H such that $\sigma(B_H) = B_H$ and $\sigma(N_H) = N_H$. In particular, we may assume that $B \subseteq B_H$ and $N \subseteq N_H$. A similar decomposition of V is also known in this case, [12]. By [40], Theorem 31.3, the unique B-invariant subspace determined by $\bar\phi$ is $eV = lin(v)$, where v is a 'heighest weight vector'. Conjugating in $M_n(F)$, we may assume that the image $\bar\phi(T_H)$ of a maximal torus T_H of H containing T is diagonal and $e = \begin{pmatrix} 1 & 0 \\ 0 & 0 \end{pmatrix}$. We have $V = \bigoplus_\lambda V_\lambda$, where the 'weight spaces' V_λ are defined by $V_\lambda = \{v \in V \mid \phi(t)v = \lambda(t)v\}$, with λ running over the set of homomorphisms $T \longrightarrow F^*$. Then $(1 - e)V$ is the direct sum of the weight spaces V_λ other than eV, and every $\bar\phi(n), n \in N$, permutes all weight spaces by left multiplication. Therefore, in the basis of V consisting of the bases of the V_λ, every $\phi(n)$ either satisfies $e\phi(n)e = 0$ or $e\phi(n) = e\phi(n)e = \phi(n)e$.

Proposition 8.15 *Let G be a finite group of Lie type. Assume that $\phi : G \longrightarrow GL_n(F)$ is an irreducible modular representation. Then there is a nonzero idempotent $e \in M_n(F)$ and a subset $I \subseteq S$ such that $P_I = \{x \in G \mid \phi(x)e = e\phi(x)e\}, P_I^- = \{x \in G \mid e\phi(x) = e\phi(x)e\}, U_I \subseteq \{x \in G \mid \phi(x)e = e\}$, and $U_I^- \subseteq \{x \in G \mid e\phi(x) = e\}$. For $g \in G, a \in M_n(F)$, let $ga = \phi(g)a, ag = a\phi(g)$. Then*

$$M_\phi = G \cup GeG \cup \{\theta\}$$

is a monoid of Lie type on G and there is a representation $\hat\phi : M_\phi \longrightarrow M_n(F)$ such that $\hat\phi(g) = \phi(g)$ for $g \in G$ and $\hat\phi(e) = e$.

Proof. We know that there exists $e = e^2 \in M_n(F)$ of rank one and a subset $I \subseteq S$ such that $eV = V^{U_I}$ and $(1 - e)V = (1 - \phi(U_I^-))V$. Moreover $P_I = \{x \in G \mid \phi(x)eV = eV\} = \{x \in G \mid \phi(x)e = e\phi(x)e\}$ and $\phi(U_I)e = e$. Then $e(1 - \phi(U_I^-))V = 0$ implies that $e\phi(U_I^-) = e$. Since $(1 - e)V$ is L_I-invariant, we also have $\phi(L_I)(1 - e) = (1 - e)\phi(L_I)(1 - e)$. Hence $e\phi(L_I) = e\phi(L_I)e$. Since $P_I^- = L_I U_I^-$ and $e\phi(U_I^-) = e$, we get $e\phi(P_I^-) = e\phi(P_I^-)e$. So, in M_ϕ we have $P_I = P(e), P_I^- \subseteq P^-(e)$ and $U_I \subseteq U(e), U_I^- \subseteq U^-(e)$.

Now, $\phi(G) \subseteq M_n(F)$ acts irreducibly by right multiplication on the row vector space W. Let $C = \{x \in G \mid e\phi(x) = e\phi(x)e\}$. Then C is a subgroup and $B^- \subseteq P_I^- \subseteq C$, so $C = P_J^-$ for some $J \subseteq S$. Clearly

$I \subseteq J$. But We is a 1-dimensional B^--invariant subspace. Dually to [16],[12] we see that $W = We \oplus W(1 - \phi(U_J))$ and the summands are $\phi(L_J)$-invariant. Since $U_J \subseteq U_I$, we have $\phi(U_J)e = e$. Therefore $W(1-\phi(U_J))e = 0$, so that $W(1-\phi(U_J)) \subseteq W(1-e)$. Hence, comparing the dimensions, we get $W(1 - \phi(U_J)) = W(1 - e)$. Now, an argument symmetric to that used in the first paragraph of the proof shows that $\phi(P_J)e = e\phi(P_J)e$. In view of the definition of P_I this implies that $J = I$. We have shown that $P_I^- = P^-(e)$ in M_ϕ.

Next, we prove that M_ϕ is closed under multiplication. Consider the representation $\overline{\phi} : H \longrightarrow GL_n(F)$ of the reductive group H corresponding to G, which is an extension of ϕ. Let N_H be the normalizer in H of a maximal torus T_H of H containing T. Note that $N = N_H \cap G$. Let $n \in N$. We have seen above that either $e\phi(n)e = 0$ or $e\phi(n) = e\phi(n)e = \phi(n)e$. But the latter is equivalent to $n \in L_I$.

Let $x \in G \backslash P_I^- P_I$. By the Bruhat decomposition $G = B^- N B$, so that $x = anb$ for some $a \in B^-, b \in B, n \in N$. Then $n \notin L_I$. So $e\phi(n)e = 0$, and hence $e\phi(x)e = e\phi(anb)e = (e\phi(a)e)\phi(n)(e\phi(b)e) = 0$. Thus, it is easy to see that $eGe = e(P_I^- P_I)e \cup \{\theta\} = eL_I \cup \{\theta\} \subseteq GeG \cup \{\theta\}$ in M_ϕ, so that M_ϕ is a monoid.

Now, as in the proof of Corollary 8.5, it follows that the rules $x \mapsto \phi(x)$ and $xe_Iy \mapsto \phi(x)e\phi(y)$, for $x, y \in G$, determine a homomorphism $\mathcal{M}_I = G \cup Ge_IG \cup \{\theta\} \longrightarrow M_n(F)$, which factors through M_ϕ. Hence, all idempotents of GeG are conjugate, because this is so in Ge_IG by Corollary 8.6. Therefore, the conditions defining a monoid of Lie type, checked above for e only, are also satisfied for the remaining idempotents of M_ϕ. It follows that M_ϕ is a monoid of Lie type. This completes the proof of the theorem. $\square$

We are now ready for the main result on modular representations, first obtained in [111]. Note that from Theorem 8.3 and the remark following Theorem 4.28 it follows that the irreducible modular representations of $\mathcal{M}$ are in one-to-one correspondence with ordered pairs (I, ϕ), where $I \subseteq S$ and ϕ is an irreducible modular representation of L_I.

Theorem 8.16 *Every irreducible modular representation of the monoid $\mathcal{M}$ restricts to an irreducible representation of the unit group G of $\mathcal{M}$. An irreducible representation of G of type (I, χ) extends to $2^{|S\backslash I|}$ non-equivalent irreducible representations of $\mathcal{M}$. If α_I denotes the number* *of homomorphisms $L_I \longrightarrow F^*$, then the number of nonequivalent irreducible representations of $\mathcal{M}$ is $\sum_{I \subseteq S} 2^{|S \setminus I|} \alpha_I$.*

Proof. Let $\phi : G \longrightarrow GL_n(F)$ be an irreducible representation of type (I, χ). From Corollary 8.5 and Proposition 8.15 it follows that there exists $e = e^2 \in M_n(F)$ of rank one and $K \subseteq S$ such that $\overline{\phi} : \mathcal{M}_K \longrightarrow M_n(F)$ given by $\overline{\phi}(xe_K y) = \phi(x)e\phi(y), \overline{\phi}(x) = \phi(x)$, for $x, y \in G$, is well defined and it is a homomorphism. Since also $P_K = \{x \in G \mid \phi(x)e = e\phi(x)e\}$, we must have $K = I$ by the definition of the pair (I, χ).

Since J_I^0 is an ideal of $\mathcal{M}_I$, the column space $V = F^n$ is also an irreducible $F_0[J_I^0]$-module. But J_I^0 is a principal factor of $\mathcal{M}$. Hence, from Section 4.3 we know that V has a structure of $F_0[\mathcal{M}]$-module such that $e_K V = 0$ for all $K \subseteq S$ with $I \not\subseteq K$. Now, let $I \subseteq K$. Then $e_K \geq e_I$ implies that $J_K V \neq 0$. Since $\mathcal{M}_K$ is a submonoid of $\mathcal{M}$ containing G, V is also an irreducible $F_0[\mathcal{M}_K]$-module. So, as above, V is also an irreducible $F_0[\mathcal{M}]$-module in a new way such that $e_K V \neq 0$ and $e_{K'} V = 0$ for $K \not\subseteq K' \subseteq S$. Thus, we have produced $2^{|S \setminus I|}$ nonequivalent irreducible $F_0[\mathcal{M}]$-modules, which are equal as $F[G]$-modules. Hence, choosing all possible $\chi : L_I \longrightarrow F^*$, we have produced $\sum_{I \subseteq S} 2^{|S \setminus I|} \alpha_I$ nonequivalent irreducible representations of $\mathcal{M}$, each restricting to an irreducible representation of G. On the other hand, we know that the number of nonequivalent irreducible representations of L_I is $\sum_{K \subseteq I} \alpha_K$. By the comment preceding the theorem, the number of irreducible representations of $\mathcal{M}$ is therefore $\sum_{I \subseteq S} \sum_{K \subseteq I} \alpha_K$. Since this is equal to $\sum_{I \subseteq S} 2^{|S \setminus I|} \alpha_I$, the assertion follows. $\square$

The following result is now a direct consequence of the remark after Theorem 4.28.

Corollary 8.17 *Let ϕ be an irreducible modular representation of G of type (I, χ), where $I \subseteq S$ and $\chi : L_I \longrightarrow F^*$ is a homomorphism. Then the degree of ϕ is equal to the rank of the matrix $\chi(Q_I)$, for the sandwich matrix Q_I of $Ge_I G$.*

Corollary 8.18 *Let M be a monoid of Lie type on G. Then any irreducible modular representation of M restricts to an irreducible representation of G.*

Proof. Let V be an irreducible $F_0[M]$-module. Then there exists a unique $\mathcal{J}$-class J of M such that $JV \neq 0$ and $M_0 V = 0$, where $M_0 =$

$\{a \in M \mid J \not\subseteq MaM\}$. So V is an irreducible $F_0[M(J)]$-module, where $M(J) = G \cup J \cup \{\theta\}$. By Corollary 8.5 it is also an irreducible $F_0[\mathcal{M}_I]$-module. Therefore, Theorem 8.16 implies that V is an irreducible $F[G]$-module. $\square$

Recall that every irreducible modular representation of G is the restriction of the representation of the corresponding reductive group $H \subseteq GL_n(F)$. The Zariski closure $\overline{H}$ of H in $M_n(F)$ is a monoid, called a reductive monoid, satisfying all axioms of a monoid of Lie type except for being finite (see [108] and the example after Lemma 8.2). The same applies to $\overline{\phi(H)}$ for any rational modular representation ϕ of H. This provides a strong link between $\overline{H}$ and monoids M of Lie type on G, in particular providing another approach to modular representations [111]. The cross section lattice of M, see Theorem 8.7, can be also recovered via the modular representations of G by considering the image of G in the semisimple algebra $F[M]/\mathcal{J}(F[M])$ and applying Theorem 8.16 and its proof.

For other results, and bibliography, on monoids of Lie type we refer to [110].

Chapter 9

Applications

One of the motivations for developing the theory of linear semigroups comes from problems on associative rings that have 'many' finite dimensional linear representations. Rings satisfying polynomial identities and Noetherian rings are the main classes of this type. However, they often lead to representations over a division algebra, which is not necessarily a field. So, an extension of our main ideas and results to the case of 'skew linear semigroups' is needed.

Throughout this chapter D denotes a division algebra. We shall be concerned with subsemigroups of $M_n(D)$. Our aim in Section 1 is to present an extension of Theorem 3.5 to this case and to discuss the differences. In particular, this leads to certain invariants for the division algebra D. Our second aim is to present in Section 3 examples of applications of this structural approach to various problems on graded rings. As an intermediate step, Section 2 provides a motivation, supported by some technical results, focusing on the case of gradations by cancellative semigroups. The main idea is to reduce problems to the case of group - graded rings, or preferably to the special case of group crossed products, for which the general theory is very well developed, [103].

9.1   Skew linear semigroups and division algebras

Because of the role of matrix algebras $M_n(D)$ over a division algebra D, a natural question that arises is whether the statement of our main

structural result, Theorem 3.5, is valid for skew linear semigroups, see
[87], Problem 35. The numerical bounds established in this theorem
for $S \subseteq M_n(K)$ were proved through an exterior power argument. One
might expect that another proof can be found, that would use basic
linear algebra only, and so it would cover the case of skew linear semigroups as well. We will show that in general this is not true, but all
non-numerical ingredients of Theorem 3.5 remain valid in this more
general setting. We will also find some necessary conditions on D for
the above structural description to be true for semigroups $S \subseteq M_n(D)$.
Most of the material of this section comes from [119].

All spaces will be right spaces over D. So the set of column vectors
D^n is a right D-space and $M_n(D)$ acts on D^n be left multiplication.
The structure of $M_n(D)$ is exactly as described in Lemma 2.1 and Theorem 2.3. Keeping the notation of Chapter 2, we denote by M_j the
subsemigroup of $M_n(D)$ consisting of all matrices of rank at most j,
and we identify the nonzero elements of the Rees factor M_j/M_{j-1} with
the corresponding elements of $M_j \setminus M_{j-1}$.

Recall from Section 2.2 that a linearly ordered set T of nonzero
idempotents of a semigroup S is called a triangular set of idempotents
if $e \prec f$ implies that $ef = \theta$ for $e, f \in T, e \prec f$.

The main structural theorem for $S \subseteq M_n(K)$ may be now extended
in the following way.

Theorem 9.1 *Let $S \subseteq M_n(D)$ be a semigroup. Put*

$$S_j = \{a \in S|\ \mathrm{rank}(a) \le j\}$$

$T_j = \{a \in S_j|\ SaS$ *does not intersect non-null $\mathcal{H}$-classes of $M_j/M_{j-1}\}$.*
*Then $S_0 \subseteq T_1 \subseteq S_1 \subseteq T_2 \subseteq \cdots \subseteq S_n = S$ are ideals of S (if nonbe-
lianempty). Moreover,*

1. *$N_j = T_j/S_{j-1}$ is a union of nilpotent ideals of S/S_{j-1} of nilpotency
   index two, and $S_j/T_j \simeq (S_j \setminus T_j) \cup \{\theta\} \subseteq S_j/S_{j-1}$ is a 0-disjoint
   union of uniform subsemigroups U_α of M_j/M_{j-1} that intersect dif-
   ferent $\mathcal{R}$-classes and different $\mathcal{L}$-classes of M_j/M_{j-1}. In particular
   $U_\alpha U_\beta \subseteq N_j$ for $\alpha \ne \beta$.*

2. *Define*

   $S'_j = \{a \in M_j/M_{j-1}|$ *there exist images b, c of elements of S*
   in M_j/M_{j-1} such that $a\mathcal{L}b$ and $a\mathcal{R}c\}$.

If every triangular set of idempotents in S'_j has at most n_j elements, where $n_j < \infty$, then N_j is nilpotent of index $\leq n_j + 2$ and S_j/T_j is a 0-disjoint union of at most n_j uniform ideals.

Proof. The proof is similar to that of Theorem 3.5. We will only say what changes should be made there.

Let $T \subseteq M_n(D)$ be a set of idempotents of rank j. Assume that $(\overline{ex}\overline{f})^2 = \theta$ for every $x \in \langle T \rangle$ and every $e, f \in T$ with $e \neq f$ (where $\overline{x}$ denotes the image of x in M_j/M_{j-1}). As in Corollary 2.13 we see that there exists a linear order $\prec$ on T such that $\overline{e}\overline{f} = \theta$ whenever $e \prec f$.

(i) $T_j/S_{j-1} \subseteq S'_j$ is a nil subsemigroup and S'_j is a semigroup of matrix type. Moreover every triangular set of idempotents in S'_j has at most n_j elements. Then Lemma 2.10 may be applied. Thus, we obtain that T_j/S_{j-1} is nilpotent of index $\leq n_j + 2$ if $n_j < \infty$.

(ii) By the proof of Theorem 3.5 we know that: if $W \neq V$ are two equivalence classes of ρ and $x \in W, y \in V$, moreover e, f are idempotents in M_j/M_{j-1} such that $e \mathcal{H} x, f \mathcal{H} y$ and $g \in M_j/M_{j-1}$ is $\mathcal{H}$-related to an element of S_j/S_{j-1}, then $(egf)^2 = \theta$. So, let T be a set of idempotents in M_j/M_{j-1} chosen from different ρ-classes. If $e, f \in T, e \neq f$ and $x \in \langle T^1 \rangle$, then $(exf) = \theta$. As explained above, T is a triangular set of idempotents. Since $T \subseteq S'_j$, $|T| \leq n_j$. Thus the number of uniform components in S_j/T_j does not exceed n_j. $\square$

In other words, if $S \subseteq M_n(D)$ for a division algebra D, then the assertion of Theorem 3.5 holds with two qualitative differences:

- there might be infinitely many uniform components consisting of matrices of a given rank j,

- the nilpotent component $N_j = T_j/S_{j-1}$ is only proved to be a union of nilpotent ideals of index two, (but we do not know whether it is nilpotent).

Assume that all n_j, defined as in Theorem 9.1, are finite. By this theorem there exists a chain of ideals $J_1 \subseteq J_2 \subseteq \cdots \subseteq J_t = S$, where $t \leq \sum_{j=1}^n n_j + n$ such that J_1 and all factors $J_i/J_{i-1}, i = 2, \ldots, t$, are uniform or nilpotent of index not exceeding $\max(n_j + 2)$. If we assume that S is π-regular, then non-nilpotent factors are completely 0-simple by Corollary 3.2. In this case the set of completely 0-simple factors of the chain is equal to the set of all $\mathcal{J}$-classes containing idempotents. Thus, the number of such classes is at most $\sum_j n_j$. The following fact, analogous to Proposition 2.14, is true: if I is an ideal of S and S/I is a nil semigroup, then S/I is nilpotent of index $\leq \prod_j(n_j + 2)$. In this case I contains all J_i/J_{i-1} which are completely 0-simple. The remaining factors are nilpotent. A simple proof of the above fact will be omitted.

We note also that the closure $\text{cl}(S)$ can be defined for any semigroup $S \subseteq M_n(D)$ and its construction is the same as in the case of linear semigroups, see Lemma 3.11 and the remark following it.

Example Assume that there exists an infinite triangular set of idempotents T in some $M_j/M_{j-1}, j < n$, (such an $M_n(D)$ will be constructed in the example before Theorem 9.16). Define $U = \{s \in M_j/M_{j-1}|\, \text{there exist } e, f \in T \text{ such that } e \preceq f, se = e \text{ and } fs = f\}$. From the remark after Lemma 2.10 we know that $S = U \cup M_{j-1}$ is a semigroup. Moreover, it has infinitely many uniform components consisting of matrices of rank j and the remaining matrices of rank j yield a nil ideal of $S/M_{j-1} = S_j/S_{j-1}$ which is not nilpotent. Therefore, in general, one cannot strengthen the statement of Theorem 9.1.

Let D be a division algebra. To D we can associate a series of numerical invariants $d_{n,j} = d_{n,j}(D)$, where $j \leq n, n \geq 1$, defined as follows: $d_{n,j}$ = the cardinality of a maximal triangular set of idempotents in the principal factor of M_j/M_{j-1} of $M_n(D)$.

The notion of a triangular set of idempotents in the case where $D = K$ is a field was studied in Lemma 2.11. Namely, if $T \subseteq M_n(K)$ is a set of elements of M_j whose images in M_j/M_{j-1} form a triangular set of idempotents. Then $|T| \leq \binom{n}{j}$. Moreover, considering diagonal idempotents we come to $d_{n,j}(K) = \binom{n}{j}$. The next step is to consider finite dimensional division algebras.

Proposition 9.2 *Assume that a division algebra D is of dimension $k < \infty$ over its centre $K = Z(D)$. Fix a basis of D over K. Let $\pi : M_n(D) \longrightarrow M_{nk}(K)$ be the representation arising from left multiplication $D^n \longrightarrow D^n$ by elements of $M_n(D)$, written in the corresponding basis of D^n. Assume that $S \subseteq M_n(D)$ is a semigroup. Then*

1. *π maps uniform (nilpotent, respectively) components of S onto uniform (nilpotent) components of $\pi(S)$,*

2. *the cardinality of maximal triangular sets of idempotents in each M_j/M_{j-1} does not exceed $\binom{nk}{jk}$, so that $d_{n,j}(D) \leq \binom{nk}{jk}$.*

Proof. First note that $\dim_D(aD^n)k = \dim_K(aD^n)$ for $a \in M_n(D)$, so that $\text{rank}(\pi(a)) = \text{rank}(a)k$. Therefore each $S_j \setminus S_{j-1}$ is mapped onto $(\pi(S))_{jk} \setminus (\pi(S))_{jk-1}$. From Theorem 9.1 it follows that the structural components of $S \subseteq M_n(D)$ are mapped onto those of $\pi(S) \subseteq M_{nk}(K)$. It is clear that any triangular set E of idempotents of S_j/S_{j-1} is mapped onto such a subset in the corresponding principal factor of $M_{nk}(K)$. The result follows. $\square$

Consequently, to get an example of a skew linear semigroup with infinitely many uniform components, one has to consider division algebras which are infinite dimensional over their centres.

Proposition 9.3 *Assume that T is a triangular set of idempotents in M_1 or M_{n-1}/M_{n-2} for $M_n(D)$. Then we have $|T| < n$.*

Proof. We will assume that T is a triangular set of idempotents of M_1. The second case may be reduced to the former by applying the mapping $a \mapsto I - a$. Suppose that $|T| > n$. Then there exist $e_1, e_2, \ldots, e_{n+1} \in T$ with $e_i e_j = 0$ for $i < j$. Let $0 \neq v_i \in \text{Im}(e_i)$. Since $v_1, v_2, \ldots, v_{n+1}$ are linearly dependent, there exist $k < n+1$ and $a_{k+1}, \ldots, a_{n+1} \in D$ such that $v_k = v_{k+1}a_{k+1} + \cdots + v_{n+1}a_{n+1}$. Thus $e_k v_k = e_k v_{k+1} a_{k+1} + \cdots + e_k v_{n+1} a_{n+1} = 0$ because $e_k e_{k+i} = 0$ for $i > 0$. This contradicts the fact that $e_k e_k = e_k$. $\square$

It follows that the first nontrivial case to consider is rank two matrices in $M_4(D)$. In this case, existence of a bound on the cardinalities of triangular sets of idempotents may be expressed in terms of chains of centralizers in D. To prove this result we will need a sequence of auxiliary lemmas. Noncommutativity of D interferes only in the proof of Lemma 9.13. The notation used in the lemmas is justified by the structure of proof of Theorem 9.14. Lower case Greek letters will denote elements of D, if not stated otherwise. If $v_1, \ldots, v_k \in D^n$, by $\text{Lin}(v_1, \ldots, v_k)$ we mean the right D-subspace of D^n spanned by these vectors.

Lemma 9.4 *Let V_1, V_2, V_3 be 2-dimensional subspaces of D^4. If V_1, V_2 and V_3 are all different and $V_i \cap V_j \neq 0$ for all i, j, then one of the following conditions holds*

1. *there exist linearly independent vectors $v_1, v_2, v_3 \in D^4$ with $V_1 = \text{Lin}(v_2, v_3), V_2 = \text{Lin}(v_1, v_3)$ and $V_3 = \text{Lin}(v_1, v_2)$,*

2. *$V_1 \cap V_2 \cap V_3 \neq 0$.*

Proof. Choose nonzero elements $v_1 \in V_2 \cap V_3$ $v_2 \in V_1 \cap V_3, v_3 \in V_1 \cap V_2$. If $V_1 \cap V_2 \cap V_3 = 0$, then v_2, v_3 are linearly independent. Since $\dim(V_1) = 2$ and $v_2, v_3 \in V_1$, it follows that $V_1 = \text{Lin}(v_2, v_3)$. Similarly $V_2 = \text{Lin}(v_1, v_3)$ and $V_3 = \text{Lin}(v_1, v_2)$. If v_1, v_2, v_3 are linearly dependent, then we must have $V_1 = V_2 = V_3$, a contradiction. The assertion follows. $\square$

Lemma 9.5 *Assume that $v_1, v_2, v_3 \in D^4$ are linearly independent. Define $V_1 = \text{Lin}(v_2, v_3), V_2 = \text{Lin}(v_1, v_3)$ and $V_3 = \text{Lin}(v_1, v_2)$. If $W \cap V_i \neq 0, i = 1, 2, 3$, for a 2-dimensional subspace W of D^4, then $W \subseteq \text{Lin}(v_1, v_2, v_3)$.*

Proof. Let $v_1\alpha_1 + v_2\alpha_2$ be a nonzero element of $W \cap V_3$. Similarly, choose nonzero $v_2\beta_2 + v_3\beta_3 \in W \cap V_1$. If $\alpha_1 \neq 0$, then $v_1\alpha_1 + v_2\alpha_2, v_2\beta_2 + v_3\beta_3$ are linearly independent. Since $\dim(W) = 2$, we get $W = \text{Lin}(v_1\alpha_1 + v_2\alpha_2, v_2\beta_2 + v_3\beta_3)$, and hence $W \subseteq \text{Lin}(v_1, v_2, v_3)$, as desired. So, suppose that $\alpha_1 = 0$. Then $0 \neq v_2\alpha_2 \in W$. Let $0 \neq v_1\gamma_1 + v_3\gamma_3 \in W \cap V_2$. Then v_2 and $v_1\gamma_1 + v_3\gamma_3$ are linearly independent elements of W. Hence $W = \text{Lin}(v_2, v_1\gamma_1 + v_3\gamma_3)$, and also $W \subseteq \text{Lin}(v_1, v_2, v_3)$. $\square$

Lemma 9.6 *Assume that $V_4, V_5, \ldots, V_n, W_4, W_5, \ldots, W_n, n \geq 4$, are 2-dimensional subspaces of D^4 such that $V_i \cap W_j \neq 0$ for $i < j$ and $V_i \cap W_i = 0$ for $4 \leq i, j \leq n$. Assume further that $W_j \subseteq W, j = 4, 5, \ldots, n$, for a 3-dimensional subspace W of D^4. Then $n \leq 6$.*

Proof. Since $\dim(V_i) + \dim(W) > 4$ and $V_i, W \subseteq D^4$ for $i \leq n$, it follows that $V_i \cap W \neq 0$. Suppose that $\dim(V_i \cap W) = 2$. Then $V_i \subseteq W$. But $W_i \subseteq W$ by the hypothesis. Since $\dim(W) = 3$, $V_i \cap W_i \neq 0$, a contradiction. Hence $\dim(V_i \cap W) = 1$. Let $0 \neq v_i \in V_i \cap W$ for $i \leq n$. If $n > 6$, then there exists $k, 4 < k \leq n$, such that $v_k \in \text{Lin}(v_4, v_5, \ldots, v_{k-1})$ because $v_4, \ldots, v_n \in W$ and $\dim(W) = 3$. Let $i < j, i, j \in \{4, \ldots, n\}$. We know that $0 \neq V_i \cap W_j \subseteq V_i \cap W$. Since $\dim(V_i \cap W) = 1$, we have $V_i \cap W_j = V_i \cap W$. Therefore $v_i \in V_i \cap W$ implies that $v_i \in W_j$. Hence $v_4, \ldots, v_{k-1} \in W_k$, so that $v_k \in W_k$ because $v_k \in \text{Lin}(v_4, \ldots, v_{k-1})$. Since $v_k \in V_k$, it follows that $V_k \cap W_k \neq 0$, a contradiction. This completes the proof. $\square$

Lemma 9.7 *Let* W, V_1, V_2, V_3 *be 2-dimensional subspaces of* D^4. *Assume that* $V_1 \cap V_2 \cap V_3 \neq 0$ *and* $V_1 \neq V_2$. *If* $\dim(V_1 + V_2 + V_3) \leq 3, W \cap V_1 \neq 0$ *and* $W \cap V_2 \neq 0$, *then also* $W \cap V_3 \neq 0$.

Proof. Let v, v_1, v_2 be nonzero elements of $V_1 \cap V_2 \cap V_3, V_1 \cap W, V_2 \cap W$ respectively. If v, v_1 are linearly dependent, then $v \in W$ because $v_1 \in W$, so that $V_3 \cap W \neq 0$. So, we may assume that v, v_1 are linearly independent. Similarly, we may assume that v, v_2 are independent. Now, if v_1, v_2 are linearly dependent, then $v_1 \in V_1 \cap V_2$. Since $v \in V_1 \cap V_2$ and v, v_1 are linearly independent, we have $\dim(V_1 \cap V_2) \geq 2$. Then $V_1 = V_2$, a contradiction. So, we may assume that v_1, v_2 are linearly independent. Since $v_1, v_2 \in W$ by our choice, $W = \text{Lin}(v_1, v_2)$. Hence $W \subseteq V_1 + V_2 + V_3$ because $v_1 \in V_1$ and $v_2 \in V_2$. Clearly $V_3 \subseteq V_1 + V_2 + V_3$. But by the hypothesis $\dim(W) = \dim(V_2) = 2$ and $\dim(V_1 + V_2 + V_3) \leq 3$. Therefore $W \cap V_3 \neq 0$, as desired. $\square$

Lemma 9.8 *Let* W, V_1, V_2, V_3 *be 2-dimensional subspaces of* D^4. *Assume that* $\dim(V_1 + V_2 + V_3) = 4$ *and* $W \cap V_i \neq 0$ *for* $i = 1, 2, 3$. *Then* $V_1 \cap V_2 \cap V_3 \subseteq W$.

Proof. We may assume that $V_1 \cap V_2 \cap V_3 \neq 0$. Let $0 \neq v \in V_1 \cap V_2 \cap V_3$. Choose $v_i \in D^4$ such that $V_i = \text{Lin}(v, v_i)$ for $i = 1, 2, 3$. Since $V_1 + V_2 + V_3 = \text{Lin}(v, v_1) + \text{Lin}(v, v_2) + \text{Lin}(v, v_3) \subseteq \text{Lin}(v, v_1, v_2, v_3)$ is of dimension 4 by the hypothesis, v, v_1, v_2, v_3 must be linearly independent. Let $\alpha_i, \beta_i \in D$ be such that $0 \neq v\alpha_i + v_i\beta_i \in W \cap V_i$ for $i = 1, 2, 3$. If there exists i such that $\beta_i = 0$, then $v \in W$, which yields the assertion. So, suppose that $\beta_i \neq 0$ for $i = 1, 2, 3$. Then $v\alpha_i + v_i\beta_i, i = 1, 2, 3$, are linearly independent elements of W. But $\dim(W) = 2$, a contradiction. This completes the proof. $\square$

Lemma 9.9 *Assume that* $V_4, V_5, \ldots, V_n, W_4, W_5, \ldots, W_n \subseteq D^4, n \geq 4$, *have dimension 2. Moreover, let* $V_i \cap W_j \neq 0$ *for* $i < j$ *and* $V_i \cap W_i = 0$, $4 \leq i, j \leq n$. *If* $\bigcap_{4 \leq j \leq n} W_j \neq 0$, *then* $n \leq 6$.

Proof. Suppose that $n \geq 7$ and $0 \neq v \in \bigcap_{4 \leq j \leq n} W_j$. Consider the factorspaces $V_i' = V_i + \text{Lin}(v), W_j' = W_j + \text{Lin}(v)$ contained in $Z' = D^4 / \text{Lin}(v) \simeq D^3$. Clearly $\dim(W_j') = 1$ because $v \in W_j$ and $\dim(V_i') = 2$ because $V_i \cap \text{Lin}(v) = 0$ (otherwise $v \in V_i \cap W_i$, a contradiction). We have $V_i' \cap W_j' \neq 0$ for $i < j$ because $v \notin V_i \cap W_j \neq 0$. So $W_j' \subseteq V_i'$

for $i < j$ because $\dim(W_j') = 1$. Suppose that $V_i' \cap W_i' \neq 0$ for some i. Then there exist $0 \neq v_i \in V_i, 0 \neq w_i \in W_i$ and $\beta_i \in D$ such that $v_i = w_i + v\beta_i$. But $w_i, v \in W_i$, so that $v_i \in W_i$. Since $v_i \in V_i$, we get $V_i \cap W_i \neq 0$, a contradiction. Hence $V_i' \cap W_i' = 0$ for $4 \leq i \leq n$. Since $n \geq 7$ and $\dim(Z') = 3$, there exists $k \in \{4, 5, \dots, n\}$ such that $W_k' \subseteq W_{k+1}' + W_{k+2}' + \cdots + W_n'$. But $W_{k+i}' \subseteq V_k'$ whenever $1 \leq i \leq n - k$. Therefore $W_k' \subseteq V_k'$. This contradicts the fact that $V_k' \cap W_k' = 0$. The result follows. $\square$

Lemma 9.10 *Let $V_{i_1}, V_{i_2}, W_{j_{n-1}}, W_{j_n}$ be 2-dimensional subspaces of D^4. Assume that $V_{i_a} \cap W_{j_b} \neq 0$ for $a = 1, 2$ and $b = n - 1, n$. If $V_{i_1} \cap V_{i_2} = 0$ and $W_{j_{n-1}} \cap W_{j_n} = 0$, then there exist linearly independent vectors $v_1, v_2, v_3, v_4 \in D^4$ such that $V_{i_1} = \mathrm{Lin}(v_1, v_2), V_{i_2} = \mathrm{Lin}(v_3, v_4), W_{j_{n-1}} = \mathrm{Lin}(v_1, v_4)$, and $W_{j_n} = \mathrm{Lin}(v_2, v_3)$.*

Proof. Let $v_1, v_2 \in D^4$ be such that $0 \neq v_1 \in V_{i_1} \cap W_{j_{n-1}}$ and $0 \neq v_2 \in V_{i_1} \cap W_{j_n}$. Then v_1, v_2 are linearly independent, because otherwise $v_1 \in W_{j_{n-1}} \cap W_{j_n}$, contradicting the hypothesis. Since $v_1, v_2 \in V_{i_1}$, we get $V_{i_1} = \mathrm{Lin}(v_1, v_2)$. Let $0 \neq v_3 \in V_{i_2} \cap W_{j_n}$ and $0 \neq v_4 \in V_{i_2} \cap W_{j_{n-1}}$. Similarly as above we check that $V_{i_2} = \mathrm{Lin}(v_3, v_4)$. Since $V_{i_1} \cap V_{i_2} = 0$ by the hypothesis, it follows that v_1, v_2, v_3, v_4 are linearly independent. Then $W_{j_n} = \mathrm{Lin}(v_2, v_3)$ because $v_2, v_3 \in W_{j_n}$, and similarly $W_{j_{n-1}} = \mathrm{Lin}(v_1, v_4)$. $\square$

Lemma 9.11 *Let v_1, v_2, v_3, v_4 be linearly independent vectors of D^4. Define spaces $V_{i_1} = \mathrm{Lin}(v_1, v_2), V_{i_2} = \mathrm{Lin}(v_3, v_4), W_{j_{n-1}} = \mathrm{Lin}(v_1, v_4), W_{j_n} = \mathrm{Lin}(v_2, v_3)$. Let V, W be 2-dimensional subspaces of D^4 such that the following condition holds: $V \neq V_{i_a}, V_{i_a} \cap W \neq 0$ for $a = 1, 2$ and $W \neq W_{j_b}, V \cap W_{j_b} \neq 0$ for $b = n - 1, n$. If moreover there exists $k \in \{1, 2, 3, 4\}$ such that $v_k \in W$ (or $v_k \in V$) then $V \cap W \neq 0$ if and only if $v_k \in V$ or $v_{k+2} \in V$ ($v_k \in W$ or $v_{k+2} \in W$, respectively), with indices taken modulo 4.*

Proof. By symmetry we may assume that $v_1 \in W$. Then we have to show that $V \cap W \neq 0$ is equivalent to $v_1 \in V$ or $v_3 \in V$. We know that $W \cap V_{i_2} \neq 0$. So there exist $\epsilon, \phi \in D$ such that $0 \neq v_4\epsilon + v_3\phi \in W \cap V_{i_2}$. By the hypothesis $v_1 \in W$, so $W = \mathrm{Lin}(v_1, v_4\epsilon + v_3\phi)$. Similarly, there exist $\alpha, \beta \in D$ such that $0 \neq v_1\alpha + v_4\beta \in V \cap W_{j_{n-1}}$ and $\delta, \gamma \in D$ such that $0 \neq v_2\gamma + v_3\delta \in V \cap W_{j_n}$. Therefore $V = \mathrm{Lin}(v_1\alpha + v_4\beta, v_2\gamma + v_3\delta)$.

Suppose first that $V \cap W \neq 0$. Then there exist $p, q, r, s \in D, (p, q, r, s) \neq (0, 0, 0, 0)$, such that $v_1 p + (v_4 \epsilon + v_3 \phi)q = (v_1 \alpha + v_4 \beta)r + (v_2 \gamma + v_3 \delta)s$. Comparing the coefficients corresponding to the vectors v_i we come to

$$\text{(1) } p = \alpha r \qquad \text{(2) } 0 = \gamma s$$
$$\text{(3) } \phi q = \delta s \qquad \text{(4) } \epsilon q = \beta r.$$

By (2) $\gamma = 0$ or $s = 0$. If $\gamma = 0$, then $v_3 \in V$, so that the right side of the desired equivalence holds. So, assume that $s = 0$. Then (3) yields $\phi q = 0$, hence $q = 0$ or $\phi = 0$. If $\phi = 0$, then $W = \text{Lin}(v_1, v_4 \epsilon) = W_{j_{n-1}}$, a contradiction. So we have $q = 0$. Then (4) implies that $\beta r = 0$, so that $r = 0$ or $\beta = 0$. If $r = 0$, then from (1) it follows that $p = 0$, so that $(p, q, r, s) = (0, 0, 0, 0)$, which is not possible. Therefore $\beta = 0$. This means that $v_1 \in V$, so again the right side of the equivalence holds.

It remains to prove the converse implication. If $v_1 \in V$, then $V \cap W \neq 0$ because $v_1 \in W$. If $v_3 \in V$, then $\gamma = 0$ and therefore $V = \text{Lin}(v_1 \alpha + v_4 \beta, v_3 \delta)$. Hence $V \subseteq \text{Lin}(v_1, v_3, v_4)$. Since $W = \text{Lin}(v_1, v_4 \epsilon + v_3 \phi)$, we also get $W \subseteq \text{Lin}(v_1, v_3, v_4)$. In view of the fact that $\dim(V) = 2, \dim(W) = 2$ and $\dim(\text{Lin}(v_1, v_3, v_4)) = 3$ this implies that $V \cap W \neq 0$, as desired. This completes the proof of the lemma. $\square$

Lemma 9.12 *Let $V_{i_1}, V_{i_2}, V_{i_3}, \overline{V}, \overline{W}, W_{j_{n-2}}, W_{j_{n-1}}, W_{j_n}$ be 2-dimensional subspaces of D^4. Assume that $V_{i_1} \cap V_{i_2} = 0$ and $W_{j_{n-1}} \cap W_{j_n} = 0$. Let $V_{i_1}, V_{i_2}, V_{i_3}, \overline{V}$ be all different and similarly let all $\overline{W}, W_{j_{n-2}}, W_{j_{n-1}}, W_{j_n}$ be different. Assume also that $V_{i_a} \cap W_{j_b} \neq 0, V_{i_a} \cap \overline{W} \neq 0$ and $\overline{V} \cap W_{j_b} \neq 0$, for $a = 1, 2, 3$ and $b = n-2, n-1, n$. If one of the spaces $V_{i_3}, \overline{V}, \overline{W}, W_{j_{n-2}}$ contains one of $V_{i_a} \cap W_{j_b}$ $(a = 1, 2, b = n - 1, n)$, then $\overline{V} \cap \overline{W} \neq 0$.*

Proof. Spaces $V_{i_1}, V_{i_2}, W_{j_{n-1}}, W_{j_n}$ satisfy the hypotheses of Lemma 9.10. Thus there exist linearly independent vectors $v_1, v_2, v_3, v_4 \in D^4$ such that $V_{i_1} = \text{Lin}(v_1, v_2), V_{i_2} = \text{Lin}(v_3, v_4), W_{j_{n-1}} = \text{Lin}(v_1, v_4)$, and $W_{j_n} = \text{Lin}(v_2, v_3)$. Consider the sequence of spaces $\overline{V}, W_{j_{n-2}}, V_{i_3}, \overline{W}$. Neighbouring spaces of this sequence have nonzero intersections. By the hypothesis we know that one of these spaces contains one of $V_{i_1} \cap W_{j_{n-1}} = \text{Lin}(v_1), V_{i_1} \cap W_{j_n} = \text{Lin}(v_2), V_{i_2} \cap W_{j_{n-1}} = \text{Lin}(v_4), V_{i_2} \cap W_{j_n} = \text{Lin}(v_3)$. This space will be denoted by Z. Thus, $v_k \in Z$ for some $k \in \{1, 2, 3, 4\}$. Let Z' be a space following or preceding Z in the sequence $\overline{V}, W_{j_{n-2}}, V_{i_3}, \overline{W}$. If we rename these spaces by putting $V = Z, W = Z'$ if $Z = V_*, Z' = W^*$ and $V = Z', W = Z$ if $Z' = V_*, Z = W^*$ (where each $*$ stands for an index or a dash), then we meet the notation and the hypotheses of Lemma 9.11. Since we know that $V \cap W \neq 0$, the "only if" part of this lemma yields $v_k \in Z'$ or $v_{k+2} \in Z'$.

If Z'' is a space neighbouring Z' in the sequence $\overline{V}, W_{j_{n-2}}, V_{i_3}, \overline{W}$, then a similar reasoning leads to $v_k \in Z''$ or $v_{k+2} \in Z''$ (note that the indices again are taken modulo 4). Repeating this reasoning we eventually get $v_k \in \overline{V}$ or $v_{k+2} \in \overline{V}$ and $v_k \in \overline{W}$ or $v_{k+2} \in \overline{W}$. By the "if" part of Lemma 9.11 this implies that $\overline{V} \cap \overline{W} \neq 0$, as desired. $\square$

We denote by $N(D)$ the supremum of the set of integers k such that there exist $a_1, a_2, \dots, a_k, b_1, b_2, \dots, b_k \in D$ with $a_i b_j = b_j a_i$ for $i < j$ and $a_i b_i \neq b_i a_i$ for $i = 1, 2, \dots, k$.

Lemma 9.13 *Let $n \geq 6$ and let a collection of 2-dimensional subspaces $V_{i_1}, V_{i_2}, V_{i_3}, V_4, \dots, V_{n-3}, W_4, W_5, \dots, W_{n-3}, W_{j_{n-2}}, W_{j_{n-1}}, W_{j_n}$ of D^4 be given. Assume that $V_{i_1} \cap V_{i_2} = 0$ and $W_{j_{n-1}} \cap W_{j_n} = 0$. Let moreover $V_{i_a} \cap W_k \neq 0$ for $a = 1, 2, 3$ and $4 \leq k \leq n - 3$, $V_{i_a} \cap W_{j_b} \neq 0$ for $a = 1, 2, 3$ and $b = n - 2, n - 1, n$, $V_k \cap W_{j_b} \neq 0$ for $4 \leq k \leq n - 3$ and $b = n - 2, n - 1, n$. Assume also that $V_i \cap W_j \neq 0$ for $4 \leq i < j \leq n - 3$ and $V_k \cap W_k = 0$ for $4 \leq k \leq n - 3$. If none of the spaces $V_{i_3}, V_4, \dots, V_{n-3}, W_4, W_5, \dots, W_{n-3}, W_{j_{n-2}}$ contains one of $V_{i_a} \cap W_{j_b}$, where $a = 1, 2, b = n - 1, n$, then $n \leq N(D) + 6$.*

Proof. By the hypothesis $V_{i_1} \cap V_{i_2} = 0$, $W_{j_{n-1}} \cap W_{j_n} = 0$, and $V_{i_a} \cap W_{j_b} \neq 0$ for $a = 1, 2$ and $b = n - 1, n$. So Lemma 9.10 may be applied to $V_{i_1}, V_{i_2}, W_{j_{n-1}}, W_{j_n}$. Then there exist linearly independent vectors $v_1, v_2, v_3, v_4 \in D^4$ such that $V_{i_1} = \text{Lin}(v_1, v_2)$, $V_{i_2} = \text{Lin}(v_3, v_4)$, $W_{j_{n-1}} = \text{Lin}(v_1, v_4)$ and $W_{j_n} = \text{Lin}(v_2, v_3)$. This implies that $V_{i_1} \cap W_{j_{n-1}} = \text{Lin}(v_1)$, $V_{i_1} \cap W_{j_n} = \text{Lin}(v_2)$, $V_{i_2} \cap W_{j_{n-1}} = \text{Lin}(v_4)$ and $V_{i_2} \cap W_{j_n} = \text{Lin}(v_3)$.

We rename the spaces by putting $V_3 = V_{i_3}$, $W_{n-2} = W_{j_{n-2}}$. By the hypothesis none of $V_3, V_4, \dots, V_{n-3}, W_4, W_5, \dots, W_{n-2}$ contains one of the vectors v_1, v_2, v_3, v_4. Define $\alpha_i, \beta_i, \gamma_i, \delta_i \in D$, $3 \leq i \leq n - 3$, such that $0 \neq v_1 \alpha_i + v_4 \gamma_i \in V_i \cap W_{j_{n-1}}$ and $0 \neq v_2 \beta_i + v_3 \delta_i \in V_i \cap W_{j_n}$. Then $V_i = \text{Lin}(v_1 \alpha_i + v_4 \gamma_i, v_2 \beta_i + v_3 \delta_i)$. Let $\overline{\alpha}_j, \overline{\beta}_j, \overline{\gamma}_j, \overline{\delta}_j \in D$, $4 \leq j \leq n - 2$, be such that $0 \neq v_1 \overline{\alpha}_j + v_2 \overline{\beta}_j \in V_{i_1} \cap W_j$ and $0 \neq v_4 \overline{\gamma}_j + v_3 \overline{\delta}_j \in V_{i_2} \cap W_j$. Hence $W_j = \text{Lin}(v_1 \overline{\alpha}_j + v_2 \overline{\beta}_j, v_4 \overline{\gamma}_j + v_3 \overline{\delta}_j)$. By the assumptions $\alpha_i, \beta_i, \gamma_i, \delta_i, \overline{\alpha}_j, \overline{\beta}_j, \overline{\gamma}_j, \overline{\delta}_j$ are all nonzero. We will find a condition for the inequality $V_i \cap W_j \neq 0$ in terms of these elements of D.

First note that the condition $V_i \cap W_j \neq 0$ is equivalent to the existence of $(p, q, r, s) \in D^4$ with $(p, q, r, s) \neq (0, 0, 0, 0)$ and

$$(v_1\overline{\alpha}_j + v_2\overline{\beta}_j)p + (v_4\overline{\gamma}_j + v_3\overline{\delta}_j)q = (v_1\alpha_i + v_4\gamma_i)r + (v_2\beta_i + v_3\delta_i)s.$$

Then we get

$$(1)\ \overline{\alpha}_j p = \alpha_i r \qquad (2)\ \overline{\beta}_j p = \beta_i s$$
$$(3)\ \overline{\delta}_j q = \delta_i s \qquad (4)\ \overline{\gamma}_j q = \gamma_i r$$

It is easy to see that $p \neq 0, q \neq 0, r \neq 0$ and $s \neq 0$. (If for instance $p = 0$ then $r = 0$ by (1). This implies that $q = 0$ by (4), so that $s = 0$ by (3). We come to $(0, 0, 0, 0) = (p, q, r, s)$, a contradiction).

From (1) we get $p = \overline{\alpha}_j^{-1}\alpha_i r$ and by (2) we have $p = \overline{\beta}_j^{-1}\beta_i s$. Then $\overline{\alpha}_j^{-1}\alpha_i r = \overline{\beta}_j^{-1}\beta_i s$. Also, eliminating q from (3) and (4) we get $\overline{\gamma}_j^{-1}\gamma_i r = \overline{\delta}_j^{-1}\delta_i s$. Therefore $rs^{-1} = \alpha_i^{-1}\overline{\alpha}_j\overline{\beta}_j^{-1}\beta_i = \gamma_i^{-1}\overline{\gamma}_j\overline{\delta}_j^{-1}\delta_i$. This may by written in the form

$$(\gamma_i\alpha_i^{-1})(\overline{\alpha}_j\overline{\beta}_j^{-1})(\beta_i\delta_i^{-1})(\overline{\delta}_j\overline{\gamma}_j^{-1}) = 1. \tag{9.1}$$

Conversely, assume that (9.1) holds. It is easy to check that $s = 1, r = \alpha_i^{-1}\overline{\alpha}_j\overline{\beta}_j^{-1}\beta_i, q = \overline{\delta}_j^{-1}\delta_i, p = \overline{\beta}_j^{-1}\beta_i$ satisfy (1) to (4). Therefore $V_i \cap W_j \neq 0$ is equivalent to (9.1).

Assume that $3 \leq i \leq n-3$ and $4 \leq j \leq n-2$. Define $a_i = \gamma_i\alpha_i^{-1}, b_j = \overline{\alpha}_j\overline{\beta}_j^{-1}, c_i = \beta_i\delta_i^{-1}, d_j = \overline{\delta}_j\overline{\gamma}_j^{-1}$. We rewrite condition (9.1) as follows

$$a_i b_j c_i d_j = 1. \tag{9.2}$$

From the hypotheses of the lemma we know that (9.2) holds for $3 \leq i < j \leq n - 2$ and does not hold for $i = j$ and $4 \leq i \leq n - 3$. In particular $a_3 b_j c_3 d_j = 1$ for $4 \leq j \leq n - 2$, so that $d_j = c_3^{-1}b_j^{-1}a_3^{-1}$. By substituting this to (9.2) we get

$$\begin{aligned} a_i b_j c_i c_3^{-1}b_j^{-1}a_3^{-1} &= 1 \quad \text{for } 4 \leq i < j \leq n - 2 \\ a_i b_j c_i c_3^{-1}b_j^{-1}a_3^{-1} &\neq 1 \quad \text{for } i = j, 4 \leq i \leq n - 3. \end{aligned} \tag{9.3}$$

In particular $a_i b_{n-2} c_i c_3^{-1}b_{n-2}^{-1}a_3^{-1} = 1$ for $4 \leq i \leq n - 3$. Thus, $a_i = a_3 b_{n-2} c_3 c_i^{-1} b_{n-2}^{-1}$. By substituting this to (9.3) we see that the element

$$a_3 b_{n-2} c_3 c_i^{-1} b_{n-2}^{-1} b_j c_i c_3^{-1} b_j^{-1} a_3^{-1}$$

is equal to 1 for $4 \leq i < j \leq n - 3$ and it is $\neq 1$ for $i = j, 4 \leq i \leq n - 3$. Transforming this equation we come to

$$b_{n-2}(c_3 c_i^{-1})(b_{n-2}^{-1} b_j)(c_i c_3^{-1}) b_j^{-1} = 1,$$

and next to

$$(b_{n-2}^{-1} b_j)(c_i c_3^{-1}) = (c_i c_3^{-1})(b_{n-2}^{-1} b_j). \tag{9.4}$$

Here (9.4) holds for $4 \leq i < j \leq n - 3$ and does not hold for $4 \leq i = j \leq n - 3$. Since $(c_i c_3^{-1})$ and $(b_{n-2}^{-1} b_j)$ are as in the definition of $N(D)$, we must have $n - 6 \leq N(D)$. This establishes the assertion of the lemma. $\square$

We are now ready for the second main result of this section. Note that $6 = \binom{4}{2} = d_{4,2}(K)$, and $N(D)$ can be considered as a noncommutativity measure of D.

Theorem 9.14 *Let T be a set of idempotents of rank 2 in $M_4(D)$ such that the image of T in M_2/M_1 is a triangular set of idempotents. Then $|T| \leq N(D) + 6$, so that $d_{4,2}(D) \leq N(D) + 6$.*

Proof. Suppose that $|T| > N(D)+6$. Then there exist $e_1, e_2, \ldots, e_n \in T$ with $n > N(D) + 6$, such that $\mathrm{rank}(e_i e_j) < 2$ for $1 \leq i < j \leq n$. Let $V_k = \ker(e_k)$ and $W_k = \mathrm{Im}(e_k)$. So we have

$$(*) \qquad V_i \cap W_j \neq 0 \text{ for } i < j \text{ and } V_i \cap W_i = 0 \text{ for } i, j = 1, 2, \ldots, n.$$

It is clear that $\dim(V_k) = \dim(W_k) = 2$. Assume that $V_i \cap W_j \neq 0$ for all $1 \leq i \neq j \leq 3$. Lemma 9.4 may be applied to V_1, V_2, V_3 (we know they are all different since, if $V_i = V_j$ for $1 \leq i < j \leq n$, then $V_j \cap W_j = V_i \cap W_j \neq 0$, a contradiction). We consider two cases (see Lemma 9.4.)

(a) There exist independent vectors v_1, v_2, v_3 with $V_1 = \mathrm{Lin}(v_2, v_3)$, $V_2 = \mathrm{Lin}(v_3, v_4)$ and $V_3 = \mathrm{Lin}(v_1, v_2)$. By condition $(*)$ we have $V_1 \cap W \neq 0$ and $V_2 \cap W \neq 0$ whenever $W = W_k$ for arbitrary $4 \leq k \leq n$. Hence Lemma 9.5 yields $W_k \subseteq \mathrm{Lin}(v_1, v_2, v_3)$ for $k = 4, 5, \ldots, n$.

Define $W = \mathrm{Lin}(v_1, v_2, v_3)$. Spaces $V_4, V_5, \ldots, V_n, W_4, W_5, \ldots, W_n$ and W satisfy the hypotheses of Lemma 9.6, so that $n \leq 6$. This contradicts the fact that $n > N(D) + 6$. Thus (a) cannot hold.

(b) $V_1 \cap V_2 \cap V_3 \neq 0$. Assume first that $\dim(V_1 + V_2 + V_3) \leq 3$. By (∗) $V_1 \cap W \neq 0$ and $V_2 \cap W \neq 0$ for $W = W_3$. By Lemma 9.7 we have $V_3 \cap W \neq 0$, thus $V_3 \cap W_3 \neq 0$. This contradicts (∗). Therefore we may assume that $\dim(V_1 + V_2 + V_3) = 4$. By condition (∗) $V_i \cap W \neq 0$ for $i = 1, 2, 3$, whenever $W = W_k$ for $4 \leq k \leq n$. Applying Lemma 9.8 to V_1, V_2, V_3 and this W we get $0 \neq V_1 \cap V_2 \cap V_3 \subseteq W$. Then $\bigcup_{4 \leq j \leq n} W_j \neq 0$. If we apply Lemma 9.9 to $V_4, V_5, \dots, V_n, W_4, W_5, \dots W_n$, then we get $n \leq 6$, a contradiction. Thus (b) cannot hold either.

Therefore there exist i_1, i_2, i_3 such that $\{i_1, i_2, i_3\} = \{1, 2, 3\}$ and $V_{i_1} \cap V_{i_2} = 0$. Our problem is symmetric with respect to the substitution $V_i \mapsto W_{m+1-i}, W_i \mapsto V_{n+1-i}$ for $i = 1, 2, \dots, n$. Hence, by a similar reasoning we see that there exist j_{n-2}, j_{n-1}, j_n with $\{j_{n-2}, j_{n-1}, j_n\} = \{n-2, n-1, n\}$ and $W_{j_{n-1}} \cap W_{j_n} = 0$.

First assume that there exists $k \in \{4, 5, \dots, n-3\}$ such that one of the spaces $V_{i_3}, V_k, W_k, V_{j_{n-2}}$ contains one of $V_{i_a} \cap W_{j_b}$, where $a = 1, 2, b = n-1, n-2$. Put $V = V_k, \overline{W} = W_k$. Now Lemma 9.12 may by applied to the sequence $V_{i_1}, V_{i_2}, V_{i_3}, \overline{V}, \overline{W}, W_{j_{n-2}}, W_{j_{n-1}}, W_{j_n}$. This yields $\overline{V} \cap \overline{W} \neq 0$, so that $V_k \cap W_k \neq 0$. This contradicts (∗). Therefore none of the spaces $V_{i_3}, V_4, \dots, V_{n-3}, W_4, W_5, \dots, W_{n-3}, W_{j_{n-2}}$ contains one of $V_{i_a} \cap W_{j_b}$, where $a = 1, 2, b = n-1, n$. So Lemma 9.13 may be applied to the sequence

$$V_{i_1}, V_{i_2}, V_{i_3}, V_4, \dots, V_{n-3}, W_4, W_5, \dots, W_{n-3}, W_{j_{n-2}}, W_{j_{n-1}}, W_{j_n}.$$

It follows that $n \leq \binom{4}{2} + \mathrm{N}(D)$, a contradiction. This completes the proof of the theorem. □

It is an open problem to find necessary and sufficient conditions for $d_{n,j}(D)$ to be finite, and to determine bounds in terms of the division algebra D, in case $n > 4$ and $j \neq 1, n-1, n$.

Remark It is easy to see that

$$\mathrm{N}(D) = \sup\{n \mid \text{there exists a chain } C_0 \supset C_1 \supset \cdots \supset C_n$$
$$\text{of centralizers of subsets of } D\}.$$

In fact, assume that $a_1, a_2, \dots a_k, b_1, b_2, \dots b_k$ are as in the definition of $\mathrm{N}(D)$. Then $C(a_0) \supset C(a_0, a_1) \supset \cdots \supset C(a_0, a_1, \dots, a_i)$, where $a_0 \in Z(D)$, because $b_i \in C(a_0, a_1, \dots, a_{i-1}) \setminus C(a_0, a_1, \dots, a_i)$ for $i = 1, 2, \dots, k$. On the other hand, assume that $C_0 \supset C_1 \supset \cdots \supset C_k$, where $C = C(A_i), A_i \subseteq D$. By induction we construct elements $a_i \in A_i$ such that $C(a_0) \supset C(a_0, a_1) \supset \cdots \supset C(a_0, \ldots, a_k)$. Choose any $a_0 \in A_0$. Assume that the elements $a_1 \in A_1, \ldots, a_{i-1} \in A_{i-1}, i < k$, have been defined. Suppose that $C(a_0, \ldots, a_{i-1}) = C(a_0, \ldots, a_{i-1}, A_i)$. Then $C(A_i) \supseteq C(a_0) \cap C(a_1) \cap \cdots \cap C(a_{i-1}) \supseteq C(A_{i-1})$, since $C_j \supseteq C_{i-1}$ for $j = 0, 1, \ldots, i-1$. This contradicts the fact that $C(A_{i-1}) \supset C(A_i)$. Therefore there exists $a_i \in A_i$ with $C(a_0, \ldots, a_{i-1}) \supset C(a_0, \ldots, a_i)$. This proves the inductive claim. Let $b_i \in C(a_0, a_1, \ldots, a_{i-1}) \setminus C(a_0, a_1, \ldots, a_i)$ for $i = 1, 2 \ldots k$. Then $a_1, \ldots, a_k, b_1, \ldots, b_k$ are as in the definition of $N(D)$. Thus $k \leq N(D)$.

Corollary 9.15 *If centralizers of noncentral elements of D are commutative, then* $N(D) \leq 2$.

Proof. Define a relation $\sim$ on $D \setminus Z(D)$ as follows: $a \sim b$ if and only if $ab = ba$. By the assumption on D, $\sim$ is an equivalence relation. Note that $C(a) = Z(D) \cup \{b \in D \setminus Z(D) \mid b \sim a\}$ whenever $a \notin Z(D)$. Thus every centralizer of a subset of D is equal to $Z(D)$, D or $Z(D) \cup X$ for a $\sim$ - class X. Therefore the length of every chain of centralizers is less than 3. By the above remark $N(D) \leq 2$. $\square$

Example Let D be the quotient ring of the skew polynomial ring $K[x, \delta]$, where δ is a derivation of the field K. It is known that if t is a noncentral element of D, then its centralizer $C(t)$ is commutative, (see [65], Lemma 1). By Corollary 9.15 $N(D) \leq 2$.

Our next example shows that there exist division algebras D with infinite triangular sets of idempotents in M_2/M_1 for $M_4(D)$. Its role is explained in the example after Theorem 9.1.

Example Define $D = D_1 \otimes_K D_2 \otimes_K \cdots$, where $D_i, i \geq 1$, are finite dimensional division algebras over their common centre K. If their ranks are pairwise relatively prime, then D is a division algebra (see [36], Lemma 4.4.8). Let α_i, β_i be any noncommuting elements of D_i. Put

$$a_i = \underbrace{1 \otimes \cdots \otimes 1}_{i-1 \text{ times}} \otimes \alpha_i \otimes 1 \otimes \cdots, \quad b_i = \underbrace{1 \otimes \cdots \otimes 1}_{i-1 \text{ times}} \otimes \beta_i \otimes 1 \otimes \cdots.$$

Thus $a_i b_j = b_j a_i$ for $i < j$ and $a_i b_i \neq b_i a_i$. Let $v_1, \ldots, v_4$ be a basis of D^4. Consider $V_i = \text{Lin}(v_1 + v_4 a_i, v_2 + v_3 a_i)$ and $W_j = \text{Lin}(v_1 + v_2 b_j, v_4 + v_3 b_j)$.

As in the proof of Lemma 9.13 it follows that $V_i \cap W_j \neq 0$ if and only if $a_i b_j = b_j a_i$ (this is a direct consequence of the equivalence of condition (9.1) with $V_i \cap W_j \neq 0$). By our choice of a_i and b_i we have $V_i \cap W_j \neq 0$ for $i < j$ and $V_i \cap W_i = 0$. Let e be the idempotents in $M_4(D)$ such that $\ker(e_i) = V_i, \operatorname{Im}(e_i) = W_i$. The images of e_i in M_2/M_1 form an infinite triangular set of idempotents ($e_i \prec e_j$ if and only if $i < j$).

We conclude with an application of Theorem 9.1 to the Burnside problem for certain skew linear semigroups. It extends the assertion of Theorem 7.3.

Theorem 9.16 *Let $S \subseteq M_n(D)$ be a π-regular semigroup. If every subgroup of S is locally finite, then S is locally finite.*

Proof. We may assume that S is finitely generated. Define S_j, S'_j, T_j as in Theorem 9.1. Let j be the least integer such that $S \subseteq M_j$. Since $S = S_j$ is finitely generated, S'_j intersects finitely many $\mathcal{L}$-classes of $M = M_j/M_{j-1}$. There exists $n_j < \infty$ such that every triangular set of idempotents in S'_j has at most n_j elements because all elements of a triangular set of idempotents belong to different $\mathcal{L}$-classes, (if $ef = \theta$ and $Me = Mf$, then there exists $x \in M$ such that $e = xf$, thus $\theta = ef = (xf)f = xf = e$, a contradiction). By Theorem 9.1 S_j/T_j is a 0-disjoint union of finitely many completely 0-simple ideals. Let $C = \mathcal{M}(G, I, M, P)$ be any of them. By assumptions on S, the group G is locally finite. So $\mathcal{M}(G, I, M, P)$ is also locally finite by Lemma 2.3 in [87]. Since C is finitely generated, C must be finite. Therefore S_j/T_j is also finite. By Lemma 2.1 of [87], T_j is finitely generated. Since T_j/S_{j-1} is nilpotent, T_j/S_{j-1} is finite. Thus S_{j-1} is finitely generated. The above reasoning may by repeated for S_{j-1} and so forth. This shows that S_j/T_j and T_j/S_{j-1} are all finite. Thus S is finite. $\square$

Corollary 9.17 *Let G be a torsion-free almost polycyclic group and D the field of fractions of the group algebra $K[G]$. If $S \subseteq M_n(D)$ is a periodic semigroup, then S is locally finite.*

Proof. We know that every periodic subgroup of $M_n(D)$ is locally finite (see [72]). Since S is a π-regular semigroup, Theorem 9.16 may by applied. Thus S is locally finite. $\square$

9.2   Graded rings: motivation
and technical results

The aim of this section is to present a general technique that can be applied, in particular, to several problems in ring theory. It is based on the structure theory of skew linear semigroups. The significance of the theory of linear groups is well known. Their structure and properties are exploited repeatedly in many important applications. On the other hand, while semigroups of endomorphisms show up often in various problems, and are a natural object to study, until recently their structure had not been exploited in applications. Our aim is to present the key ingredients of such an approach and to illustrate it with some applications to graded rings. We start with preliminary motivating remarks on graded rings, leading to semigroups of matrices.

We will consider rings R graded by a semigroup S. That is, $R = \bigoplus_{s \in S} R_s$ as additive groups, and $R_s R_t \subseteq R_{st}$ for every $s, t \in S$. $\mathrm{H}(R) = \bigcup_{s \in S} R_s$ is the set of homogeneous elements of R. Clearly, $\mathrm{H}(R)$ is a multiplicative subsemigroup. If $X \subseteq \mathrm{H}(R)$, then by $\mathbf{Z}\{X\}$ we denote the (homogeneous) subring of R generated by X. For $r \in R, r \neq 0$, we put $\mathrm{supp}(r) = \{s \in S \,|\, r_s \neq 0\}$. If $I \subseteq R$ is a subset, then $\mathrm{supp}(I) = \{\mathrm{supp}(r) \mid 0 \neq r \in I\}$, while the graded part of I is defined by $I_{gr} = \bigoplus_{s \in S}(I \cap R_s)$. The degree map $\deg : \mathrm{H}(R) \setminus \{0\} \longrightarrow S$ is defined by $\deg(h) = s$ for $h \in R_s$.

Group crossed products $R = A * G$, for a ring A and a group G, form the simplest class of graded rings, [103]. Here $R_g = Ag$ for $g \in G$, so each component R_g contains a unit of R provided that A is a unital ring. On the other hand, if for a graded ring R the set $H = \mathrm{H}(R) \setminus \{0\}$ consists of regular elements (that is, non - zero divisors in R), then under many classical finiteness conditions on R, H is an Ore subset of R and one can form the homogeneous localization RH^{-1}. If additionally S is cancellative, then the image $\deg(H)$ of H under the degree map is an Ore semigroup, so RH^{-1} has a natural structure of a group - graded ring with components $\bigcup_{s,t \in S, s=gt} R_s(H \cap R_t)^{-1}$, for $g \in G = \deg(H)\deg(H)^{-1}$. Since each component contains a unit, the following result shows that $RH^{-1} = A * G$, a group crossed product of the group G.

Lemma 9.18 *Assume that every component of R contains an invertible element. Then $S = \mathrm{supp}(R)$ is a group and $R \simeq R_e * S$, a group crossed* *product.*

Proof. Let $u_s \in R_s, s \in S$, be an invertible element. Let G be the group generated by all u_s. It is clear that $\deg(G) = S$ is a group. Next $R_{s^{-1}} u_s \subseteq R_e$, so that $R_{s^{-1}} \subseteq R_e u_s^{-1}$. Hence $R = \sum_{s \in S} R_e u_s^{-1}$ and this is a direct sum. Write $g_s = u_s^{-1}$. Clearly $g_s g_t = u_{st} g_{st}$ for some $u_{st} \in G \cap R_e$. Therefore, for $r, r' \in R_e$ and $s, s' \in S$ we get $(rg_s)(r'g_{s'}) = rg_s r' g_s^{-1} g_s g_{s'} = (rg_s r' g_s^{-1} u_{ss'}) g_{ss'}$. Also $(r')^{(s)} = g_s r' g_s^{-1} \in R_e$. Hence u_{st} yield the desired cocycle and conjugations by g_s are the desired automorphisms of R_e. $\square$

Thus, a natural idea is to study R via RH^{-1}, or more generally to reduce problems on an arbitrary graded ring R to certain crossed products derived from R. Such an approach is quite natural and it turns out to be very fruitful even in the simplest case where R is a domain and $S = \mathbf{Z}$, see [3],[137]. What are the most general circumstances this can be applied? For example, if $H(R) \subseteq M_n(D)$ for a division algebra D, and F is the group of units of the monoid $eM_n(D)e$ for some idempotent $e \in M_n(D)$, then $H(R) \cap F \subseteq F \simeq GL_j(D)$, where $j = \text{rank}(e)$, (if nonempty) is a natural candidate for a 'nice' set of homogeneous elements of the graded subring $\mathbf{Z}\{H(R) \cap F\}$. Our main motivating ideas are built along these lines, aiming at applications to rings with many homomorphic images embeddable into simple Artinian rings.

Lemma 9.19 *Let P be a prime ideal of a graded ring R such that R/P is a right Goldie ring. Then*

1. *the image H' of $H(R)$ in R/P has a unique ideal uniform component I, consisting of all matrices of the least nonzero rank in the classical quotient ring $Q_{cl}(R/P)$ and the zero matrix. Furthermore, if $\widehat{I}$ is the completely 0-simple closure of I, then $\widehat{H'} = H' \cup \widehat{I}$ is a semigroup in which $\widehat{I}$ is an ideal,*

2. *if $a \in H(R)$ is such that its image a' in R/P belongs to I and is nonnilpotent, then $a'R'a' \subseteq R/P$ is the subring generated by $T' = a'H'a' \setminus \{0\}$, it is a prime right Goldie ring and T' is a cancellative semigroup.*

Proof. 1) Let $Q_{cl}(R') = M_n(D)$ be the classical ring of quotients of $R' = R/P$. Consider the set I of elements q in H' such that the matrix q is equal to zero or has a minimal positive rank in $M_n(D)$. Clearly $I \neq 0$, because $R' \neq 0$.

Suppose that J is a nonzero nil ideal of H'. Then J is nilpotent by Theorem 9.1 and Proposition 9.3, (see also [24, 17.20, 17.22]). Hence J generates a nilpotent subring N of R'. Therefore N is a nonzero nilpotent ideal of R'. This contradicts the primeness of R' and shows that H' has no nonzero nil ideals.

Furthermore, a similar argument shows that for any nonzero ideals I_1 and I_2 of H' we have $I_1 I_2 \neq \{0\}$. Hence applying the structure theorem for skew linear semigroups, Theorem 9.1, to the subsemigroup H of $Q_{cl}(R)$ it follows that I is the unique ideal uniform component of H'. The first part of the lemma now follows as in Lemma 3.11.

2) The $\mathcal{H}$ - class of a' in $\widehat{I} \subseteq Q_{cl}(R')$ contains an idempotent e, so that $a'Q_{cl}(R')a' = eQ_{cl}(R')e$ is again a simple Artinian ring and a' is invertible in this ring. Similarly, each nonzero $b \in a'H'a' \subseteq a'\widehat{I}a'$ is a regular element of $a'R'a'$, so T' is a cancellative semigroup. Suppose $0 \neq x \in (r_1a'R' + \cdots + r_ka'R') \cap ra'R'$, for some $r, r_1, \ldots, r_k \in a'R'a'$. Then, because R' is prime, $xR'a' \neq \{0\}$. Therefore $(r_1a'R'a' + \cdots + r_na'R'a') \cap ra'R'a' \neq \{0\}$. Since by assumption R' is right Goldie, this implies that $a'R'a'$ has finite right Goldie dimension.

Suppose that J, L are ideals of $a'R'a'$ such that $JL = \{0\}$. Clearly $Ja'R'a'L = \{0\}$ implies that $Ja' = \{0\}$ or $a'L = \{0\}$. Thus $J = \{0\}$ or $L = \{0\}$ because a' is a unit in $a'Q_{cl}(R)a'$. Therefore $a'R'a'$ is prime.

Since $a'R'a'$ is a subring of $Q_{cl}(R)$, it also satisfies the ascending chain condition on right annihilators. It follows that $a'R'a'$ is a prime right Goldie ring. $\square$

The above lemma applies in particular to prime homomorphic images of a graded ring R which satisfies a polynomial identity or is right Noetherian.

For the rest of this section we fix a prime ideal P of R such that R/P is a right Goldie ring. Let $Q_{cl}(R/P) = M_n(D)$ for a division algebra D. Our aim is to study R/P via a homogeneous subring of R with a nicer graded structure. For any $x \in R$ and $X \subseteq R$, the images of x and X in the ring R/P will be denoted by $\overline{x}$ and $\overline{T}$, respectively.

Choose a minimal nonzero idempotent $e \in M_n(D)$ such that $F =$

$\{a \in eM_n(D)e \mid \mathrm{rank}(a) = \mathrm{rank}(e)\}$ intersects $\overline{\mathrm{H}(R)}$. Let T be the inverse image in $\mathrm{H}(R)$ of $\overline{T} = F \cap \overline{\mathrm{H}(R)}$. Note that F is a (maximal) subgroup of the multiplicative semigroup $M_n(D)$. The degree map $\deg : \mathrm{H}(R) \setminus \{0\} \longrightarrow S$ restricted to T is a homomorphism $T \longrightarrow \deg(T) \subseteq S$. Consider the subring $R_T = \mathbf{Z}\{T\}$ generated by T. It is an $\deg(T)$ - graded ring. In many cases one can show that T is a right Ore subset of R_T and the homogeneous localization $R' = R_T T^{-1}$ is graded by the group G of quotients of $\deg(T)$. Hence $R' = A * G$, a group crossed product. Moreover, as we will see in Section 9.3, the image $\overline{R_T}$ of R_T in R/P to a large extent is responsible for the properties of R/P. However, in general, one also needs to consider the intersection of $\mathrm{H}(R)$ with other maximal subgroups of $M_n(D)$, and not only those coming from the matrices of minimal nonzero rank. Then, in order to implement the above motivating ideas, one has to use the full strength of the structural approach to subsemigroups of $M_n(D)$.

For the remainder of this section we focus on the most important case where S is a cancellative semigroup.

If $a \in \mathrm{H}(R)$, then clearly $a\,\mathrm{H}(R)a \subseteq \mathrm{H}(R)$ and $a\,\mathrm{H}(R)a$ is a graded subring of R. Moreover, if S is cancellative, then $sts = st's, s, t, t' \in S$ implies that $t = t'$, so that $a\,\mathrm{H}(R)a = \mathrm{H}(aRa)$.

We will fix an element a in $T \subseteq \mathrm{H}(R)$. Let $H_a = a\,\mathrm{H}(R)a \setminus P$. Then $\overline{a}$ belongs to a subgroup F of $M_n(D)$. Moreover, $\overline{a\,\mathrm{H}(R)a} \subseteq F \cup \{0\}$. Hence H_a is the set of all elements q in $a\,\mathrm{H}(R)a$ such that $\overline{q} \in F$. It follows that $H_a \subseteq T$ is a multiplicative subsemigroup of R. Since H_a does not contain zero, evidently H_a has no zero divisors, that is, $xy \neq 0$ for any $x, y \in H_a$.

The additive subgroup $R_{(a)}$ generated in R by H_a is a subring of $\mathbf{Z}\{T\}$. Obviously, the image $\overline{R_{(a)}} \subseteq R/P$ is equal to $\overline{a}\overline{R}\overline{a}$. Since H_a consists of homogeneous elements, $R_{(a)}$ is a homogeneous subring of aRa.

Note that R/P_{gr} is graded and $\mathrm{H}(R/P_{gr}) = \bigcup_{s \in S} R_s/(R_s \cap P)$. Hence, the semigroup of homogeneous elements of R/P_{gr} trivially intersects P/P_{gr}. Clearly P/P_{gr} is a prime ideal of R/P_{gr}. So, studying prime homomorphic images of R we may replace R by R/P_{gr} and hence consider the special case where P has no nonzero homogeneous elements, that is $P \cap \mathrm{H}(R) = 0$. Then we get $H_a = a\,\mathrm{H}(R)a \setminus \{0\}$, $R_{(a)} = aRa$ and $\mathrm{H}(R_{(a)}) = a\,\mathrm{H}(R)a$. Hence $r_s \in (H_a)_s = H_a \cap R_s$, for any element $r \in R_{(a)}$ and any $s \in \mathrm{supp}(r)$.

Lemma 9.20 *If P has no nonzero homogeneous elements, then all elements of H_a are regular in $R_{(a)}$. In particular, H_a is a cancellative semigroup.*

Proof. Suppose that $rh = 0$ for some elements $h \in H_a$, $r \in R_{(a)}$. The cancellativity of S implies that $r_s h = 0$ for each $s \in S$. We know that $r_s \in H_a$. Since H_a has no zero divisors, this yields $r_s = 0$ for all s; whence $r = 0$. We have shown that all elements of H_a are right regular in $R_{(a)}$.

If we take any elements x, y, z in H_a such that $yx = zx$, then $(y - z)x = 0$ and $y - z \in R_{(a)}$ imply $y = z$. Thus H_a is right cancellative. Similarly, H_a consists of left regular elements of $R_{(a)}$ and it is left cancellative. $\square$

A multiplicative subsemigroup T of of regular elements of R is said to be a right (left) Ore subset if, for each $r \in R$ and $t \in T$, there exist $r' \in R$ and $t' \in T$ such that $rt' = tr'$ (respectively, $t'r = r't$). By $\mathcal{B}(R)$ we denote the prime radical of R.

Lemma 9.21 *Assume that P has no nonzero homogeneous elements, and one of the following conditions holds*

1. *R is an algebra with no free subalgebras of rank two,*

2. *R has finite right Goldie dimension.*

Then H_a is a right Ore semigroup and it is a right Ore subset of the ring $R_{(a)}$. If 1) holds, then it is also a left Ore subset of $R_{(a)}$.

Proof. Take any elements $x, y \in H_a$.

1) We may assume that x, y are not invertible in H_a since otherwise $xH_a \cap yH_a \neq \emptyset$. By the hypothesis there exists a nonzero polynomial f with zero constant term and with coefficients in the base field K such that $f(x, y) = 0$. Assume that f is of minimal degree. Clearly, $f(x, y) = xf_1(x, y) + yf_2(x, y)$ for some polynomials f_1 and f_2. By symmetry we may assume that f_1 is nonzero. Here $c = f_1(x, y)$ and $d = f_2(x, y)$ are elements of the algebra R^1 obtained by adjoining an identity to R in case R has no identity element, and $R^1 = R$ otherwise. Assume first that $xc \neq 0$. In view of Lemma 9.20 we may replace xc, yd by xcx, ydx respectively. Then $c \in R$. Choose $s \in S$ such that $(xc)_s \neq 0$. The

cancellativity of S implies that $(xc)_s = x(c_t)$ for some $t \in S$. Similarly, $(yd)_s = y(d_u)$ for some $u \in S$. We get $x(c_t) + y(d_u) = f(x, y)_s = 0$, whence $0 \neq x(c_t) = y(-d_u)$.

On the other hand, suppose that $xc = 0$. If the constant term of f_1 is zero, then $c \in R$ and from Lemma 9.20 it follows that $c = f_1(x, y) = 0$, contradicting the choice of f. So suppose that the constant term of f_1 is nonzero, so $f_1 = g + \lambda$ for some $\lambda \in K$ and a polynomial g with zero constant term. Then $xc = 0$ can be rewritten as $x = xe$, where $c = g(x, y) + \lambda$ and $e = -g(x, y)\lambda^{-1} \in R_{(a)}$. This leads to $x = xe_v$ for some $e_v \in H_a$. By Lemma 9.20 $e = e_v$ is the identity of H_a. But then $g(x, y) = -\lambda \neq 0$ and we have $xg_1(x, y) + yg_2(x, y) = -\lambda$ for some polynomials g_1, g_2. An argument as in the first paragraph of the proof shows that one of the elements x, y is invertible in H_a, again leading to a contradiction.

Since $x, y \in H_a$, the definition of c and d implies that $c_t \in H_a \cap R_t$ and $-d_u \in H_a \cap R_u$. Thus $xH_a \cap yH_a \neq \emptyset$. This means that H_a is a right Ore semigroup. Similarly, H_a is a left Ore semigroup.

2) By the hypothesis the right ideals $x^j yaR, j = 1, 2, \ldots$, are dependent. So $x^k yar_k + \cdots + x^n yar_n = 0$ for some $r_i \in R$ and $n > k \geq 1$ with $x^n yar_n \neq 0$. Choosing maximal such k we have $x^k yar_k \neq 0$. So we have $b' + b'' = 0$, where $b' = x^k yc, b'' = x^{k+1}d$ for some $c, d \in aR$. Also $b' \neq 0$ by the choice of k. As in 1) we see that $x^k yc_s = x^{k+1}d_t \neq 0$ for some $s \in \mathrm{supp}(c)$ and $t \in \mathrm{supp}(d)$. So $u = x^k yc_s \notin P$ because $P \cap \mathrm{H}(R) = 0$. Hence there exists $z \in \mathrm{H}(R)$ such that $uza \notin P$. So $x^k y(c_s za) = x^{k+1}(d_t za) \in H_a$ and both elements in the parentheses lie in aRa. Hence, Lemma 9.20 allows us to cancel x^k on the left. Therefore $xH_a \cap yH_a \neq \emptyset$, as desired.

So, we have shown that H_a is a right Ore semigroup. Therefore it has the group of right quotients $H_a H_a^{-1}$. Since $R_{(a)}$ is the subring generated by H_a, it follows easily that H_a is a right Ore subset in $R_{(a)}$. Namely, if $h, h_1, \ldots, h_n \in H_a$, then choosing a common denominator $g \in H_a$ we can write $h^{-1}h_i = g_i g^{-1}$ for some $g_i \in H_a$. So, for $r = h_1 + \cdots + h_n$ we get $rg = h(g_1 + \cdots + g_n)$, as desired. Similarly $H_{(a)}$ is a left Ore subset of $R_{(a)}$ if 1) holds. $\square$

Denote by S_a the image $\deg(H_a) \subseteq S$. Since $\deg$ is a homomorphism when restricted to H_a, under the hypotheses of Lemma 9.21 it follows that S_a is a (cancellative) right Ore semigroup. It has a group of quotients $G_a = S_a S_a^{-1}$. Clearly, $R_{(a)}$ is G_a - graded with components $(R_{(a)})_x = 0$ for $x \in G_a \setminus S_a$ and $(R_{(a)})_x = aR_t a$ for $x \in S_a$ with $x = \deg(a)t \deg(a)$. Also $S_a \subseteq \deg(T)$ and the latter is a subsemigroup of S. Since $\deg(a) \deg(T) \deg(a) \subseteq S_a$, it follows that T is also right Ore and the groups of quotients of T, S_a are equal.

Since H_a is a right Ore subset, we can consider localization $R_{(a)} H_a^{-1}$. Obviously, $\tilde{R}_a = R_{(a)} H_a^{-1}$ is $G_a = S_a S_a^{-1}$ - graded, because H_a^{-1} consists of homogeneous elements. Clearly, the identity component D_a of $\tilde{R}_a$ is equal to $\sum_{s \in S_a} (R_{(a)})_s (H_a)_s^{-1}$. Consider $r_t, h_t \in (H_a)_t, r_s, h_s \in (H_a)_s$. Then $r_t h_t^{-1} = r_t h_u (h_t h_u)^{-1}$ and $r_s h_s^{-1} = r_s h_v (h_s h_v)^{-1}$, where $h_u \in (H_a)_u, h_v \in (H_a)_v$ are such that $h_t h_u = h_s h_v$. Since $tu = sv$, we see that $r_s h_s^{-1}, r_t h_t^{-1} \in (H_a)_{tu}(H_a)_{tu}^{-1}$. This implies easily that $D_a = \bigcup_{s \in S_a} (H_a)_s (H_a)_s^{-1} \cup \{0\}$. Therefore D_a is a division algebra. Besides, every homogeneous component of $\tilde{R}_a$ contains an invertible element from $H_a H_a^{-1}$. This means that $\tilde{R}_a \simeq D_a * G_a$ a crossed product of D_a and G_a, as Lemma 9.18 shows.

Note that the natural map from $R_{(a)}$ onto $\overline{aRa} \subseteq R/P$ extends to $\tilde{R}_a = D_a * G_a \longrightarrow \overline{a} Q_{cl}(R/P)\overline{a}$. In order to use the above observations, one needs to establish some relations between the properties of $R, R_{(a)}$, and $\tilde{R}_a$.

Lemma 9.22 *Assume that P has no nonzero homogeneous elements and S_a is a right Ore semigroup. Then*

1. *If $M \subset M'$ are right ideals of $\tilde{R}_{(a)}$, then also $(M \cap R_{(a)})aR \subset (M' \cap R_{(a)})aR$. In particular, if R has right Krull dimension, or is right Noetherian, then $\tilde{R}_{(a)}$ has right Krull dimension, not exceeding that of R, respectively $\tilde{R}_{(a)}$ is right Noetherian.*

2. *Assume that prime homomorphic images of $\tilde{R}_{(a)}$ are right Goldie. If $Q \subset Q'$ are prime ideals of $\tilde{R}_{(a)}$, then there exist prime ideals $W \subset W'$ of R such that $W \cap R_{(a)} = Q \cap R_{(a)}, W' \cap R_{(a)} = Q' \cap R_{(a)}$. Moreover, the classical Krull dimension of $\tilde{R}_{(a)}$ does not exceed that of R.*

Proof. Assume that $M \subset M'$ are right ideals of $\tilde{R}_{(a)}$. Then $(M \cap R_{(a)})aR \subseteq (M' \cap R_{(a)})aR$. Suppose that $(M \cap R_{(a)})aR = (M' \cap R_{(a)})aR$. Then $(M \cap R_{(a)})R_{(a)} = (M \cap R_{(a)})aRa = (M' \cap R_{(a)})aRa = (M' \cap$

$R_{(a)})R_{(a)}$. Since $\tilde{R}_{(a)} = R_{(a)}H_a^{-1}$, we come to $M = (M \cap R_{(a)})\tilde{R}_{(a)} \cap R_{(a)} = (M' \cap R_{(a)})\tilde{R}_a \cap R_{(a)} = M'$, a contradiction. This shows that a strictly decreasing (increasing) chain of right ideals of $\tilde{R}_{(a)}$ yields a strictly decreasing (increasing) chain of right ideals of R. So the first assertion follows.

$Q \cap R_{(a)} \subseteq Q' \cap R_{(a)}$ are prime ideals of $R_{(a)}$ by the hypothesis on $\tilde{R}_{(a)}$. Since $(Q \cap R_{(a)})\tilde{R}_{(a)} = Q$, the inclusion above is proper. Let J be the ideal generated by $Q \cap R_{(a)}$ in R. Since $R(Q \cap R_{(a)})R \cap R_{(a)} \subseteq Q \cap R_{(a)}$, we must have $J \cap R_{(a)} = Q \cap R_{(a)}$. So, let W be an ideal of R which is maximal with respect to the condition $W \cap R_{(a)} = Q \cap R_{(a)}$. It is clear that W is a prime ideal because $Q \cap R_{(a)}$ is a prime ideal of $R_{(a)}$. Similarly, let J' be the ideal of R generated by $Q' \cap R_{(a)}$. Write $I = (W + J') \cap R_{(a)}$. Then $aIa \subseteq aWa + aJ'a \subseteq W \cap R_{(a)} + J' \cap R_{(a)} = Q \cap R_{(a)} + Q' \cap R_{(a)} = Q' \cap R_{(a)}$. Then $I \subseteq Q' \cap R_{(a)}$ since the latter is prime. So we get $I = Q' \cap R_{(a)}$ because $J' \cap R_{(a)} = Q' \cap R_{(a)}$. Therefore there exists an ideal W' of R which is maximal with respect to $W \subseteq W'$ and $W' \cap R_{(a)} = Q' \cap R_{(a)}$. Again it is clear that W' is a prime ideal of R. Since $Q \cap R_{(a)} \subset Q' \cap R_{(a)}$, we must have $W \neq W'$. The above can also be applied to finite ascending chains of prime ideals of $\tilde{R}_{(a)}$, so the remaining assertion follows. $\square$

We note that the additional hypothesis in assertion 2) of Lemma 9.22 is satisfied whenever R is a PI - algebra or, by 1) of this lemma, if R has right Krull dimension, [81].

The theory of group - graded rings is especially well developed for gradations by groups of some special types. Our next aim is to show that certain important ring theoretic properties lead to gradations by such groups. We say that R is a graded algebra if R is an algebra over a field K and all components are K-subspaces.

Lemma 9.23 Assume that R is an algebra of finite Gelfand - Kirillov dimension. Then G_a is locally almost nilpotent.

Proof. Choose a subalgebra B generated by finitely many nonzero homogeneous elements $r_i \in R_{s_i}, i = 1, \ldots, k$, of $R_{(a)}$. Let $w_1, \ldots, w_n \in H_a$ be some words in r_i such that $\deg(w_1), \ldots, \deg(w_n) \in S_a$ are different. Since the components of R are subspaces, $w_1, \ldots, w_n$ are linearly independent over K. Therefore, the growth function of the semigroup

$S' = \langle s_1, \dots, s_k \rangle \subseteq S_a$ is bounded above by the growth function of B. So S' has polynomial growth. Since it is cancellative, it generates an almost nilpotent subgroup of the group $G_a = S_a S_a^{-1}$ by Theorem 7.1 and Theorem 7.2. $\square$

Lemma 9.24 *Let R be an S - graded algebra satisfying a polynomial identity and let A be a multiplicative subsemigroup of* $\mathrm{H}(R)$. *If A does not contain zero, then the subsemigroup* $\deg(A)$ *of S has the permutation property. In particular, G_a has a subgroup of finite index whose commutator subgroup is finite.*

Proof.   Let $H = \deg(A)$. Every PI - algebra satisfies a multilinear identity, that is, an identity of the form

$$(*)\qquad x_1 \cdots x_n + \sum_{1 \neq \sigma \in S_n} k_\sigma x_{\sigma(1)} \cdots x_{\sigma(n)} = 0,$$

where S_n is the symmetric group, k_σ are elements of the base field, [117].

Take any elements $a_1, a_2, \dots, a_n$ in A. Suppose that $a_i \in R_{s_i}$, for $i = 1, \dots, n$. Applying $(*)$ to the elements $a_1, \dots, a_n$ we get

$$a_1 \cdots a_n \in R_{s_1 \cdots s_n} \cap \sum_{1 \neq \sigma \in S_n} R_{s_{\sigma(1)} \cdots s_{\sigma(n)}}.$$

Given that A does not contain 0 it follows that $a_1 \cdots a_n \neq 0$. Therefore

$$0 \neq a_1 \cdots a_n \in R_{s_1 \cdots s_n} \cap R_{s_{\sigma(1)} \cdots s_{\sigma(n)}}$$

for some $\sigma \neq 1$. Hence $s_1 \cdots s_n = s_{\sigma(1)} \cdots s_{\sigma(n)}$. This means that H has the permutation property, as claimed. Finally, the assertion on G_a follows from Lemma 5.25 since we can take $A = H_a$. $\square$

9.3   Applications to graded rings

This section is devoted to examples of applications of the approach presented before. We shall discuss four different topics: prime Goldie semigroup algebras, homogeneity of the Jacobson radical of a graded ring, graded rings with finiteness conditions and semigroup algebras that are principal ideal rings. Here the semigroup algebra $R = K[S]$, and the contracted semigroup algebra $K_0[S]$, of a semigroup S over a field K are viewed as graded rings with components $R_s = Ks$ for all $s \in S$, respectively for all nonzero $s \in S$. So, it is natural to consider the subsemigroup S of $H(R)$ rather than entire $H(R)$.

A number of other important and easy-to-state problems have been recently treated via representations in simple Artinian rings. In particular, several problems concerning presentability of idempotents as a sum of nilpotents in an algebra over a field of characteristic zero have been solved within the algebras $M_4(D)$ [121]. Clearly, the methods go beyond the multiplicative structure, but they carry a lot of the flavour of this chapter. This, together with the material of Section 9.1, magnifies the feeling of an abundance of algebraic structures within the algebras $M_n(D)$ (compared to the case where D is a field), providing more motivation for a study of skew linear representations of associative algebras, [122].

Since a complete bibliography would have to be very long, only the key references are provided, from which the main results come and from which an extensive bibliography can be traced. Our general references to ring theory and to graded rings are [81] and [103]. For basic facts on group rings and semigroup rings we refer to [102],[87]. In many cases we give a sketch of the proof only, the emphasis being on the way semigroups of matrices come to the picture.

9.3.1 Prime Goldie semigroup algebras

In [136] Zelmanov described prime semigroup algebras that satisfy a polynomial identity. Our first aim is to extend his result to a description of prime Goldie semigroup algebras. We do this in the more general context of contracted semigroup algebras. The results come from [48]. The formalism of Munn algebras will be used, see Section 4.2. In particular, in this context ○ will stand for the usual matrix multiplication. We begin with a technical result, which is an extension of Lemma 3.27.

Lemma 9.25 *Let J be a uniform subsemigroup of a completely 0-simple semigroup with closure $\hat{J} = \mathcal{M}(G, X, Y, P)$ such that $|Y| = r < \infty$. Assume that for any maximal subgroup H of $\hat{J}$, the semigroup $J \cap H$ satisfies the right Ore condition and $K[G]$ is semiprime right Goldie. Then*

1. *G is the group of right quotients of a subsemigroup T such that J has a subsemigroup I isomorphic to*

$$\mathcal{M}(T, X, Y, P') \subseteq \mathcal{M}(G, X, Y, P') \simeq \hat{J},$$

for a sandwich matrix P',

2. *if $K_0[J]$ is semiprime, then $|X| \leq |Y|$,*

3. *if $|X| = r$ and $P \circ a \neq 0$ for every nonzero $a \in K_0[J]$, then the rings*

$$K_0[I] \subseteq K_0[J] \subseteq K_0[\hat{J}]$$

have the same classical ring of right quotients which is isomorphic to the classical ring of quotients of $M_r(K[G])$.

Proof. (1) To construct T we introduce some notation. For $x \in X, y \in Y$, denote by J_{xy} the set of $(g, x, y) \in J$, where $g \in G$. Choose $u \in X, v \in Y$ such that J_{uv} is contained in a maximal subgroup H of $\hat{J}$. Let

$$T = \bigcup_{x \in X, y \in Y} J_{uy} J_{xv} \subseteq J_{uv} \cup \{\theta\}.$$

For each $x \in X, y \in Y$, choose $z_{xv} \in J_{xv}$ and $r_{uy} \in J_{uy}$. Define

$$I = \left(\bigcup_{x \in X, y \in Y} z_{xv} T r_{uy} \right) \cup \{\theta\} \subseteq J.$$

As in [136] (or see the proof of Theorem 22.5 in [87]), I is a semigroup which is isomorphic to the semigroup of matrix type $\mathcal{M}(T, X, Y, P')$, where P' is the $r \times r$ matrix with (y, x)-entry $r_{uy} z_{xv}$. Since

$$\left(\bigcup_{x \in X, y \in Y} z_{xv} \hat{J}_{uv} r_{uy} \right) \cup \{\theta\} = \hat{J},$$

we get that $\hat{J}$ can be identified with $\mathcal{M}(H, X, Y, P')$. By assumption, H is the group of right quotients of $J \cap H$. So, it is easily verified that H is as well the group of right quotients of T.

(2) Suppose $|X| > r$. Let Q be the (semisimple Artinian) ring of quotients of $K[G]$. There exists a nonzero $(r + 1) \times r$ matrix b over Q

such that $P_1 \circ b = 0$, where P_1 is a fixed $r \times (r+1)$ submatrix of P. Since Q is the ring of right quotients of $K[T]$, there exists an $r \times r$ matrix c over $K[T]$ such that $b \circ c \neq 0$ is a matrix over $K[T]$. Then $b \circ c$ can be viewed as an element of $K_0[I]$ which annihilates $K_0[J]$ on the right. This contradicts the semiprimeness of $K_0[J]$.

(3) Since, by assumption, $K[H]$ is semiprime right Goldie, the matrix ring $M_r(K[H])$ has a right classical ring of quotients, say Q'. Hence, for any $\alpha \in M_r(K[H])$ there exists a regular diagonal matrix $\delta \in M_r(K[T])$ such that $\alpha \circ \delta \in M_r(K[T])$. The assertion on the sandwich matrix P is equivalent with the right annihilator of $K_0[J]$ being trivial. Therefore $P' \circ a \neq 0$ for every nonzero $a \in K_0[I]$. It follows that $P' \circ \alpha \neq 0$ for every nonzero $\alpha \in M_r(K[H])$, so P' is a left regular element of $M_r(K[H])$. Therefore it is also right regular because $M_r(K[H])$ is semiprime Goldie. Further

$$\alpha \circ \delta \circ P' \in M_r(K[T]) \circ P'$$

and the matrix $\delta \circ P'$ is regular in $M_r(K[H])$. Since

$$K_0[I] \circ P' = M_r(K[T]) \circ P' \subseteq K_0[J] \circ P' \subseteq K_0[\hat{J}] \circ P' \subseteq M_r(K[H]) \subseteq Q',$$

we obtain that Q' is the ring of right quotients of all these rings. As P' is invertible in Q', the natural homomorphism $K_0[\hat{J}] \longrightarrow K_0[\hat{J}] \circ P'$ is an isomorphism. Hence,

$$K_0[I] \subseteq K_0[J] \subseteq K_0[\hat{J}]$$

have a common ring of right quotients isomorphic to Q'. $\square$

Theorem 9.26 *Let K be a field and S a semigroup. Consider the following conditions:*

1. *the contracted semigroup algebra $K_0[S]$ is prime left and right Goldie,*

2. *S does not contain noncommutative free subsemigroups and $K_0[S]$ is prime left Goldie,*

3. *S contains a right ideal A and a left ideal B such that AB and BA are both uniform subsemigroups in a common completely 0-simple semigroup with equal closures, $AB = \mathcal{M}(T, r, r, P)$, where T is a cancellative semigroup such that $K[T]$ is prime left and right Goldie, and the right and left annihilators of the contracted semigroup ring $K_0[AB]$ in $K_0[S]$ are trivial.*

Then, (1) and (3) are equivalent, and (2) implies (3).

Proof. Assume $R = K_0[S]$ is prime left Goldie. It is easy to see that
the assertions of Lemma 9.19 remain valid for the subsemigroup S of
$H(R) = KS = \{\lambda s \mid \lambda \in K, s \in S\}$. This is because $s\mathcal{H}\lambda s$, if $\lambda \neq 0$, in
the left quotient ring Q of R. Hence S has an ideal I which is uniform
in a completely 0-simple semigroup $\hat{I} = \mathcal{M}(G, X, Y, P)$ contained in Q.
It follows also that, for any maximal subgroup H of $\hat{I}$, H is a group of
left quotients of $T = H \cap I$ and $K[H \cap I]$ is prime left Goldie. Note that
then also the ring $K[H]$ is prime left Goldie. Because the 'columns' of
I define, in a natural way, left ideals in the left Goldie ring $K_0[S]$, it is
clear that Y is finite. For simplicity we write $Y = \{1, \ldots, r\}$.

As an ideal of a prime left Goldie ring, $K_0[I]$ is prime left Goldie
as well. Assume now that $K_0[S]$ is also right Goldie or that S does
not contain noncommutative free subsemigroups, that is, we assume
condition (1) or (2) of the statement of the theorem is satisfied. We
claim that then $K_0[I]$ is also right Goldie. The claim clearly holds
if $K_0[S]$ is right Goldie. So, we deal with the case that S does not
contain noncommutative free subsemigroups. From Lemma 9.25 and
its right-left dual we know that $|X| = |Y| = r$ and $K_0[I]$ is right Goldie.
Therefore $K[H \cap I]$ also is right Goldie. Moreover, we may assume that
$X = Y = \{1, \ldots, r\}$.

Using the notation of the proof of Lemma 9.25, let

$$A = \bigcup_{x \in X, y \in Y} z_{xv} I_{uy} \cup \{\theta\}, \quad B = \bigcup_{x \in X, y \in Y} I_{xv} r_{uy} \cup \{\theta\}.$$

Then the right (left) annihilator of $R = AB$ in $K_0[S]$ is trivial because
it coincides with the right (left) annihilator of $K_0[I]$. Thus, (3) follows.

We now prove that (3) implies (1). Let A be a right ideal of S and
B a left ideal of S satisfying the assumptions stated in (3). Since $K[T]$
is prime left and right Goldie, the cancellative semigroup T has a left
and right group of quotients, say G. It follows that AB and BA are
uniform in the completely 0-simple semigroup $\hat{J} = \mathcal{M}(G, r, r, P)$.

Now $K_0[AB]$ is isomorphic to the Munn algebra $\mathcal{M}(K[T], r, r, P)$,
and also $K_0[\hat{J}] \simeq \mathcal{M}(K[G], r, r, P)$. Since, by assumption, $K_0[AB]$ has
zero right and left annihilators in $K_0[S]$, and thus also in $K_0[AB]$, it is
easily seen that $P \circ a \neq 0$ and $a \circ P \neq 0$ for any nonzero $a \in K_0[\hat{J}]$.
Hence, by Lemma 9.25, $K_0[AB] \subseteq K_0[\hat{J}]$ have the same prime (left and

right) classical ring of quotients. So both rings are prime left and right Goldie. Clearly we then also obtain that $K_0[(AB)^2]$ is prime (left and right) Goldie.

Let L be a nonzero ideal of $K_0[S]$. Then ALB is an ideal of AB and clearly $ALB \subseteq L$. If $ALB = 0$, then $BALBA = 0$, and therefore $K_0[(AB)^2]LK_0[(AB)^2] = 0$. The hypothesis on the annihilator of $K_0[AB]$ implies that $L = 0$, a contradiction. Therefore, every nonzero ideal of $K_0[S]$ intersects $K_0[AB]$ nontrivially. The primeness of $K_0[AB]$ implies that $K_0[S]$ is prime.

For each $1 \leq i \leq r$, choose $e_i = (c_i, i, 1) \in BA$, that is we pick an element from each row of the first column of $BA \subseteq J \subseteq \hat{J}$. Then the free right $K[G]$-module $M = \bigoplus_{i=1}^r K[G]$ with basis e_i, $1 \leq i \leq r$, can be identified with $K_0[\hat{J}]e_1$. Let

$$\varphi : K_0[S] \longrightarrow \mathrm{End}_{K[G]}(M)$$

be the ring morphism defined via left multiplication, that is, for $\alpha \in K_0[S]$, $\varphi(\alpha)$ is the matrix of the left multiplication by α in the above basis of M. Since $K_0[S]$ is prime, φ is an embedding. Clearly, in the same way one defines a ring isomorphism $K_0[\hat{J}] \longrightarrow \mathrm{End}_{K[G]}(M)$. Abusing notation we will denote this map also by φ. Hence

$$\varphi(K_0[AB]) \subseteq \varphi(K_0[S]) \subseteq \varphi(K_0[\hat{J}]) = \mathrm{End}_{K[G]}(M) \simeq M_r(K[G]).$$

By the previous we know that $\varphi(K_0[AB])$ and $\varphi(K_0[\hat{J}])$ are prime (left and right) Goldie, and they both have the same classical ring of quotients. It follows that $\varphi(K_0[S])$, and thus also $K_0[S]$, is prime (left and right) Goldie. $\square$

9.3.2 Semilocal and perfect group - graded rings

Following [46], we consider rings satisfying certain classical finiteness conditions. Such conditions have been studied by many authors, but in most cases not from a structural point of view. However, most of the known results follow from the structure theorem presented below. The first result shows that, roughly speaking, the case where H(R) is a completely 0-simple semigroup is crucial. Recall that a semigroup N with zero θ is left T-nilpotent if for every $x_1, x_2, \ldots$ in N there exists $n \geq 1$ such that $x_1 x_2 \cdots x_n = \theta$. A ring R is semilocal if $R/\mathcal{J}(R)$ is semisimple Artinian. R is called left perfect if it is semilocal and $\mathcal{J}(R)$ is left T-nilpotent.

Proposition 9.27 *Let R be a left perfect (semilocal, respectively) ring graded by a semigroup S. Then $\mathrm{H}(R)$ has a finite ideal chain whose factors are either left T-nilpotent (nil) or completely 0-simple.*

The idea of the proof:
    i) first show that $\mathrm{H}(R)$ is π-regular, that is, a power of every $a \in R$ lies in a subgroup of the semigroup $\mathrm{H}(R)$. (This is an extension of the well-known fact that a simple Artinian ring has this property, extending Proposition 1.3). The proof uses the following useful lemma:

    (∗)   If a ring R graded by a semigroup S is semilocal and $a \in R_s$ is not nilpotent, then s is periodic in S.

    ii) next, use the assumption on R together with the fact that a 0-simple semigroup which is π-regular must be completely 0-simple. $\square$

We use the above proposition to reduce the study of R to that of S-graded rings whose homogeneous elements form a semigroup with a nicer structure. The largest homogeneous ideal contained in the Jacobson radical $\mathcal{J}(R)$ of $R = \bigoplus_{s \in S} R_s$ will be denoted by $\mathcal{J}_{gr}(R)$. Let N be the maximal nil ideal of $\mathrm{H}(R)$ for a left perfect S-graded ring R. Then N is left T-nilpotent. So the subring $\mathbf{Z}\{N\}$ generated by N in R is a left T-nilpotent homogeneous ideal of R. We denote this ring by J_1. It follows that $J_1 = \mathcal{J}_{gr}(R)$. By the proposition, the semigroup $\mathrm{H}(R)/N$ has a minimal completely 0-simple ideal I. Let $I_1 = (\mathbf{Z}\{I\} + \mathcal{J}_{gr}(R))/\mathcal{J}_{gr}(R)$ be the image of $\mathbf{Z}\{I\}$ in the S-graded ring $R' = R/\mathcal{J}_{gr}(R)$. Then the image of I in I_1/J_1 is a completely 0-simple ideal in $\mathrm{H}(I_1/J_1)$. Continuing this process we obtain a chain

$$I_0 = \{0\} \subseteq J_1 \subseteq I_1 \subseteq J_2 \subseteq I_2 \subseteq \cdots \subseteq J_m \subseteq I_m \subseteq J_{m+1} = R$$

of homogeneous ideals of R, with m less than or equal to the length of the R-module $R/\mathcal{J}(R)$, and such that each J_i/I_i is left T-nilpotent and each $\mathrm{H}(I_i/J_i)$ has a completely 0-simple ideal L that generates I_i/J_i as an additive group, and in particular $\mathrm{supp}(I_i/J_i) = \mathrm{supp}(L)$.

    Next, we need the following technical lemma.

Lemma 9.28 *Let G be a group with identity e and let R be a semilocal G-graded ring with $J(R_e)$ nil. If L is a homogeneous left ideal of R with $L \cap R_e \subseteq J(R_e)$, then $L \subseteq J(R)$. In particular, if $R_{g^{-1}} r_g \subseteq J(R_e)$ for some $r_g \in R_g$, then $r_g \in J(R)$. Hence $J(R_e) \subseteq J(R)$ and $J(R_e) = J(R) \cap R_e$.*

Proof: Using (∗), first show that $H(R)$ is nil modulo R_e. Then the assumptions imply that $\bigcup_{g \in G} L_g$ must be nil, and consequently $L \subseteq J(R)$.

If $R_{g^{-1}} r_g \subseteq J(R_e)$, then the above applied to $L = R r_g$ implies that $r_g \in J(R)$. This yields $J(R_e) \subseteq J(R) \cap R_e$ (the converse is well known). □

The main result in the group - graded case shows that the rings in question are built of group crossed products.

Theorem 9.29 ([46]) *Let G be a group and let R be a G-graded ring with $J(R_e) \subseteq J_{gr}(R) \neq R$. Then the G-graded ring $R/J_{gr}(R)$ is semilocal (respectively left perfect, semiprimary, left Artinian) if and only if the following conditions are satisfied*

1. *$R_e/J(R_e) \simeq M_{n_1}(D_1) \times \cdots \times M_{n_r}(D_r)$, where each D_i is a division algebra,*

2. *for any complete set of orthogonal idempotents e_u of $R_e/J(R_e)$, $1 \leq u \leq q = n_1 + \cdots + n_r$, the ring $R' = R/J_{gr}(R)$ is the direct product of matrix rings over crossed products over some periodic subgroups H_i of G*

$$M_{m_1}(D_{(1)} * H_1) \times \cdots \times M_{m_l}(D_{(l)} * H_l),$$

with $q = m_1 + \cdots + m_l$, and for each $1 \leq i \leq l$,

$$M_{m_i}(D_{(i)} * H_i) = R' e_j R', \qquad D_{(i)} = e_j(R_e/J(R_e))e_j$$

for some $1 \leq j \leq q$. In particular these matrix rings are homogeneous subrings.

3. *each crossed product $D_{(i)} * H_i$ is semilocal (respectively left perfect, semiprimary, left Artinian).*

The idea of the proof: we can assume that $\mathcal{J}_{gr}(R) = 0$. Then R_e has a unity and one verifies that this is the unity of R as well. Write $1 = \sum e_i$ for primitive orthogonal idempotents e_i of R_e. Each $e_i R_e e_i$ is a division algebra. But $H(e_i R e_i) \setminus \{0\}$ is cancellative and Lemma 9.28 implies that the grading is non-degenerate ($R_{g^{-1}} r_g \neq 0 \neq r_g R_{g^{-1}}$ for $0 \neq r_g \in R_g, g \in G$). Via Lemma 9.18 this allows us to show that $e_i R e_i \simeq e_i R_e e_i * H_i$ for a group H_i. R semilocal implies that H_i is periodic. Glue the pieces together, writing first R as a generalised matrix ring $\sum_{i,j} e_i R e_j$, to get the structure of R. $\square$

The hypothesis $\mathcal{J}(R_e) \subseteq \mathcal{J}_{gr}(R)$ is always satisfied for a left perfect ring R, because of Lemma 9.28 and Lemma 9.30 below.

Clearly, the condition $\mathcal{J}_{gr}(R) \neq R$ is satisfied for a group - graded ring with unity. For an arbitrary group - graded ring R one obtains from the theorem that R is left perfect if and only if $\mathcal{J}_{gr}(R)$ is left T-nilpotent and $R/\mathcal{J}_{gr}(R)$ satisfies conditions (1) - (3) if $R \neq \mathcal{J}_{gr}(R)$.

An extension of Theorem 9.29 to rings graded by a semigroup S is also known, [46]. It turns out that R modulo $\mathcal{J}(R)$ can be covered by finitely many subrings of the form $a R_G b$, where $a, b \in H(R)$ and $R_G = \bigoplus_{g \in G} R_g$ is a G-graded (semilocal) ring, for a subgroup G of S. The following lemma is easy in case of rings with unity, but creates unexpected problems in the general case, which in particular is needed to handle semigroup gradations.

Lemma 9.30 *Let H be a subgroup of an S-graded ring R. If R is semilocal (left perfect), then so is the ring $R_H = \bigoplus_{h \in H} R_h$.*

9.3.3 The radical of a graded ring

The homogeneity problem for the radical of R asks whether $\mathcal{J}_{gr}(R) = \mathcal{J}(R)$ under some conditions on R and on S. The theorems of Bergman on **Z**-graded rings and of Cohen and Montgomery on algebras over a field of characteristic zero graded by a finite group G are the most basic results in this area. For these and various generalizations we refer to [59].

S is said to be a u.p. (unique product) semigroup if for every non-empty finite subsets $A, B \subseteq S$ there exists an element of AB which has a unique presentation in the form ab, where $a \in A, b \in B$, [87]. Such an

S is cancellative. If it has a group of quotients, then this group must be torsion - free. It is an open problem whether the radical of any S-graded ring is homogeneous.

Our main result reads as follows.

Theorem 9.31 ([61]) *The Jacobson radical of an S-graded ring R is homogeneous if either of the following conditions holds*

1. *S is a cancellative semigroup and R is a PI-algebra over a field of characteristic zero.*

2. *S is a u.p.-semigroup and at least one of the following conditions is satisfied:*

   (a) all nil subsemigroups of $\mathrm{H}(R/\mathcal{J}_{gr}(R))$ are locally nilpotent,

   (b) every nil subsemigroup of every right primitive homomorphic image of R is locally nilpotent,

   (c) for every minimal prime ideal P of R, the ring R/P is a domain or embeds into a matrix ring over a division algebra.

Conditions (a),(b),(c) cover in particular the case where R is PI, semilocal, or it is right Goldie modulo the prime radical. The same proof shows also that the prime radical of R is homogeneous in any of the above cases.

We will outline some ideas of the proof. For an element $x \in R$ by $\mathrm{H}(x)$ we denote the subsemigroup of $\mathrm{H}(R)$ generated by the set of homogeneous components of x. The first lemma, roughly speaking, allows us to consider two cases only: the case of a group gradation and the case where the study of quasi inverses of elements $x \in \mathcal{J}(R)$ requires only the ring generated by $\mathrm{H}(x)$.

Lemma 9.32 *Let S be a cancellative semigroup, R an S-graded ring, I the ideal of nonunits of S and $G = S \setminus I$ (if S has no identity, then $S = I$). Assume that $x + y = yx$ for some $x \in R_I, y \in R$. Then $y \in A$, where $A = Z\{\mathrm{H}(x)\}$ is the subring generated by $\mathrm{H}(x)$.*

An element $a \in R$ is called rigid if $xay = 0$ implies $xa_ty = 0$ for every $t \in \mathrm{supp}(a)$ and $x, y \in \mathrm{H}(R)$. This is a useful notion, derived from the often exploited properties of nonzero elements of shortest length in a given ideal of R.

Lemma 9.33 *Let S be a cancellative semigroup, R an S-graded ring, r a rigid element of R. If $\mathrm{H}(r)$ contains 0, then $\mathrm{H}(r)$ is nilpotent.*

So, when studying the 'local' properties of elements $r \in \mathcal{J}(R)$ (those expressible in terms of the ring $\mathbf{Z}\{\mathrm{H}(r)\}$), the only other case to consider is when $\mathrm{H}(r)$ has no zero divisors, and hence deg is a homomorphism on $\mathrm{H}(r)$. This, together with a reduction to semigroups of matrices (possible because of the assumptions on R) allows us to apply the ideas of Sections 9.1,9.2. The case of a PI - algebra R is the easiest to explain. Namely, Lemma 9.24 leads to the case of a gradation by a group which is locally almost abelian. So known results on gradations by such groups (extending the theorems of Bergman and of Cohen and Montgomery, [59]) can be applied. If $\mathcal{J}(R/\mathcal{J}_{gr}(R)) \neq 0$, this yields a nonzero nil ideal of the semigroup $\mathrm{H}(R/\mathcal{J}(R))$ constructed from certain $0 \neq r \in \mathcal{J}(R)$, which leads to a contradiction. On the other hand, the proof in case (c) requires an induction that involves all 'layers' of the skew linear semigroups which are the images of $\mathrm{H}(R)$ in the quotient rings of the prime rings R/P, for minimal primes P of R.

9.3.4   Principal ideal semigroup algebras

Our last application is again concerned with semigroup algebras $K[S]$ viewed as S-graded rings with components $Ks, s \in S$. We assume that $K[S]$ has a unity and that every left ideal of $K[S]$ is principal. The main problem is to find a structural description of such algebras and of the underlying semigroups. The group ring case is our point of departure.

Theorem 9.34 ([101]) *Let G be a group and K a field. The following conditions are equivalent:*

1. *$K[G]$ is a principal left ideal ring,*

2. *if char $K = 0$ then G is finite or finite-by-infinite cyclic,*
   if char $K = p > 0$ then G is (finite p')-by-(cyclic p) or G is (finite p')-by-infinite cyclic.

Note that such rings are finitely generated K-algebras, they satisfy a polynomial identity and have Gelfand-Kirillov dimension 1. The following extension to cancellative semigroups has been obtained in [49].

Theorem 9.35 *Let T be a cancellative monoid and K a field of characteristic p (not necessarily nonzero). The following conditions are equivalent:*

1. $K[T]$ *is a principal left ideal ring,*

2. T *is a semigroup satisfying one of the following conditions:*

   (a) T is a group satisfying the conditions of Theorem 9.34,

   (b) T contains a finite p'-subgroup H and a nonperiodic element x such that $xH = Hx$, $T = \bigcup_{i \in \mathbf{N}} Hx^i$ and the central idempotents of $K[H]$ are central in $K[T]$.

The first major step of our approach to the general case is the following result.

Theorem 9.36 ([47]) *Let $K[S]$ be a principal left ideal ring. Then $K[S]$ satisfies a polynomial identity.*

The idea of the proof: for simplicity we discuss only the case where $S \subseteq M_n(D)$ and the K-subalgebra A generated by S is an order in $M_n(D)$. Consider a maximal subgroup $G \simeq GL_j(D)$ of $M_n(D)$. General techniques allow to lift (right) ideals of $K[\mathrm{gp}(S \cap G)]$ to (right) ideals of $K[S]$, [87], Lemma 7.21 (compare Lemma 9.22). The assumption on $K[S]$ implies then that $K[\mathrm{gp}(S \cap G)]$ is right Noetherian. The main difficulty is in showing that this group ring actually is a principal ideal ring. Then it is PI by Theorem 9.34. Finally, the structure of $S \subseteq M_n(D)$ together with the Goldie condition for A allow us to show that $K[S]$ is PI as well. $\square$

Corollary 9.37 *If $K[S]$ is a principal left ideal ring, then S is finitely generated and $K[S]$ embeds into a matrix ring $M_n(F)$ over a field extension F of K. In particular, S is a linear semigroup. The groups associated to S are either finite or finite-by-infinite cyclic.*

Proof. $K[S]$ is finitely generated because it is a left Noetherian PI-algebra, [87], Theorem 19.14. A theorem of Ananin implies that a finitely generated left Noetherian PI - algebra embeds into a matrix ring over a (finitely generated) commutative algebra A, [2]. A classical result of Malcev then implies that $K[S]$ embeds into matrices over a field, [76]. Theorem 9.34 is used to get the remaining assertion. $\square$

The following is an easy consequence.

Corollary 9.38 *If $K[S]$ is a principal left ideal ring, then the Gelfand-Kirillov dimension of $K[S]$ is equal to its classical Krull dimension and it is 0 or 1. In the former case S is finite. Moreover, every prime Artinian homomorphic image of $K[S]$ is finite dimensional over K.*

The following construction plays a crucial role in our final result.

Example 1 *Assume that H is a finite group whose order is not divisible by the characteristic of K. Let T be a monoid with group of units H such that $T = \bigcup_{i \geq 0} Hx^i$ for some $x \in T$, and either this union is disjoint or $x^n = \theta$ for some $n \geq 1$. Assume also that $Hx = xH$ and the central idempotents of $K[H]$ commute with x. Then $K[T]$ is a principal ideal ring. In fact, one can check that $K[T] \simeq A \oplus B$, where A is a semisimple Artinian ring and B is a finite direct sum of algebras of the type $C[x, \sigma]/(x^k)$ for a semisimple Artinian ring C, an automorphism σ of C and some $k \geq 1$. Note that this extends the construction of 2 b) in Theorem 9.35.*

Let C be a prime homomorphic image of a principal left ideal ring $K[S]$. We use the fact that C is a finitely generated prime PI - algebra of classical Krull dimension 0 or 1. First, we claim that, if an ideal J of C is idempotent, then either $J = 0$ or $J = C$. Let $J = Ca$. Then $CaCa = Ca^2$, so that $J = Ca^2 = Ja$. Therefore $ba = a$ for some $b \in J$. Thus $bax = ax$ for all $x \in C$ and consequently b is a left identity of J. Therefore it is an identity of J because C is prime, so $J = 0$ or $J = C$. Secondly, if J is a nonzero ideal of C, then C/J is of finite dimension over K. Indeed, the classical Krull dimension of C/J is 0. Hence $\mathrm{GK}(C/J) = \mathrm{clKdim}\ (C/J) = 0$.

The former observation will allow us to decompose certain algebras into a direct product of its simpler blocks. The latter leads to the strategy of considering the finite dimensional case first, and then applying this to the general case.

It is well known that finite dimensional algebras which are principal left ideal rings also are principal right ideal rings. Moreover, they are exactly finite direct sums of matrix rings over local algebras whose radical is a principal ideal, [21], Theorem 9.4.1. This decomposition of $K[S]$ turns out to be a refinement of a decomposition on the semigroup level, and leads to a complete description of $K[S]$ as follows.

Proposition 9.39 *Let S be a finite semigroup and N a nilpotent ideal of S, or $N = \emptyset$, such that S/N is completely 0-simple. Then the contracted semigroup algebra $K_0[S]$ is a principal ideal ring if and only if one of the following conditions is satisfied*

1. *$S \simeq \mathcal{M}(G, n, n, Q)$ for a finite group G satisfying the conditions of Theorem 9.34 and for a sandwich matrix Q that is invertible over $K[G]$,*

2. *$S \simeq \mathcal{M}(T, n, n, Q)$ for a finite monoid T satisfying the conditions of Example 1 and for a sandwich matrix Q that is invertible over $K_0[T]$.*

Proposition 9.40 *Let S be a finite semigroup. Then $K_0[S]$ is a principal ideal ring if and only if S has a chain of ideals $I_1 \subseteq I_2 \subseteq \cdots \subseteq I_r = S$ such that I_1 and every factor I_j/I_{j-1} is of the type described in Proposition 9.39.*

The remaining step is to show that principal ideal rings $K_0[S]$ can be approached as inverse limits of finite dimensional semigroup algebras. This leads to our main result.

Theorem 9.41 ([47]) *The following conditions are equivalent:*

1. *$K_0[S]$ is a principal (left an right) ideal ring,*

2. *there exists an ideal chain*

$$I_1 \subset \cdots \subset I_t = S$$

such that I_1 and every factor I_j/I_{j-1} is of the form $\mathcal{M}(T, n, n, P)$ for an invertible over $K_0[T]$ sandwich matrix P, and one of the following conditions holds:

(a) *T is a group of the type described in Theorem 9.34,*

(b) T is a monoid of the type described in Example 1.

In case the equivalent conditions are satisfied it follows that

$$K_0[S] \simeq K_0[I_1] \oplus K_0[I_2/I_1] \oplus \cdots \oplus K_0[I_t/I_{t-1}]$$

and each component is isomorphic to some $M_n(K[T])$. Moreover, $K_0[S]$ is a finite module over its centre, which is finitely generated.

It may be checked that each of the semigroups I_1, I_j/I_{j-1}, yielding the structural blocks in the above theorem, is of the form $\mathcal{M}(T', n, n, Q)$, where T' is a homomorphic image of a cancellative monoid T satisfying the conditions of Theorem 9.35. In case T is not a group, one of the conditions for $K[T]$ to be of such a type is that the central idempotents of the semisimple ring $K[H]$ are central in $K[T]$, where H is a finite subgroup of $T = \bigcup_{i \geq 0} Hx^i$. In case $K[H]$ is split (that is, a direct product of matrix rings over K) this means that the centre of $K[H]$ is contained in the centre of $K[G]$, where G is the group of quotients of S. Therefore, this condition can be given an intrinsic formulation in terms of S only: the S - conjugacy class of every element h of H coincides with the conjugacy class of h in H.

The methods used in the proof of Theorem 9.41 allow us to establish the left - right symmetry of the principal ideal condition for any semiprime algebra $K[S]$. This leads to the following result.

Theorem 9.42 *Let S be a semigroup and K a field. Then $K_0[S]$ is a semiprime principal left ideal ring if and only if there exists an ideal chain*

$$I_1 \subset \cdots \subset I_t = S$$

such that I_1 and every factor I_j/I_{j-1} is of the form $\mathcal{M}(T, n, n, P)$ for an invertible over $K_0[T]$ sandwich matrix P and a monoid T such that

1. *either T is an infinite group as in Theorem 9.34,*

2. *or $T = \bigcup_i Hx^i$ is of the type described in Example 1 and such that for every primitive central idempotent $e \in K[H]$, either $K[H]ex = 0$ or $K[H]ex^i \neq 0$ for all $i \geq 1$.*

Moreover, if the equivalent conditions are satisfied, then $K_0[S]$ is a principal right ideal ring and the direct sum decomposition of Theorem 9.41 occurs.

If G is as in Theorem 9.34 and $K[G]$ is prime, then G is trivial or infinite cyclic. If T is as in Example 1 and $K_0[T]$ is prime, then H is trivial. Therefore, the following is an easy consequence of Theorem 9.42.

Corollary 9.43 $K_0[S]$ *is a prime principal left ideal ring if and only if*

$$S \simeq \mathcal{M}(\{1\}, n, n, Q), \quad S \simeq \mathcal{M}(\langle x \rangle^1, n, n, Q)$$

or

$$S \simeq \mathcal{M}(\langle x, x^{-1} \rangle, n, n, Q)$$

where the matrix Q is invertible in $M_n(K)$, $M_n(K[x])$ or $M_n(K[x, x^{-1}])$ respectively. Hence, $K_0[S] \simeq M_n(K)$, $M_n(K[x])$, or $M_n(K[x, x^{-1}])$.

Bibliography

[1] Amitsur S.A. and Small L.W. GK – dimensions of corners and ideals, Israel J. Math. 69(1990),152-160.

[2] Anan'in A.Z. An intriguing story about representable algebras, in: Ring Theory, pp.31-38, Weizmann Science Press, Jerusalem, 1989.

[3] Artin M. and Stafford J.T. Noncommutative graded domains with quadratic growth, Invent. Math. 122(1995),231-276.

[4] Bass H. The degree of polynomial growth of finitely generated nilpotent groups, Proc. London Math. Soc. 25(4)(1972),603-614.

[5] Boffa M. and Bryant R.M. Les groupes linéaires vérifiant une identité monoidale, C. R. Acad. Sci. Paris, t.308, Ser.I (1989),127-128.

[6] M.Boffa and F.Point Identités de Thue – Morse dans les groupes, C. R. Acad. Sci. Paris, t.312, Ser.I (1991),667-670.

[7] Borel A. Linear Algebraic Groups, Springer, New York, 1991.

[8] Borevic Z.I. and Shafarevic I.R. Number Theory, Academic Press, New York, 1966.

[9] Bourbaki N. Algèbre, ch. 3 Algèbre Multilinéaire, Hermann, Paris, 1958.

[10] Bourbaki N. Algèbre, ch. 6 Valuations, Hermann, Paris, 1964.

[11] Brown T.C. On van der Waerden's theorem on arithmetic progressions, Notices Amer. Math. Soc. 16(1969),245.

[12] Cabanes M. Irreducible modules and Levi supplements, J. Algebra 90(1984),84-97.

[13] Carter R.W. Finite Groups of Lie Type – Conjugacy Classes and Complex Characters, Wiley, New York, 1985.

[14] Chevalley C. Sur certains groupes simples, Tohoku Math. J. 7(1955),14-66.

[15] Clifford A.H. and Preston G.B. Algebraic Theory of Semigroups, American Mathematical Society, Providence, vol.1, 1961; vol.2, 1967.

[16] Curtis C.W. Modular representations of finite groups with split (B,N)–pair, Lect. Notes in Math. 131, pp.57-95, Springer, New York, 1970.

[17] Curtis C.W. and Reiner I. Representation Theory of Finite Groups and Associative Algebras, Wiley, New York, 1962.

[18] Curtis C.W and Reiner I. Methods of Representation Theory, with applications to finite groups and orders, Wiley, New York, vol.1 1981, vol.2 1987.

[19] Curzio M., Longobardi P., May M. and Robinson D.J.S. A permutational property of groups, Arch. Math. 44(1985),385-389.

[20] de Luca A. and Varricchio S. A finiteness condition for semigroups generalizing a theorem of Coudrain and Schutzenberger, Adv. Math. 108(1994),91-103.

[21] Drozd Ju.A. and Kirichenko V.V. Finite Dimensional Algebras, Springer, Berlin, 1994.

[22] Edigarjan B.M. and Faddeev D.K. On the nonsingularity of certain matrices connected with representations of a semigroup of matrices over a finite field, Izv. Akad. Nauk Armjan. SSR Ser. Mat. 9(1974),421-426 (in Russian).

[23] Faddeev D.K. On representations of a full matrix semigroup over a finite field, Doklady Akad. Nauk SSSR 230(1976),1290-1293 (in Russian).

[24] Faith C. Algebra II, Ring Theory, Springer, New York, 1976.

[25] Fountain J. and Gould V.A.R. Completely 0-simple semigroups of quotients II, in: Contributions to General Algebra 3, pp. 115-124, Verlag Holder-Pichler-Temperly, 1985.

[26] Fountain J. and Petrich M. Completely 0-simple semigroups of quotients III, Math. Proc. Cambridge Phil. Soc. 105(1989),263-275.

[27] Garzon M. and Zalcstein Y. Linear semigroups with permutation properties, Semigroup Forum 35(1987),369-371.

[28] Glover D.J. A study of certain modular representations, J. Algebra 51(1978),425-475.

[29] Gluskin L.M. On matrix semigroups, Izv. AN SSSR 22(1958),439-448 (in Russian).

[30] Green J.A. The characters of the finite general linear group, Trans. Amer. Math. Soc. 80(1955),402-447.

[31] Grigorchuk R.I. Cancellative semigroups of power growth, Mat. Zametki 43(1988),305-319 (in Russian).

[32] Gromov M. Groups of polynomial growth and expanding maps, Publ. Math. IHES 53(1) (1981),53-73.

[33] Hall M. Combinatorial Theory, Wiley, New York, 1986.

[34] Harris J.C. and Kuhn N.J. Stable decompositions of classifying spaces of finite abelian p-groups, Math. Proc. Cambridge Phil. Soc. 103(1988),427-449.

[35] Hartley B. and Pickel P.F. Free subgroups in the unit group of integral group rings, Canad. J. Math. 32(1980),1342-1352.

[36] Herstein I.N. Noncommutative Rings, Math. Assoc. of America, 1968.

[37] Hofmann K. and Skryago A.M. Finite dimensional continuous representations of compact regular semigroups, Semigroup Forum 28(1984),199-234.

[38] Howie J.M. Fundamentals of Semigroup Theory, Oxford University Press, 1995.

[39] Huang W. Nilpotent algebraic monoids, J. Algebra 179(1996),720-731.

[40] Humphreys J.E. Linear Algebraic Groups, Springer, New York, 1975.

[41] Iwahori N. On the structure of a Hecke ring of a Chevalley group over a finite field, J. Fac. Sci. Univ. Tokyo, Sec. I, 10(1964),215-236.

[42] Jacob G. Séries rationelles. Problèmes de finitude et de décidabilité, in: Séries Formelles en variables non commutatives et applications, pp.135-178, LITP, 1977.

[43] Jacob G. La finitude des représentations linéaires des semi-groupes est décidable, J. Algebra 52(1978),437-459.

[44] Jacobson N. Structure of Rings, Amer. Math. Soc., Providence, 1968.

[45] Jespers E. and Okniński J. Nilpotent semigroups and semigroup algebras, J. Algebra 169(1994),984-1011.

[46] Jespers E. and Okniński J. Descending chain conditions and graded rings, J. Algebra 178(1995),458-479.

[47] Jespers E. and Okniński J. Semigroup algebras that are principal ideal rings, J. Algebra 183(1996),837-863.

[48] Jespers E. and Okniński J. Noetherian semigroup algebras, to appear.

[49] Jespers E. and Wauters P. Principal ideal semigroup rings, Comm. Algebra 23(1995),5057-5076.

[50] Jones P.R. Analogues of the bicyclic semigroup in simple semigroups without idempotents, Proc. Roy. Soc. Edinburgh 106A (1987),11-24.

[51] Justin J. Propriétés combinatoires de certains semi-groupes, C. R. Acad. Sci. Paris 269(1969),1113-1115.

[52] Justin J. Groupes et semi-groupes à croissance linéaire, C. R. Acad. Sci. Paris 273(1971),212-214.

[53] Justin J. Généralisation du théorème de van der Waerden sur les semi-groupes répétitifs, J. Combinatorial Theory. Ser. A. 12 (1972),357-367.

[54] Justin J. Characterization of the repetitive commutative semigroups, J. Algebra 21(1972),87-90.

[55] Justin J. Groupes linéaires répétitifs, C. R. Acad. Sci. Paris 292(1981),349-350.

[56] Kaplansky I. The Engel – Kolchin theorem revisited, in: Contributions to Algebra, pp.233-237, Academic Press, New York, 1977.

[57] Kargapolov M.I. and Merzljakov Ju.I. Fundamentals of the Theory of Groups, Springer, New York, 1979.

[58] Karpilovsky G. Unit Groups of Group Rings, Longman, 1989.

[59] Karpilovsky G. The Jacobson Radical of Classical Rings, Longman, 1991.

[60] Kelarev A.V. A simple matrix semigroup which is not completely simple, Semigroup Forum 37(1988),123-125.

[61] Kelarev A.V. and Okniński J. The Jacobson radical of graded PI – rings and related classes of rings, J. Algebra 186(1996),818-830.

[62] Kelarev A.V. and Okniński J. A combinatorial property and growth for semigroups of matrices, Comm. Algebra, to appear.

[63] Kelarev A.V. and Shumyatsky P.V. Soluble and linear repetitive groups, Bull. Austral. Math. Soc. 52(1995),253-261.

[64] Klein A. Free subsemigroups of domains, Proc. Amer. Math. Soc. 116(1992),339-341.

[65] Klein A.A and Makar-Limanov L. Skew fields of differential operators, Israel J. Math. 72(1990), 281-287.

[66] Kovacs L. Semigroup algebras of the full matrix semigroup over a finite field, Proc. Amer. Math. Soc. 116(1992),911-919.

[67] Krause G.R. and Lenagan T.H. Growth of Algebras and Gelfand – Kirillov dimension, Pitman, London, 1985.

[68] Krop L. On the representations of the full matrix semigroup on homogeneous polynomials, II, J. Algebra 102(1986),284-300.

[69] Lallement G. On nilpotency in semigroups, Pacific J. Math. 42(1972),693-700.

[70] Lallement G. Semigroups and Combinatorial Applications, Wiley, New York, 1979.

[71] Lambrou M., Longstaff W.E. and Radjavi H. Spectral conditions and reducibility of operator semigroups, Indiana Univ. Math. J. 41(1992),449-464.

[72] Lichtman A.I. On linear groups over a field of fractions of a polycyclic group ring, Israel J. Math. 42(1982),318-326.

[73] Lothaire M. Combinatorics on Words, Addison-Wesley, Reading, Mass., 1983.

[74] Makar-Limanov L. On free subsemigroups of skew fields, Proc. Amer. Math. Soc. 91(1984),189-191.

[75] Makar-Limanov L. On group rings of nilpotent groups, Israel J. Math. 48(1984),244-248.

[76] Malcev A.I. On representation of infinite algebras, Mat. Sb. 13(1943),263-285 (in Russian).

[77] Malcev A.I. Nilpotent semigroups, Uchen. Zap. Ivanovsk. Ped. Inst. 4(1953),107-111 (in Russian).

[78] Margulis G.A. Discrete Subgroups of Semisimple Lie Groups, Springer, Berlin, 1991.

[79] McAlister D.B. Rings related to completely 0-simple semigroups, J. Austral. Math. Soc. 12(1971),257-274.

[80] McAlister D.B. Representations of semigroups by linear transformations, Semigroup Forum 2(1971),189-263; 283-320.

[81] McConnell J.C. and Robson J.C. Noncommutative Noetherian Rings, Wiley, New York, 1987.

[82] McNaughton R. and Zalcstein Y. The Burnside problem for semigroups, J. Algebra 34(1975),292-299.

[83] Merzljakov Ju.I. Rational Groups, Nauka, Moscow, 1986 (in Russian).

[84] Montgomery S. Prime ideals and group actions in noncommutative algebras, Contemp. Math. 88(1989),103-124.

[85] Nagahara T. and Yokota S. On traces of separable simple subalgebras in matrix rings, Canad. Math. Bull. 104(1995),104-111.

[86] Neumann B.H. and Taylor T. Subsemigroups of nilpotent groups, Proc. Roy. Soc., Ser. A 274(1963),1-4.

[87] Okniński J. Semigroup Algebras, Marcel Dekker, New York, 1991.

[88] Okniński J. Linear representations of semigroups, in: Monoids and Semigroups with Applications, pp.257-277, World Sci., 1991.

[89] Okniński J. Linear semigroups with identities, in: Semigroups – Theory and Applications to Formal Languages and Codes, pp.201-211, World Sci., 1993.

[90] Okniński J. Gelfand – Kirillov dimension of noetherian semigroup algebras, J. Algebra 162(1993),302-316.

[91] Okniński J. Growth of linear semigroups, J. Austral. Math. Soc. 60(1996),18-30.

[92] Okniński J. Linear semigroups of polynomial growth in positive characteristic, J. Pure Appl. Algebra 107(1996),253-261.

[93] Okniński J. Growth of linear semigroups, II, J. Algebra 178 (1995),561-580; Corrigendum 181(1996),660-661.

[94] Okniński J. Nilpotent semigroups of matrices, Math. Proc. Cambridge Phil. Soc. 120(1996),617-630.

[95] Okniński J. Triangularizable semigroups of matrices, Linear Algebra Appl. 262(1997),111-118.

[96] Okniński J. and Ponizovskii J.S. A new matrix representation theorem for semigroups, Semigroup Forum 52(1996),293-305.

[97] Okniński J. and Putcha M.S. PI semigroup algebras of linear semigroups, Proc. Amer. Math. Soc. 109(1990),39-46.

[98] Okniński J. and Putcha M.S. Complex representations of matrix semigroups, Trans. Amer. Math. Soc. 323(1991),563-581.

[99] Okniński J. and Putcha M.S. Semigroup algebras of linear semigroups, J. Algebra 151(1992),304-321.

[100] Okniński J. and Salwa A. Generalised Tits alternative for linear semigroups, J. Pure Appl. Algebra 103(1995),211-220.

[101] Passman D.S. Observations on group rings, Comm. Algebra 5(1977),1119-1162.

[102] Passman D.S. The Algebraic Structure of Group Rings, Wiley, New York, 1977.

[103] Passman D.S. Infinite Crossed Products, Academic Press, 1989.

[104] Petrich M. Rings and Semigroups, Lect. Notes in Math. 380, Springer, New York, 1974.

[105] Point F. Conditions of quasi-nilpotency in certain varieties of groups, Comm. Algebra 22(1)(1994),355-370.

[106] Ponizovskii J.S. On irreducible matrix semigroups, Semigroup Forum 24(1982),117-148.

[107] Ponizovskii J.S. Matrix representations of semigroups, to appear.

[108] Putcha M.S. Linear Algebraic Monoids, London Math. Soc. Lecture Note Ser.133, Cambridge, 1988.

[109] Putcha M.S. Monoids on groups with BN – pairs, J. Algebra 120(1989),139-169.

[110] Putcha M.S. Monoids of Lie type, in: Semigroups, Formal Languages and Groups, pp.353-367, NATO ASI Series, Kluwer, 1995.

[111] Putcha M.S. and Renner L. The canonical compactification of a finite group of Lie type, Trans. Amer. Math. Soc. 337(1993),305-319.

[112] Radjavi H. On the reduction and triangularization of semigroups of operators, J. Operator Theory 13(1985),63-71.

[113] Radjavi H. A trace condition equivalent to simultaneous triangularizability, Canad. J. Math. 38(1986),376-386.

[114] Renner L. Modular representations of finite monoids of Lie type, in: Semigroups, Formal Languages and Groups, pp.381-390, NATO ASI Series, Kluwer, Dordrecht, 1995.

[115] Renner L. Analogue of the Bruhat decomposition for algebraic monoids II. The length function and the trichotomy, J.Algebra 175(1995),697-714.

[116] Reutenauer C. An Ogden – like iteration lemma for rational languages, Acta Informatica 13(1980),189-197.

[117] Rowen L.H. Polynomial Identities in Ring Theory, Academic Press, New York, 1980.

[118] Rosenblatt J.M. Invariant measures and growth conditions, Trans. Amer. Math. Soc. 193(1974),33-53.

[119] Salwa A. Structure of skew linear semigroups, Int. J. Algebra and Comput. 3(1993),101-113.

[120] Salwa A. Representations of semigroups arising from finite groups of Lie type, Trans. Amer. Math. Soc. 348(1996),2931-2945.

[121] Salwa A. Representing idempotents as a sum of two nilpotents – an approach via matrices over division rings, to appear.

[122] Schofield A.H. Representations of Rings over Skew Fields, London Math. Soc. Lect. Note Ser. 92, Cambridge Univ. Press, 1985.

[123] Shalev A. Combinatorial applications in residually finite groups II, J. Algebra 157(1993),51-61.

[124] Shirshov A.I. On certain near – Engel groups, Algebra i Logika 2(5)(1963),5-18 (in Russian).

[125] Shirvani M. and Wehrfritz B.A.F. Skew Linear Groups. London Math. Soc. Lect. Note Ser. 118, Cambridge, 1986.

[126] Skryago A.M. Simple semigroups of matrices, VINITI, 1981 (in Russian).

[127] Skryago A.M. On a property of matrices, Vestsi Akad. Nauk BSSR 5(1982),32-35 (in Russian).

[128] Small L.W., Stafford J.T. and Warfield R.B. Affine algebras of Gelfand – Kirillov one are PI, Math. Proc. Cambridge Phil. Soc. 97(1995),407-414.

[129] Solomon L. The Bruhat decomposition, Tits system and Iwahori ring for the monoid of matrices over a finite field, Geom. Ded. 30(1990),15-49.

[130] Solomon L. An introduction to reductive monoids, in: Semigroups, Formal Languages and Groups, pp.295-352, NATO ASI Series, Kluwer, 1995.

[131] Steinberg R. Endomorphisms of linear algebraic groups, Mem. Amer. Math. Soc. 80, 1968.

[132] Tits J. Free subgroups of linear groups, J. Algebra 20(1972),250-270.

[133] Watters J.F. Block triangularization of algebras of matrices, Linear Algebra Appl. 32(1980),3-7.

[134] Wehrfritz B.A.F. Infinite Linear Groups, Springer, 1973.

[135] Zariski O. and Samuel P. Commutative Algebra, vol.1, 1958; vol.2, 1960.

[136] Zelmanov E.I. Semigroup algebras with identities, Sib. Math. J. 18(1977),787-798.

[137] Zhang J.J. On Gelfand – Kirillov transcendence degree, Trans. Amer. Math. Soc. 348(1996),2867-2899.

Index

algebra
    Artinian, 267, 281, 286
    division, 110, 129, 251, 264
    Hecke, 55
    locally finite, 184
    Munn, 5, 50, 98, 111, 164, 275, 278
    Noetherian, 166, 272, 285
    PI, 157, 163, 166, 184, 187, 274, 283, 285
    principal ideal, 284, 287
    semigroup, 4, 50, 98, 157, 163, 237, 284, 287
        contracted, 4, 111, 150, 277
    semisimple, 50, 105, 237
    separable, 110
almost nilpotent semigroup, 172
almost semisimple semigroup, 209, 219
almost unipotent semigroup, 138, 141, 188
associated group, 59, 67, 71, 82, 83, 93, 95, 133, 137, 142, 143, 151, 173, 179, 190, 192, 199, 219, 285

basic ideal, 116
BN-pair, 225
Borel subgroup, 25, 34, 55, 226

bottom layer, 123, 125
Bruhat decomposition, 25, 226, 235

closed set, 5
closure
    completely 0-simple, 59, 77
    π-regular, 8, 78, 80, 81, 105, 145, 147, 148, 180, 192, 254
    Zariski, 5, 8, 80, 95, 131, 137, 158, 161, 180
commutator subgroup, 158, 168
completely 0-simple closure, 59, 77
completely 0-simple semigroup, 2, 30, 62, 68, 86, 120, 162, 237, 280
completely reducible semigroup, 54, 102
connected component, 8, 168, 174
constructible set, 169
Coxeter generators, 225, 230
cross section lattice, 235, 250
crossed product, 266, 281

degree map, 266
dimension
    Gelfand-Kirillov, 184, 190, 203, 211, 273, 286

Main notation

Chapter 1

$\langle A \rangle$	subsemigroup generated by A
$\text{gp}(A)$	subgroup generated by A
S^1	semigroup with identity adjoined
S^0	semigroup with zero adjoined
θ	the zero element of a semigroup
$M_n(K)$	$n \times n$ full linear monoid over a field K
$GL_n(K)$	full linear group
$|A|$	the cardinality of A
$\mathcal{R}, \mathcal{L}, \mathcal{J}, \mathcal{H}$	Green's relations
S/I	Rees factor semigroup
$K\{A\}$	K-linear span of A
$K[S]$	semigroup algebra
$K_0[S]$	contracted semigroup algebra
$\text{supp}(a)$	the support of $a \in K[S]$
$A \circ B$	ordinary product of (generalized) matrices
$S = \mathcal{M}(G, X, Y, P)$	semigroup of matrix type over a group G
(g, x, y)	typical element of $\mathcal{M}(G, X, Y, P)$
$S_{(x)}$	row determined by $x \in X$
$S^{(y)}$	column determined by $y \in Y$
$S^{(y)}_{(x)}$	intersection of $S_{(x)}, S^{(y)}$ in $\mathcal{M}(G, X, Y, P)$
$\mathcal{M}(K[G], X, Y, P)$	Munn algebra
$\overline{K}$	algebraic closure of K
$\text{rank}(a)$	rank of a matrix a
K^n	the space of column vectors
$\text{ker}(a), \text{Im}(a)$	kernel, image of a linear map
$\overline{S}$	Zariski closure of S

cl(S)	π-regular closure of S
$\mathbf{Z}, \mathbf{Q}, \mathbf{R}, \mathbf{C}$	integers, rationals, reals, complexes
$\mathbf{F}_q$	field of q elements
ch(K)	characteristic of K
G^c	the connected component of G
Λ^j	j-th exterior power
$\text{End}_R(V)$	endomorphisms of an R-module V
$\text{Aut}_R(V)$	automorphisms of an R-module V
M_j	matrices of rank $\leq j$ in $M_n(K)$

Chapter 2

$\mathcal{M}(H_j, X_j, Y_j, Q_j)$	a Rees matrix presentation of M_j/M_{j-1}
$\text{Lin}_K(v_1, \ldots, v_k)$	K-span of vectors $v_1, \ldots, v_k$
P_j, P_j^-	maximal parabolic subgroups of $GL_n(K)$
U_j, U_j^-	unipotent radicals of P_j, P_j^-
L_j	Levi factor of P_j, P_j^-
$GL_n(K)/P_j$	left coset representatives of P_j
$GL_n(K)/P_j^-$	right coset representatives of P_j^-
R	Renner monoid for $M_n(K)$
W	Weyl group for $GL_n(K)$
B	upper triangular matrices in $GL_n(K)$
U	unipotent matrices in B
T	diagonal matrices in $GL_n(K)$
W_j	matrices of rank $\leq j$ in W
E_{ij}	(i,j)-th matrix unit
a^*	transpose of a
$\delta(a, b)$	distance on R
$\sigma(d(a, b))$	rank sequence associated to a, b

Chapter 3

$\hat{U}$	completely 0-simple closure of U
S_j	matrices of rank $\leq j$ in S
T_j	a distinguished ideal of S_j
U_α	uniform components of S
K^*	multiplicative group of K

S_G	the semigroup generated by S, G
$\mathrm{cl}(S)$	π-regular closure of S

Chapter 4

$C(A), R(A)$	column space, row space of a subset $A \subseteq M_n(K)$
$\mathrm{tr}(s)$	trace of s
$[D : K]$	dimension of D over K
$\tilde{R} = \mathcal{M}(R, X, Y, P)$	Munn algebra over R
$\tilde{R}_{(x)}^{(y)}, \tilde{R}_{X'}^{Y'}$	subalgebras of $\tilde{R}$ determined by $x \in X, y \in Y$ and by $X' \subseteq X, Y' \subseteq Y$
$M_X^{row}(R)$	$X \times X$ matrices over R with finitely many nonzero rows
$l_R(Z), r_R(Z)$	left, right annihilators of Z in R
$\mathrm{Row}(P)$	left R-module generated by rows of P
$\mathrm{Col}(P)$	right R-module generated by columns of P
$\mathcal{B}(J)$	basic ideal determined by J
$V(P), V_0(P)$	$\tilde{R}$-modules associated to R-module V
$X.r$	r copies of the set X
$\overline{P}$	P viewed as $Y.r \times X.r$ matrix
$\mathrm{rank}(P)$	rank of the sandwich matrix P
$\mathcal{J}(K[S])$	Jacobson radical of $K[S]$
$I(S)$	idealizer of S
$\Omega(S)$	translational hull of S

Chapter 5

$K[x_{ij}]$	the coordinate ring of $M_n(K)$
$x_m(x, y, u_1, \ldots, u_m),$ $y_m(x, y, u_1, \ldots, u_m)$	Malcev words
$X_m = Y_m$	Malcev identity
G_u	unipotent elements of G
$E(A)$	set of idempotents in A
$\mathcal{P}_m, \mathcal{P}$	permutation properties
$[H, H]$	commutator subgroup of H

Chapter 6

$\mathcal{R}(G)$	solvable radical of G
G_s	semisimple elements of G
$\|g\|$	norm of g
$\chi(x)$	characteristic polynomial
$A(g), A'(g)$	attracting space and repulsing space
$PGL_n(K)$	projective linear group
$\hat{X}$	topological closure of X

Chapter 7

$\mathrm{GK}(S)$	Gelfand - Kirillov dimension of S
$d_S(m)$	growth function
$Z(H)$	centre of H
$L(S, X, k)$	repetitivity constant
$|w|$	length of a word w
$f_A(m)$	growth function with respect to a subset A of S
$G(L/E)$	Galois group of L over E
$\mathrm{rk}(S)$	rank of a semigroup S

Chapter 8

B, N	BN-pair for a group G
W	Weyl subgroup
S	system of simple reflections
$l(a)$	length of a
W_I	parabolic subgroup of W
B^-	opposite Borel subgroup
P_I, P_I^-	standard opposite parabolics
U_I, U_I^-	unipotent radicals of P_I, P_I^-
L_I	Levi factor
$R_u(P)$	unipotent radical of P
$X_\alpha, X_{i,j}$	root subgroups
Φ, Φ^+, Φ^-	index sets for root subgroups
η_I	a map $G \longrightarrow L_I \cup \{\theta\}$
$P(e), P^-(e)$	parabolics associated to an idempotent

$U(e), U^-(e), L(e)$	distinguished subgroups of $P(e), P^-(e)$
$C_G(A), N_G(A)$	centralizer, normalizer of A in G
M_σ	finite reductive monoid
$\mathcal{M}_{(G)}, \mathcal{M}$	universal monoid of Lie type on G
e_I	standard idempotent of $\mathcal{M}$
J_I^0	principal factor of $\mathcal{M}$
$\mathcal{M}_I$	$G \cup J_I^0$
$M(J)$	three $\mathcal{J}$-class monoid of Lie type
$\mathcal{U}(M)$	set of $\mathcal{J}$-classes of M
$\Lambda(M)$	cross section of idempotents
$GL(V)$	automorphisms of the space V
π_K^L	induced representation

Chapter 9

$M_n(D)$	full skew linear monoid over D
$d_{n,j}(D), \mathrm{N}(D)$	numerical invariants for D
$\oplus_{s \in S} R_s$	S-graded ring
$\mathrm{H}(R)$	semigroup of homogeneous elements of a graded ring R
$\mathrm{supp}(r)$	support of r in R
$\mathbf{Z}\{X\}$	additive subgroup generated by X
I_{gr}	graded part of I
$\deg(h)$	degree of a homogeneous element h
$A * G$	crossed product
$Q_{cl}(R)$	classical quotient ring of R
H_a	a subsemigroup of $\mathrm{H}(R)$ related to a
$R_{(a)}$	additive group generated by H_a
S_a	image of H_a under the degree map
G_a	group of quotients of S_a
$\tilde{R}_{(a)}$	localization of $R_{(a)}$ with respect to H_a
$\mathcal{J}(R)$	Jacobson radical of R
$\mathcal{J}_{gr}(R)$	graded part of $\mathcal{J}(R)$
$\mathrm{clKdim}(R)$	classical Krull dimension of R

www.ingramcontent.com/pod-product-compliance
Lightning Source LLC
Chambersburg PA
CBHW050635190326
41458CB00008B/2277